Physiological Principles in Medicine

General Editors

Dr R. N. Hardy
Physiological Laboratory, Cambridge

Professor M. Hobsley
Department of Surgical Studies, The Middlesex Hospital and the Middlesex Hospital Medical School, London

Professor K. B. Saunders
Department of Medicine, St. George's Hospital and St. George's Hospital Medical School, London

Neurophysiology

Physiological Principles in Medicine

Books are published in linked pairs—the preclinical volume linked to its clinical counterpart, as follows:

Endocrine Physiology by Richard N. Hardy
Clinical Endocrinology by Peter Daggett

Digestive System Physiology by Paul A. Sanford
Disorders of the Digestive System by Michael Hobsley

Respiratory Physiology by John Widdicombe and Andrew Davies
Respiratory Disorders by Ian R. Cameron and Nigel T. Bateman

Neurophysiology by R. H. S. Carpenter
Clinical Neurology by C. D. Marsden

In preparation:

Clinical Neurology by C. D. Marsden

Reproduction and the Fetus by Alan L. R. Findlay
Gynaecology, Obstetrics and the Neonate by S. J. Steele

Neurophysiology

R. H. S. Carpenter

Lecturer in Physiology, University of Cambridge;
Fellow and Tutor, Gonville and Caius College, Cambridge

Edward Arnold

First published 1984
by Edward Arnold (Publishers) Ltd
41 Bedford Square
London WC1B 3DQ

British Library Cataloguing in Publication Data

Carpenter, R. H. S.
Neurophysiology.—(Physiological principles in medicine, ISSN 0260-2946)
1. Neurophysiology
I. Title II. Series
612'.8 QP361

ISBN 0-7131-4431-9

Text set in 10/11 pt Baskerville
by CK Typesetters, 26 Mulgrave Road, Sutton, Surrey
Printed and Bound in Great Britain by Thomson Litho Ltd, East Kilbride

General preface to series

Student textbooks of medicine seek to present the subject of human diseases and their treatment in a manner that is not only informative, but interesting and readily assimilable. It is also important, in a field where knowledge advances rapidly, that the principles are emphasized rather than details, so that information remains valid for as long as possible.

These factors all favour an approach which concentrates on each disease as a disturbance of normal structure and function. Therapy, in principle, follows logically from a knowledge of the disturbance, though it is in this field that the most rapid changes in information occur.

A disturbance of normal structure without any disturbance of function is not important to the patient except for cosmetic or psychological considerations. Therefore, it is the disturbance in function which should be stressed. Preclinical students must get a firm grasp of physiology in a way that shows them how it is related to disease, while clinical students must be presented with descriptions of disease which stress the basic disturbance of function that is responsible for symptoms and signs. This approach should increase interest, reduce the burden on the student's memory and remain valid despite alterations in the details of treatment, so long as the fundamental physiological concepts remain unchallenged.

In the present Series, the major physiological systems are each covered by a pair of books, one preclinical and one clinical, in which the authors have attempted to meet the requirements discussed above. A particular feature is the provision of cross-references between the two members of a pair of books to facilitate the blending of basic science and clinical expertise that is the goal of this Series.

RNH
MH
KBS

Dark mysteries are here—old pathways, secret places
Under the tangled cortex, grown snugly thick now—
Intricately synapsed: electrode-proof.

Preface

The supervision system practised at Cambridge and elsewhere brings many benefits both to teacher and taught: not least, that lecturers are brought face to face with the results of deficiencies in their own teaching in a peculiarly immediate and painful way. What has seemed to many supervisors a most worrying trend over the last ten years or so is the extent to which a student may come away from a series of lectures on (let us say) the circulation, with an impressive amount of detailed information, including perhaps the minutiae of experiments published only a month or two previously, yet with little sense of what might be called *function*: of what the circulation really does, of how it responds to actual examples of changed external conditions, and how it relates to other major systems. And in the case of the central nervous system things seem even worse: a student may acquire an immensely detailed knowledge of the anatomical intricacies of the motor system, yet not be able to tell you even in the broadest terms what the cerebellum actually does, or have the slightest feel for what kinds of processes must be involved in such an act as throwing a cricket ball. The result is much knowledge, but little understanding, and very little sense of ignorance.

I believe this to be the result of two factors. The first is, paradoxically, that over the last decade or so, Universities and Teaching Hospitals have quite rightly begun to take teaching much more seriously than once was the case, and consequently a perfectly laudable sense of competition has developed amongst lecturers to gain the approval of their audiences. But students—at least in the short term—tend to form judgements rather on the basis of the number of 'facts' that they have succeeded in copying down in the course of a lecture: the more recent these facts are, the better they are pleased. Lecturers naturally respond to this by filling their lectures with increasing amounts of detail, at the expense of fundamental principles. The students' notebooks swell with quantities of undigested information, but they are bewildered—even resentful—when asked simple but basic questions like 'how does a man stand upright?'. This change in emphasis has made physiology less enjoyable either to study or teach than it used to be, as well as less educational in the broadest sense: there is no time and little motivation to ask questions of oneself, and all is reduced, in the end, to rote-learning.

The second factor that has debased the intellectual quality of much of our teaching is the increasing emphasis that is put on mechanism instead of function. More time is often spent in talking about the detailed physics of nerve conduction than in discussing exactly what information is being carried by nerves, how it is coded, and how the nervous system is actually used. Again, lecturers' fear of instant student opinion is perhaps partly the cause: most students get immediate and easy satisfaction (of a limited kind) by seeing the detailed steps that cause a particular phenomenon; and if all can be reduced to a series of biochemical reactions, then so much the better. To understand whole systems and their interactions requires rather more effort of thought, and one can never be sure one is right. But in the long run, and most particularly for medical students, it is precisely the large-scale functioning of physiological systems that is important. A doctor needs to have a feel for what is likely to be the consequence of chronic heart failure in terms of problems of fluid balance, or for what may happen if his asthmatic patient decides on a holiday in the Andes. Whether the cardiac action potential is due mainly to calcium or to sodium, and whether or not the substantia nigra projects to the red nucleus, are for him matters of singularly little interest or significance.

This book is an attempt to go counter to this trend by starting from the premise that a more satisfactory way to teach physiology is to build a scaffold of general principles on which factual details may later be hung as the need arises, and to prefer to consider *what* systems do rather than *how* they do it. However, this is largely a matter of emphasis and organization rather than of content, and the reader will find details of mechanism if they are required. Above all, the aim has been to recreate something of the intellectual excitement of the study of physiology that has been lost sight of in recent years, and to encourage the student to think and to question. If it is at all successful in this, the thanks should go not to me but rather to those past and present students of mine for whose intellectual stimulation I am—as all teachers must surely be—deeply indebted.

Cambridge, 1984 R. H. S. Carpenter

Acknowledgements

It is a pleasure to acknowledge my indebtedness to Dr Susan Aufgaerdem, Professor George L. Engel, Mr Austin Hockaday, Dr J. Keast-Butler, Dr Richard Kessel, Dr Peter Lewis, Dr J. Purdon Martin, Dr N. R. C. Roberton, Dr T. D. M. Roberts and Mr Peter Starling for their help in providing illustrations; to the editors and to the staff of Edward Arnold for their criticisms and support; to various authors and publishers where mentioned for permission to reproduce material; and above all to the late Dr R. N. Hardy, who died so tragically while the book was in its final stages: his friendly encouragement, and his qualities of wisdom and humanity are sadly missed by all who knew him.

Contents

	General preface to series	**v**
	Preface	**vii**
1	**Studying the brain**	**1**
	The history of the brain	1
	Central neurones	7
	Anatomical methods	10
	Physiological methods	15
2	**Nerves**	**19**
	The flow of electrical current along nerves	20
	The cyclical regeneration of action potentials	22
	The dependence of membrane potential on ionic permeabilities	24
	The dependence of ionic permeability on membrane potential	29
	Conduction velocity	33
	The compound action potential	35
	Threshold properties	38
	The all-or-none law	42
	Neural codes	44
3	**Receptors and synapses**	**47**
	Sensory receptors	47
	Transduction in the Pacinian corpuscle	47
	The initiation of impulses	51
	Functions of adaptation	55
	Synaptic transmission	58
	Electrical transmission	59
	Chemical transmission at the neuromuscular junction	61
	Central excitatory synapses	64
	Inhibitory synapses	68
	Voltage and current inhibition	72
	Presynaptic inhibition	73
	Long-term changes in excitability	76
	The development of synaptic connections	78

4 Skin sense **82**
Sensory modalities 82
The receptors and their central connections 84
Neural responses to cutaneous stimulation 89
Lateral inhibition 91
Responses from smaller afferents 94
Central responses 96
Pain 97

5 Proprioception **100**
Muscle proprioceptors 100
Joint receptors 106
The vestibular apparatus 107
The sensory cells 108
Otolith organs (utricle and saccule) 109
The semicircular canals 111

6 Hearing **115**
The nature of sound 115
Sound spectra 118
The structure of the ear 122
Fourier analysis by the cochlea 126
Responses from auditory fibres 128
Central pathways and responses 131
Spatial localization of sound 133

7 Vision **138**
Photometry 139
Light and dark adaptation 141
Image-forming by the eye 142
The control of the pupil 146
The retina 147
The receptors 150
Electrical events in retinal receptors 152
Retinal interneurones 154
Central visual pathways 156
Mechanisms of adaptation 161
Visual acuity 166
Lateral inhibition 173
Flicker sensitivity 174
Colour vision 176
Colour mixing 177
Chromatic adaptation 181
Clinical disturbances of colour vision 182
The use of vision 183

8 Smell and taste **191**
Olfaction: the receptors and their connections 191

Psychophysics of smell 200
Theories of the olfactory mechanism 202
Gustation: the receptors 205
Theories of gustation 207

9 Types of motor control 210
Difficulties in studying motor systems 210
The use of feedback 212
The hierarchy of control 220

10 The spinal level 224
Descending pathways 225
Sensory feedback from muscles 231
The servohypothesis 234

11 The control of posture 241
The importance of support 241
Vestibular contribution to posture 245
Visual contributions to posture 248
How does the eye tell us about head movement? 250
Vestibular and visual interactions 251
Neck reflexes 253

12 Higher levels of motor control 256
Motor cortex 256
Corticospinal concepts 259
Co-ordination of somatosensory input with motor output 261
Cerebellum 264
Disorders of the cerebellum 268
Theories of cerebellar action 270
The basal ganglia 273

13 Analysis and storage of information by the cerebral cortex 277
Frontal cortex 280
The parietal lobe 282
Left-right asymmetry in the brain 287
The temporal lobe: neo- and archi-cortex 291
The neural mechanism of memory 297

14 The highest levels of control 303
Motivational maps 303
Emotion 306
Sleep 312
'Mind' and consciousness 316

References and further reading 320

Index 331

1

Studying the brain

This book is about trying to understand the brain; we might begin by asking whether such an aim is not in fact hopelessly ambitious.

The human brain is a machine whose complexity far exceeds anything made by Man. It is made up of units called *neurones* that provide both the pathways by which information is transmitted within it and also the computing machinery with which its decisions are made. There are quite a lot of these neurones: about thirty times as many as the total number of men, women and children on this planet. Each neurone communicates with a large number of its neighbours, perhaps several thousand of them, and it is the pattern of these connections that determines what the brain does. Studying the brain is thus like studying human society: for a society can only be fully understood in terms of the interactions that each of its individuals makes with the circle of people with whom he is in contact. Consequently the understanding of the brain is a task as daunting as trying to comprehend the behaviour of the entire human race, its politics, its economics, and all other aspects of what it does; in fact, about thirty times more difficult. For this reason, the study of the brain has in some respects a closer affinity with 'arts' subjects like history than it does with much conventional science, and this for many people is part of its attraction. Our brains *need* to be complex: part of what they do is to embody a kind of working model of the outside world, that enables us to imagine in advance what would be the result of different courses of action. It follows that the brain must be at least as complicated as the world we experience. How has such complexity come about? The evolutionary history of the brain is not well understood, but was probably something like the account that follows.

The history of the brain

The co-ordination of a single-celled organism such as an amoeba is essentially chemical: its brain is its nucleus, acting in conjunction with its other organelles. But the proper co-operation of the cells of a multicellular organism clearly demands some kind of communication between them, particularly when, as in Hydra, there is specialization of cells into different functions: secretion, movement, nutrition, defence and so on. In small and slow creatures, such communication may still be chemical; but as an organism gets bigger, it takes a disproportionately longer time for a chemical signal released at one end of it to reach the other, because the time taken for diffusion is proportional to the square of the distance travelled. If speed of response is not particularly important, and if in addition there is some kind of fluid circulation that will increase the rate of dispersion of chemical transmitters, this kind of communication may still be satisfactory even in very large organisms. Our own hormonal control systems are of course precisely of this kind. But such systems are not only slow, they are also imprecise. When a sudden fright leads to release of adrenaline into our blood, it acts indiscriminately on the whole body. For fast and localized action, we need an arrangement that will release chemical transmitter as rapidly as possible at the site where it is required, and nowhere else (Fig. 1.1). Without spatial selectivity of this kind, we would need as many different transmitter substances as target organs: in fact, as we shall see, all the thousands of millions of skeletal muscle cells in our bodies are controlled by just one chemical transmitter: acetylcholine.

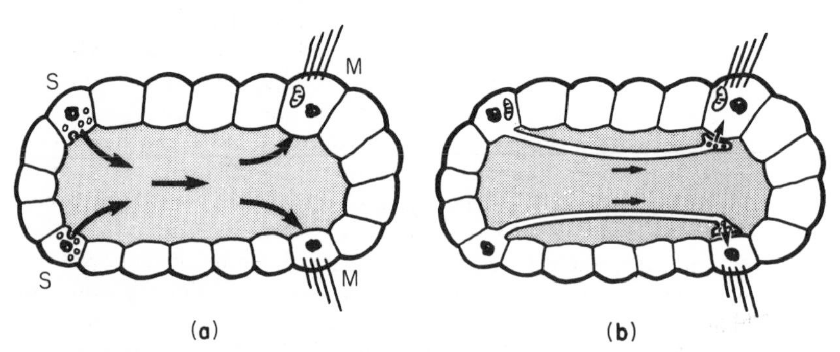

Fig. 1.1 (**a**) A hypothetical multicellular organism with sensory cells (S) that control motor cells (M) by releasing a chemical transmitter or hormone into the common fluid space. (**b**) Direct connections between sensory and motor cells by means of nerve axons, providing communication that is both quicker and more specific. Their ultimate action on the motor cells is still chemical.

This function of localized secretion is carried out by the nerve cells or *neurones*. Neurones are of ectodermal origin, and some remain in epithelia as *sensory receptors* that are sensitive to mechanical or chemical stimuli, to temperature or to electromagnetic radiation. Others have migrated inward, and have become specialized as *interneurones*, responding only to the chemicals released locally by sensory neurones or other interneurones, and in turn releasing transmitter at their *terminals* which form junctions called *synapses* either with interneurones or with effectors such as muscles or secretory cells. They thus provide the communication channels by which information is passed rapidly from one part of the central nervous system to another, through mechanisms that form the subject of Chapters 2 and 3. In Hydra, for example, we find a network of such intercommunicating neurones, making contact on the one hand with sensory cells on its surface that respond to touch and chemical stimuli, and on the other with muscle cells and secretory glands. Hydra's brain is thus spread more-or-less uniformly throughout its body, with only a slight increase in density in the region of its mouth: yet even such a relatively undifferentiated structure is capable of well co-ordinated, even 'purposeful' behaviour.

The next step in the evolution of the nervous system comes with the increasing specialization of sensory organs, particularly of *teloreceptors* such as eyes and olfactory receptors. For an animal that normally moves in one particular direction, such organs tend to develop at the front end, and the result of the consequent extra flux of sensory information to a localized region is an increased proliferation of interneurones in the head. In Planaria we have the first true brain of this kind, a dense concentration of neurones close to the eyes and sensory lobes of the head, giving rise to a pair of nerve cords that run down the body and send off side branches connecting with other neurones and effector cells. In segmented animals like the earthworm, the nerve cords show a series of swellings or *ganglia*, one to each segment; each of them can be thought of as a kind of sub-brain, and a decapitated earthworm is still capable of many kinds of segmental and intersegmental co-ordination. Though our bodies are not of course segmented, our nerve cord, the *spinal cord*, still shows some segmental properties, particularly in the organization of the incoming and outgoing fibres, and in the existence of corresponding chains of ganglia along each side (Fig. 1.2). We shall see later, in Chapter 10, that our spinal

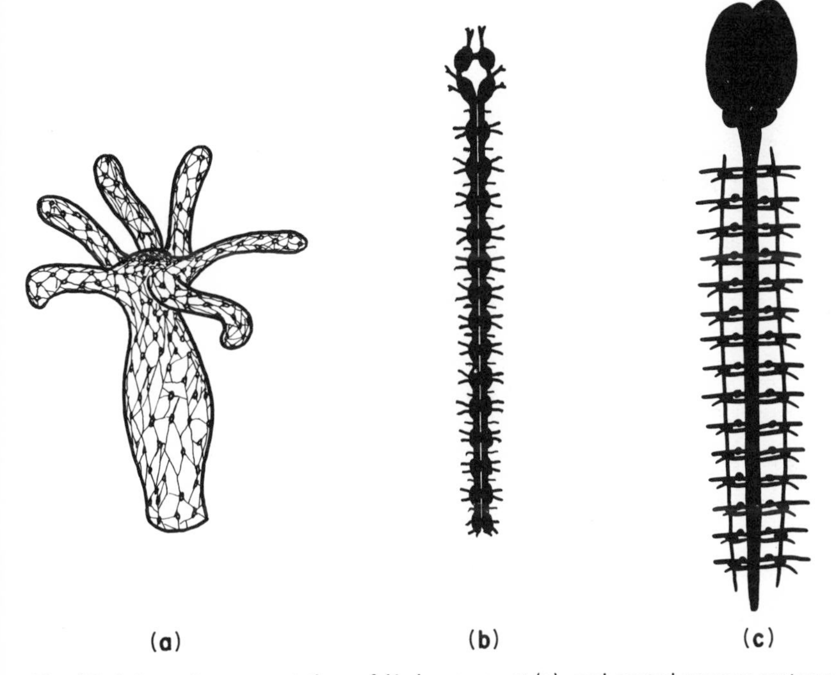

Fig. 1.2 Schematic representations of *Hydra* nerve-net (**a**), and central nervous systems of earthworm (**b**) and Man (**c**). (Partly after Buchsbaum, 1971).

cord is also capable of a limited degree of brain-like activity. The primitive nerve net has not been altogether superseded, but survives as an adjunct to the central nervous system in the diffuse networks near the viscera that control gut movement and some other visceral functions.

The subsequent development of the brain is rather more complex, and not well understood. By looking at its evolutionary history in conjunction with the sequence of its growth in fetal development, one can postulate a framework that may help to relate the primitive nervous system to the more intricate structure of the adult human brain.

The central nervous system is derived from a narrow strip of ectoderm, the *neural plate*, which runs down the middle of the vertebrate embryo's back. The

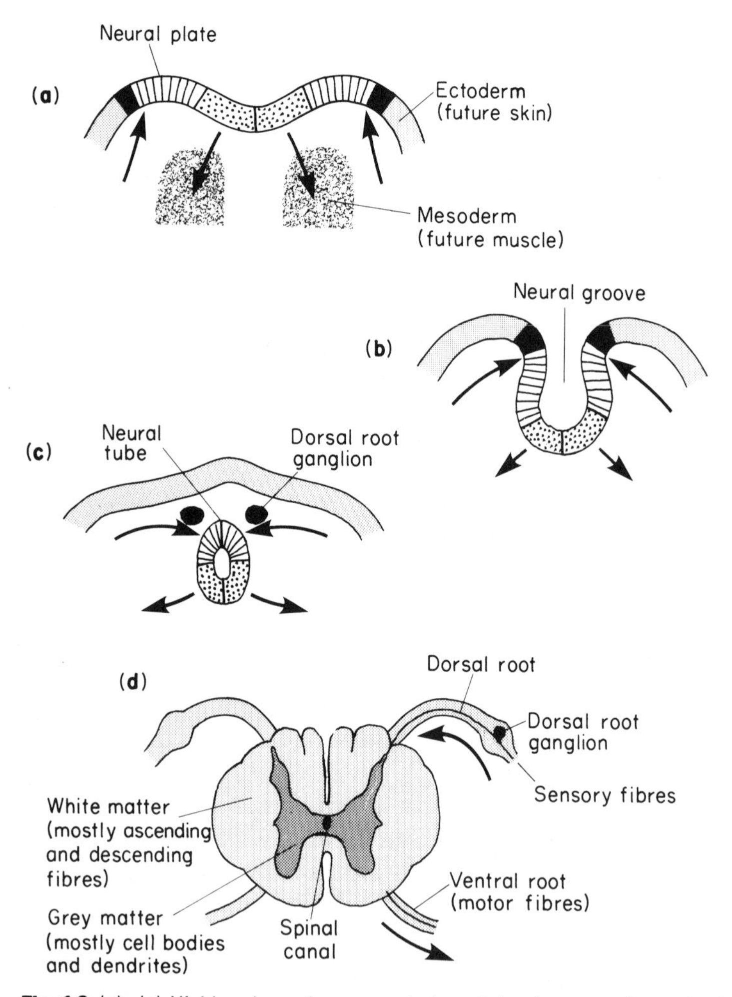

Fig. 1.3 (**a**)–(**c**) Highly schematic representation of development of neural tube from neural plate, showing relative positions of sensory (striped) and motor (stippled) regions. (**d**) Cross-section of adult human spinal cord, at the level of the second thoracic vertebra

centre of this strip becomes depressed into a trough or groove, and eventually its edges come to meet in the middle to form a closed structure, the *neural tube* (Fig. 1.3). It is natural for sensory fibres from the skin to enter at the margins of the neural plate, and for motor fibres to the more medial musculature to leave the plate nearer the midline, and as a consequence one finds that it is in the dorsal half of the neural tube that the sensory fibres terminate (their cell bodies lying in the *dorsal root ganglia* on each side of the tube), while the cell bodies of the efferent motor fibres lie in the ventrolateral part of the neural tube, and this arrangement is evident in the adult spinal cord (Fig. 1.3**d**). Here one can see in cross-section the *ventral* and *dorsal horns*, consisting of masses of *grey matter* (mostly cell bodies), a less prominent central region concerned with the neural control of visceral function, and a surrounding sheath consisting of *white matter*, mainly bundles of nerve fibres running longitudinally up and down the cord.

At the cephalic end of the neural tube, a modification of this basic plan occurs. The central fluid-filled canal, which is very small in the spinal cord, widens out at two separate points to form hollow chambers or *ventricles*: at the same time it migrates back to the dorsal surface of the neural tube, so that the ventricles are open on their dorsal side. This surface is covered by the *choroid membrane*, the site of production of the cerebrospinal fluid that fills the canals and ventricles of the brain. The more caudal of the ventricles is called the fourth ventricle, and the region around it is the *hindbrain* or rhombencephalon; it is connected to the more rostral third ventricle by the *cerebral aqueduct*. The region round the aqueduct is called the *midbrain* or mesencephalon and that round the third ventricle is the *forebrain* or prosencephalon (Fig. 1.4). Subsequently, the third ventricle produces a pair of swellings at the front end, which become inflated into the *lateral ventricles*: the neural tissue surrounding them forms the *cerebral hemispheres* (telencephalon) while the rest of the forebrain is called the diencephalon. The hindbrain is likewise divided into two regions: the caudal part is called the *medulla*, and the rostral part of it (metencephalon) is marked by the outgrowth of the *cerebellum* over the dorsal

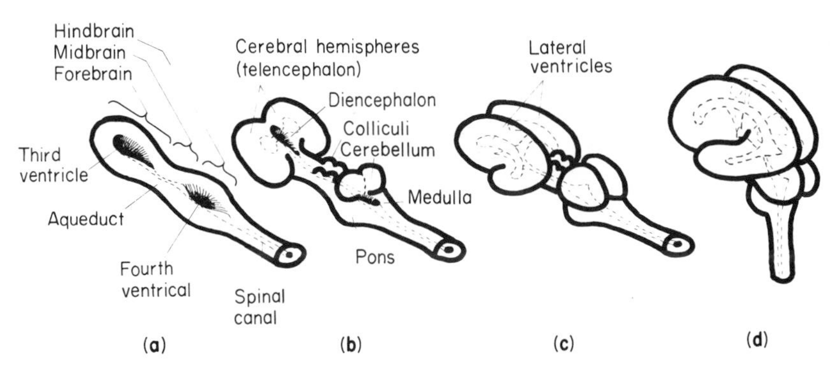

Fig. 1.4 The notional steps leading from neural tube to human brain. (**a**) The opening of the canal to and from the third and fourth ventricles. (**b,c**) The growth of the cerebellum, and of the cerebral hemispheres with their associated lateral ventricles. (**d**) Human brain, showing greatly enlarged cerebral hemispheres, and flexion of the neural tube

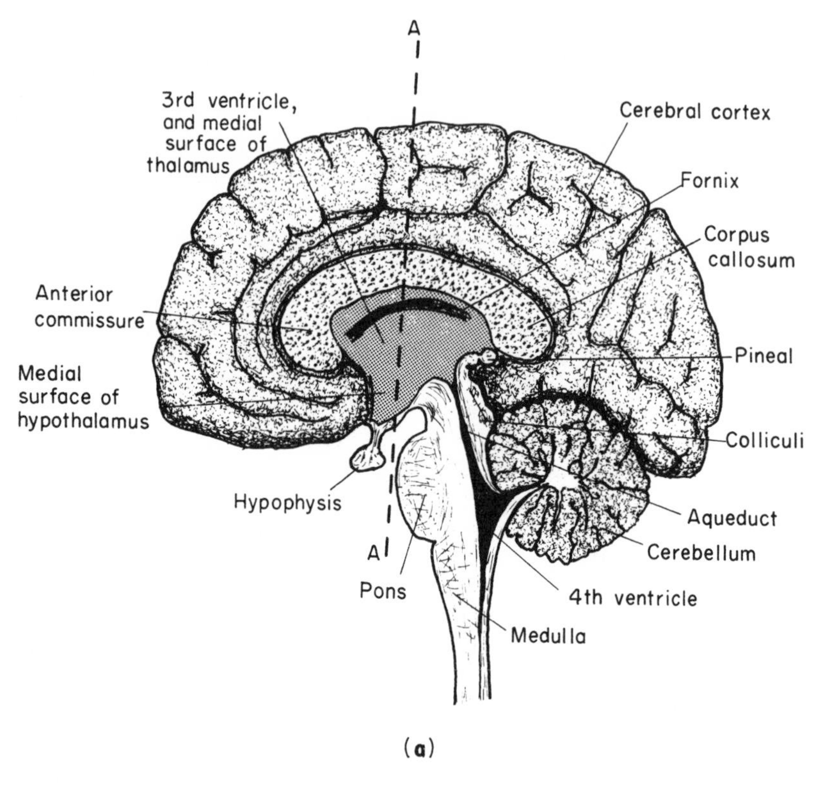

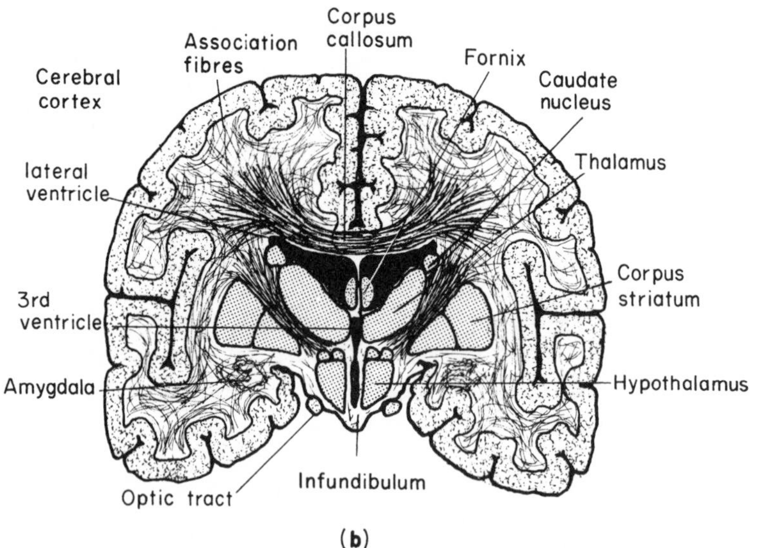

Fig. 1.5 Semi-schematic sections of the human brain. (**a**) Sagittal, showing medial aspect; (**b**) transverse, in the plane A–A

surface, and a massive bundle of fibres associated with it, the *pons*, on the ventral surface. In less developed species such as fish all these structures are easily recognizable without dissection, but in Man the extraordinary ballooning growth of the cerebral hemispheres has not only engulfed the other surface features (leaving only the cerebellum and medulla peeping out at the back), but the massive fibre tracts needed to connect the cerebral hemispheres to each other and to the rest of the brain have tended to elbow older structures out of the way and have often considerably distorted their shape. Another factor that makes the anatomy of the human brain somewhat confusing is that the neural tube has become bent forward: while the axis of the hindbrain is near vertical, that of the forebrain is horizontal (Fig. 1.4).

The most important areas of the human brain are shown in the sagittal and transverse sections of Fig. 1.5. The *cerebellum* is an important co-ordinating system for posture and for motor movements in general, that arose originally as an adjunct to the vestibular apparatus, a sensory organ concerned with balance and the detection of movement (Chapter 5). Important landmarks on the dorsal surface of the midbrain are the four bumps (corpora quadrigemina) formed by the *superior* and *inferior colliculi*, primitive sensory integrating areas for vision and hearing respectively; in higher vertebrates their function is mainly that of organizing orienting reactions and other semi-reflex responses to visual and auditory stimuli. In the diencephalon, on each side of the third ventricle, lie the two halves of the *thalamus*, a dense group of nuclei whose neurones partly act as relays for fibres that project upward to the cerebral hemispheres. Close to them but more lateral is the *corpus striatum*, an old area that is concerned in the control of movements (Chapter 12). Also in the diencephalon, but lying on the floor of the third ventricle, is the *hypothalamus*, the brain's interface with the hormonal and autonomic systems that control the body's internal homoeostasis, and as such lying for the most part outside the scope of this book. More laterally, various nuclei, fibre tracts and other areas form a loosely defined system called the *limbic system*, which connects both with the hypothalamus, with the olfactory areas, and with many other regions of the brain; they are thought to be concerned with such functions as emotion, motivation and certain kinds of memory (Chapter 14). Finally, there is the *cerebral cortex*, which covers the lateral ventricles and is deeply convoluted and furrowed in higher vertebrates, enabling a large superficial area of tissue to be crammed into a relatively small volume.

The divisions of the brain described so far are very gross ones, and for the most part quite obvious to the naked eye. Finer anatomical distinctions can only be made by looking at the neurones themselves, and the way their populations vary from one region to another.

Central neurones

Neurones from different parts of the nervous system show a wide range of shapes and sizes (Fig. 1.6). What they have in common is a compact cell body containing the nucleus, and a number of projecting filaments that generally show extensive branching: these projections form the pathways by which

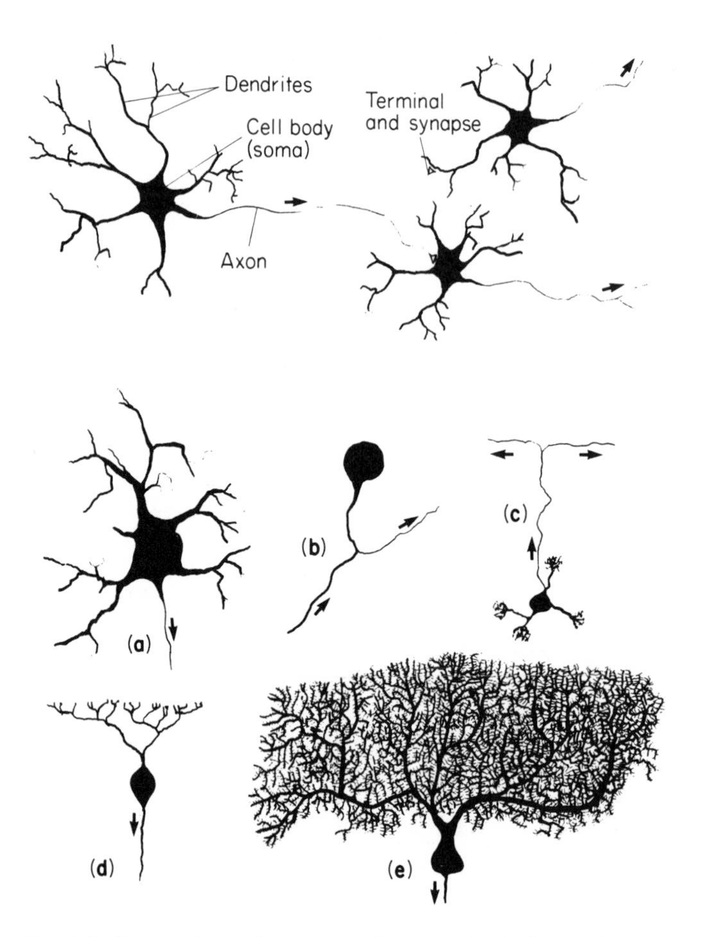

Fig. 1.6 Above, schematic representation of a 'classical' central neurone synapsing with two others, in one case on a dendrite and in the other on the cell body or soma. Below, some typical and somewhat idealized central neurones: (**a**) motor neurone; (**b**) dorsal root (unipolar) neurone; (**c**) granule cell, (cerebellum); (**d**) bipolar cell, retina; (**e**) Purkinje cell, cerebellum. The arrows indicate the axon

information from different sources is gathered together by the neurone, and then transmitted in turn to some other region. In a 'classical' neurone, one of them (the *axon*) forms the output of the cell, and the others (*dendrites*) form the input. However, there are many exceptions to this rule: the dorsal root ganglion cells for instance have no dendrites at all and the cell body simply lies to one side of a single continuous axon (Fig. 1.6**b**); and in many sites within the brain the dendrites are known to act as outputs as well as inputs. The axon is specialized for carrying information rapidly over long distances, and may often be very long indeed—as for example those that carry muscle commands all the way from the cerebral cortex to the bottom of the spinal cord. As a rule they carry their information in the form of propagated *action* *potentials,* discussed in Chapter 2. The larger axons are frequently swathed in layers of *myelin*, a lipid substance that speeds up the conduction of action potentials by acting as an electrical insulator: these layers are the result of accessory *Schwann cells* sending out myelin-rich processes that wrap themselves round and round the axons to form a kind of swiss-roll (Fig. 1.7); the myelin is interrupted at regular intervals, at the *nodes of Ranvier*.

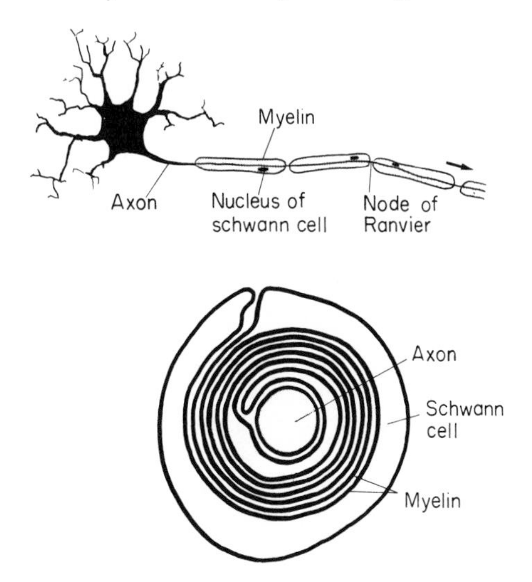

Fig. 1.7 Above, diagram of a neurone with myelinated axon (not to scale: the distance between the nodes of Ranvier is typically of the order of 1 mm, while the cell body might be 20–80μm in diameter). Below, schematic cross-section of myelinated axon, showing the layers of myelin formed by the Schwann cell wrapping itself round and round

The far end of the axon is usually branched, and its terminals make synaptic contact either with the dendrites or cell bodies of other neurones, with secretory cells, or (in the case of motor neurones) with muscle cells. At this point the signal—previously essentially electrical in nature—causes the release from the terminal of a tiny quantity of chemical transmitter, which then acts on the post-synaptic cell. We shall see later that it is the *pattern* of excitation—in time or space—arriving at the dendrites and body of a neurone that determines whether or not in turn it sends an action potential down its axon. In that respect, a neurone is a kind of miniature brain in its own right, that makes decisions on the basis of the pattern of activity at its inputs. Thanks to the work of biophysicists who have studied the way in which different synaptic inputs to a neurone interact with one another, we have a fair idea of the general rules that determine whether or not a cell will respond to a given pattern of excitation. In particular, we know that these rules depend very critically on the shape of the dendritic tree and the distribution of synapses upon it; two synapses interact in a very different way if they lie near one another on a single branch, rather than far apart on separate branches. It follows from this that a full understanding of the function of a single neurone requires detailed knowledge of the size and shape of its dendrites, and of the origin of all the thousands of afferent fibres synapsing with it. This is an impossibly daunting task, even for one neurone, let alone the hundred thousand million or so that constitute the brain. In practice, one cannot hope to do much more than to try to identify neurones that are typical of a particular area, and form some idea of their 'average' properties. We are helped in this by the fact that neurones are to a large extent grouped in homogeneous communities called *nuclei*, groups of similar cells projecting to the same area of the nervous system and whose afferent fibres likewise have common origins. The existence of nuclei implies that of the *tracts* that join them, and much of the work of neuroanatomists is to identify the inputs and outputs associated with particular nuclei.

Anatomical methods

Just looking with the naked eye at sections of the brain, one can make out the grosser nuclei and tracts; but a microscope is needed to make out the details of neuronal connections. Since neuronal tissue is virtually transparent, some kind of stain is needed: the problem then is that if we stain *all* the nerve cells, the brain is such a densely knotted structure that the whole thing will simply come out black and we shall be no better off: what is needed is a *selective* stain. One such is the *Golgi silver stain*, which has the odd property that it is only taken up by a very small percentage of the neurones in the tissue to which it is applied, apparently at random; those cells that do take it up do so completely, resulting in a complete and often very beautiful delineation of their dendritic structure (see for example Fig. 12.7). Another technique is to inject a dye such as Procion Yellow into a neurone through a micropipette inserted into it; this dye then diffuses into most parts of the cell but not outside it, and provides a good way of marking a cell whose electrical responses have previously been recorded with the same micropipette. Some marker substances are *transported* along axons, either from the cell body to the terminals (orthograde) or in the opposite direction (retrograde). One example is *horseradish peroxidase* (HRP); after extracellular injection at a particular site, it is taken up by axon terminals and carried back to the cell body. In this way one may identify the origin of efferents to a particular region of the brain. Similarly, labelled amino acids (tritiated leucine, for example) are taken up by cell bodies and transported towards the terminals, enabling one to identify the areas to which a particular nucleus projects. A related technique for tracing axonal pathways is to study the *degeneration* that results from injury to a nerve fibre. Two kinds of degeneration follow damage to an axon: orthograde or Wallerian degeneration, distal to the cut; and retrograde degeneration, in the direction of the cell body. In the first case, one may use the Nauta stain that identifies certain of the degeneration products; in retrograde degeneration one may see various characteristic changes in the cell body (Fig. 1.8). Sometimes these degenerative changes may actually extend beyond the synapse to affect the next neurone along, and are then described as *transneuronal* orthograde or

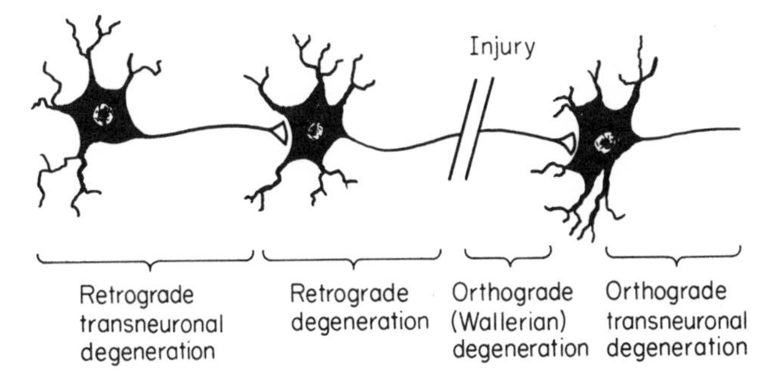

Fig. 1.8 Schematic representation of the types of degeneration that may follow injury to a neurone

retrograde degeneration. In monkeys, for instance, removal of the eye results in shrinkage of the neurones in the thalamus with which the fibres from the eye make contact. Thus by making a lesion in a particular nucleus one may in principle trace both the afferent and efferent pathways associated with it, and in some cases one may identify the second-order cells with which these fibres synapse as well. A problem with degeneration studies, but not with HRP or labelled amino acids, is that they do not distinguish between fibres that genuinely begin or end in a particular region, and those *fibres of passage* that merely happen to pass through it.

A wiring diagram not enough

If these anatomical techniques were perfect, and if we had the patience to identify each of the 10^{13} or so individual pathways in the central nervous system, we would end up with something like a wiring diagram of the brain. Would we really be much the wiser as a result? With our knowledge of the biophysical properties of each individual neurone we could then in principle calculate how the whole assembly of neurones would react to any given pattern of stimulation at the sensory receptors: the problem is of course the almost inconceivable difficulty of ever carrying out such a calculation. With modern computers, the accurate simulation of the behaviour of even a hundred neurones connected together in a realistic network is about the limit of what can be achieved within a reasonable period of time. Thus it is not so much that we know too little of the behaviour of individual neurones, but rather that we lack the conceptual techniques for analysing their behaviour as a whole. The brain is in fact composed of essentially rather simple elements joined together in extremely complex ways, and in that respect has certain similarities with a digital computer, Figure 1.9 shows the circuit diagram of a tiny part of a very small computer, of the circuits that perform the relatively trivial task of getting data from the keyboard. Without knowing very much about logic circuits, one can see at once that there is a relatively limited number of types of element in the circuit (the logic gates), and it is clear that the function of the whole circuit must reside in the way that these standard units are connected to one another.

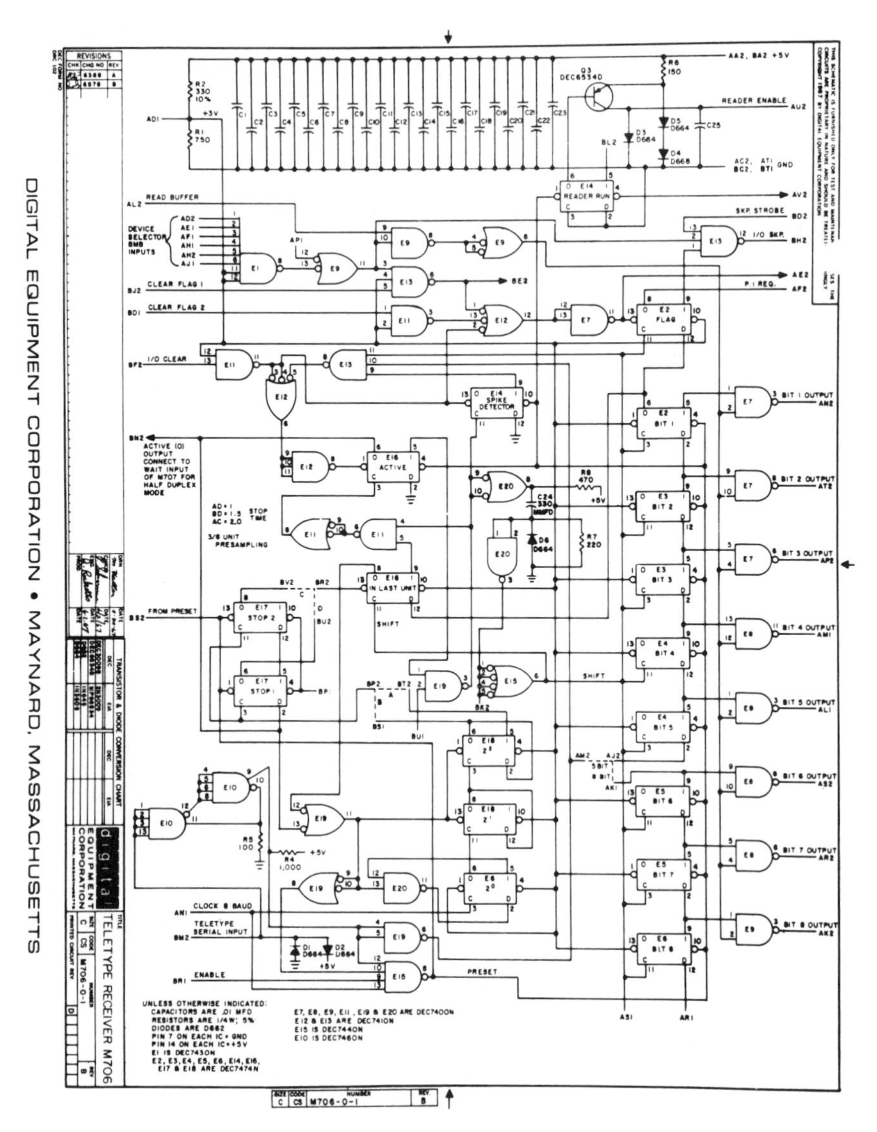

Fig. 1.9 Logic diagram of a very minor part (the keyboard receiver) of a very small computer. (Courtesy of Digital Equipment Corporation)

Now a full understanding of the details of this circuit's operation requires a considerable degree of intellectual effort; to master the computer of which it forms a part might take years of study; yet the brain is thousands of millions of times larger in scale.

Another example of a highly complex system made out of vast numbers of quite simple elements is provided by substances like foam rubber (Fig. 1.10). Here too, the shape and behaviour of each individual element—each

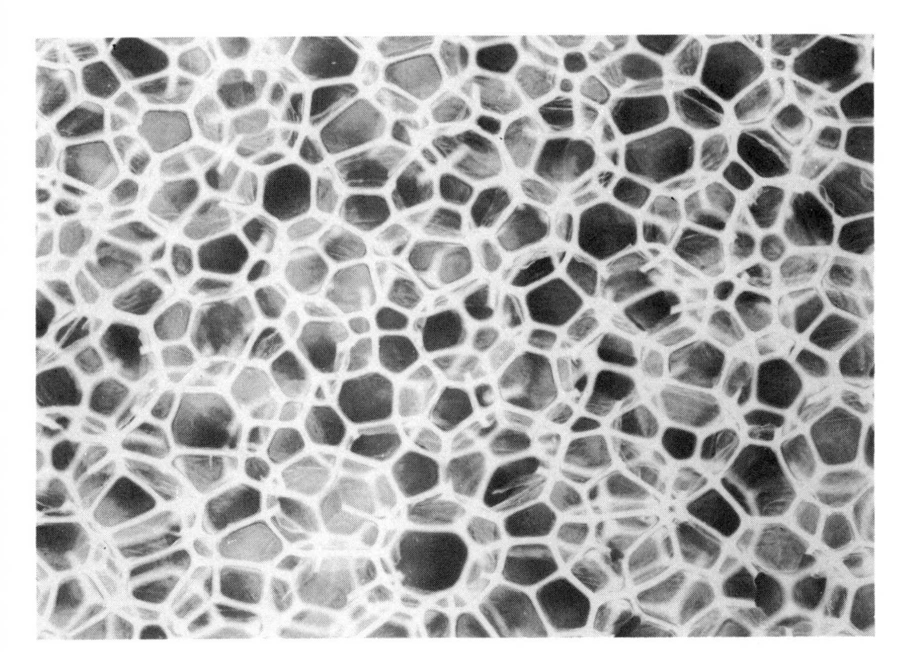

Fig. 1.10 Microphotograph of foam rubber

bubble—is unique, yet determined by relatively simple physical laws. But it would clearly be a hopeless task to attempt to work out the overall properties of a slab of such foam—its elasticity, for example—by painstakingly considering one bubble at a time and calculating its effects on each of its neighbours. An alternative to a *reductionist* approach of this sort is to step back a little from the system, until the differences between each of its elements become blurred, and then try to derive some description of their average behaviour *en masse* without the necessity of considering each one in detail. A well known example of this kind of *holistic* approach is the study of the physical properties of gases. Gases exert a pressure because their molecules are constantly rushing about and bumping against the walls of their containers, generating an average force that depends on their velocity or temperature, and on how many of them there are. If one knew the exact position and velocity of every single molecule of such a gas, then in principle one could compute each of their individual trajectories and hence predict their collisions with the walls and with each other: one would then have a perfect understanding of the gas. But this is clearly out of the question. A more tractable approach is to turn the large numbers involved into an actual advantage, by considering only the 'lumped' properties of the molecules, their average behaviour as a whole, and then derive statistical descriptions of the way in which pressure depends on volume and temperature, through the methods of *statistical mechanics*.

One might hope that by analogy one might be able to develop a sort of statistical mechanics for neurones. Unfortunately, however, the one thing we cannot do with neurones is to take averages: it is precisely the *differences* between them—the patterns of their activity—that are significant. It is as unhelpful to talk about the average behaviour of a sheet of cells in the retina or cerebral cortex as it would be to talk about an average page of this book, or an average Beethoven piano sonata. In fact we shall see again and again that at every level of the brain there are special mechanisms of *lateral inhibition* whose specific function is to exaggerate differences of activity between neighbouring neurones and thus to enhance or amplify any patterns that may exist, and such mechanisms make the whole system more uncomputable than ever. A simple example may illustrate why this is. Suppose we take a piece of card, dribble a line of ink across it, and then up-end it (Fig. 1.11). The ink starts to run down,

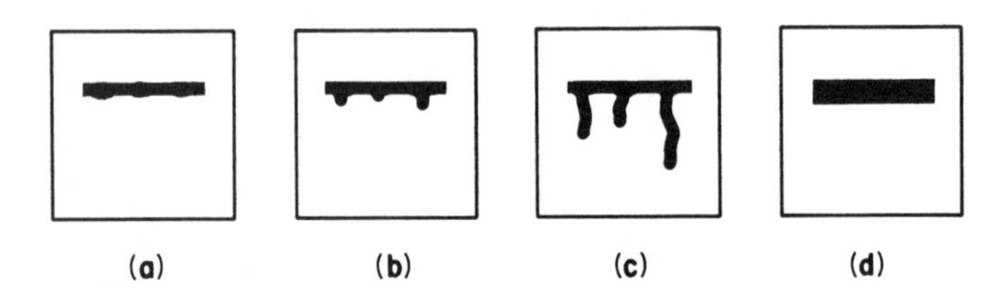

Fig. 1.11 A system whch amplifies patterns. We make a horizontal dribble of ink on a vertical card (**a**). Drops begin to form at points where there happens to be more ink (**b**), and in doing so they draw ink away from neighbouring regions, thus exaggerating the original non-uniformity (**c**). A prediction of the behaviour of the ink based only on averages (**d**) is the one result that is *never* observed!

in an irregular way that is partly a function of the amount of ink at different points along the line. But as soon as a drop starts to form at the lower edge, it begins to draw ink off from neighbouring regions, thus reinforcing itself at their expense. We end up with a series of discrete trickles, in which the pattern of 'activity' (quantity of ink at different distances) has been automatically amplified. A description in terms of average behaviour would be meaningless: on average, the ink descends uniformly along all its length, the one thing that is *never* actually observed!

Another reason why we cannot take averages is that one person's neuronal connections are different from another's, both at the level of description of individual nerve cells, but also to a certain extent at grosser levels. In a later chapter we shall see that the movements of various parts of the body are represented in a systematic way on the surface of a part of the cerebral cortex called the motor area. These 'motor maps' not only differ quite markedly from one individual to another, but even in the same individual when measured on different occasions. This is very far from being an isolated example: we shall see later that connections in many parts of the brain are to a large extent determined by experience, and that it is precisely this *plasticity* of function that enables us to learn to adapt our behaviour to circumstances. So the functioning of the brain is probably as much a matter of how it has been *programmed* as of its hardware—of the general outline of its wiring specified by genetic instructions—and these programs are necessarily as varied as the experiences of the brain's owner. Thus it is also by no means clear, even if we did manage to make some sense of the wiring diagram of A's brain, that it would throw much light on the functioning of B's.

There are thus many levels at which one may try to investigate the brain, and corresponding to them a number of distinct branches of the neurosciences have emerged. To pursue the computer analogy a little further, biophysicists investigate the properties of the individual logic gates; neuroanatomists trace the connections that link one unit to another; psychologists describe the programs in the machine and how they got there; and neuropharmacologists examine the colours of the wires. The task of neurophysiologists has generally been to try to bridge the levels at which the other disciplines perform their investigations, by trying to correlate anatomical structure with patterns of neuronal activity, and neural events with overt behaviour and sensation.

Physiological methods

There are essentially three techniques that are used to try to correlate the activity of particular neurones with particular functions: they are *recording, stimulation,* and *lesions.*

All neurones generate electrical currents when they are active, and these electrical effects may be picked up by means of *electrodes*: either gross electrodes that look at the average responses of many hundreds of neurones at once, or microelectrodes that are small enough to impale single nerve cells and record their activity in isolation from whatever else may be going on round them. Though 'micro', intracellular electrodes are not much smaller than the neurones themselves, and there is a danger of bias in making such recordings since populations of smaller cells may be missed entirely. Some workers have experimented with arrays of electrodes, which have the advantage that one may then be able to observe something of the spatial pattern of activity: trying to deduce the function of a whole nucleus by recording from a single cell is rather like reading a book by looking at just one letter on each page. Electrical recording has proved most successful at the sensory or input side of the brain, since one may then find out what a particular cell responds to by presenting it with a variety of different sensory stimuli: the difficulties of recording from the motor side are considered in Chapter 9.

Local metabolic effects may also be used to pinpoint neural activity. Two such methods have been developed recently. One is to treat an animal with labelled *deoxyglucose* (DOG), and then present it with a particular task, such as looking at a specific kind of visual stimulus. Those cells that are most active during that period take up the DOG preferentially, and their localization becomes apparent when the brain is subsequently sectioned. Another technique is to examine local changes in blood flow, which may be monitored outside the skull by using an array of radiation detectors, and labelling the blood with something like radioactive xenon. Though it cannot provide accurate localization, it has the advantage that changes in the pattern of activity can be observed in conscious human subjects: some examples are described on p.288.

Experimental *stimulation* of the brain may be electrical, using gross or microelectrodes, or sometimes by local microinjection of pharmacological agents or other substances. Stimulation is a natural way to try to investigate the motor system, though for reasons discussed more fully in Chapter 9 it has not often proved a very helpful technique. Stimulation of single cells is usually inadequate to achieve any overt response at all, while simultaneous stimulation of large numbers of them with gross electrodes and heavy currents has proved in general to be too 'unphysiological' to give meaningful results. However, stimulation at one site while recording from another is often a good way to establish the existence of functional connecting pathways.

Finally *lesions*: apart from their use in demonstrating anatomical pathways through degeneration, they may also be used to try to associate particular functions with particular regions of the brain. If destruction of a specific area X of the brain results in the loss of some function Y, then it might seem reasonable to conclude that Y is localized in X. The trouble is, however, that there are many other ways in which such a result might be explained. The function Y might in fact be localized somewhere else altogether, but either with connecting fibres that merely happen to run through the region X, or perhaps requiring some kind of tonic permissive influence from X; or a lesion at X may simply interfere with the blood supply to the true area of localization of Y (Fig. 1.12). There is an apocryphal story of the man who demonstrated his findings about the auditory system of the grasshopper at a meeting of the

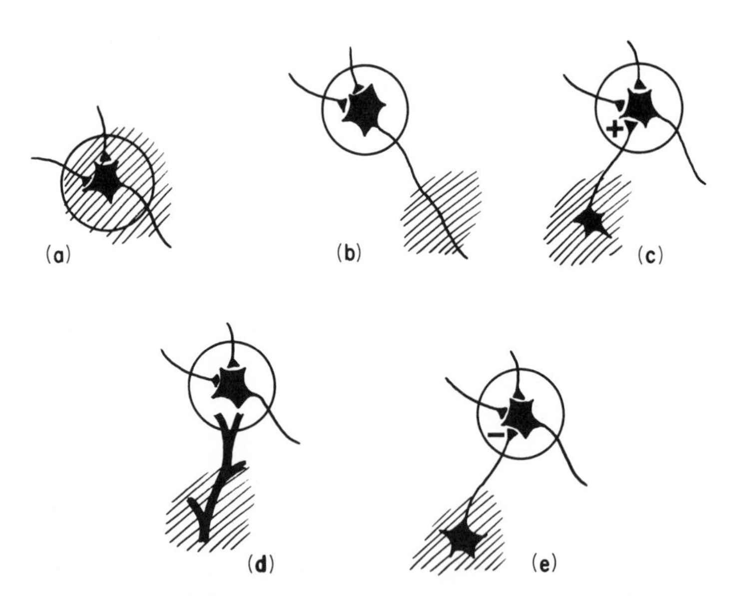

Fig. 1.12 If a lesion causes the loss of a particular function, it may indeed be because it has directly interfered with the area responsible for that function (**a**); but equally, it may be that it has merely interrupted fibres of passage (**b**), or has abolished a tonic 'permissive' input (**c**), or has interfered with blood supply (**d**). A lesion that abolishes a source of tonic inhibition may give rise to 'release'—the appearance of new and abnormal reactions (**e**).

Royal Society. 'Gentlemen' he said, 'I have here a grasshopper that I have trained to jump whenever I clap my hands.' He claps, and the grasshopper jumps. 'Now, gentlemen, I shall cut off its legs—so—and you will now observe that when I clap my hands it no longer jumps. It is therefore clear that grasshoppers hear with their legs.' Perhaps the logical flaw is obvious here—the fact that grasshoppers *do* hear with their legs is beside the point! Yet hardly less glaringly flawed deductions have quite often found their way into the literature. Paradoxically, one is on safer ground if one finds that a lesion in area X has absolutely *no* effect whatever on function Y: for then one can be fairly sure (unless the function is localized at two independent sites that somehow work in parallel) that Y is *not* localized in X.

Sometimes the result of a lesion is not the abolition of a function, but the sudden appearance of new behaviour not previously seen, a phenomenon called *release* (Fig. 1.12). Again, it is easy to jump to erroneous conclusions. If you remove one of the circuit boards from your hi-fi and it starts to make a whistling noise, is it fair to conclude that the function of the board you removed was to inhibit whistling? Yet it is still commonly said that because the effect of lesions in certain parts of the corpus striatum is a jerkiness in moving, and tremor at rest, the function of those areas is to smooth out movements and inhibit tremor!

Part of the problem is undoubtedly that the whole notion of localization of function in centres is too crude and simple-minded, a hangover from the days of phrenology, with its picture of the brain divided up into a number of discrete little compartments with highly specialized functions (Fig. 1.13). Some classical experiments performed on rats by the psychologist Karl Lashley cast doubt on such a view of localization. He made lesions of different sizes in the cerebral cortex, avoiding the regions of it that are specialized as primary areas of input and output, and found that the defect of performance in a task like learning to run a maze depended more or less on the quantity of cortex removed, and rather little on the region where it was taken from. Large areas of the brain seem in fact not to be rigidly committed to specialized tasks, but may change their function as the result of experience, or of damage to neighbouring areas. The situation is reminiscent of that in a telephone exchange, where automatic switches called selectors have the job of connecting one subscriber to another in response to signals from his telephone dial. But there is not one set of selectors for each subscriber: if there were, the total number of selectors required would become astronomical. Instead, there is a common pool of uncommitted selectors: when you lift your receiver, a device called a line-finder hunts through the available bank of equipment until it finds one free, and then gives you the dialling tone. This means that in the course of a day, any one selector may be used by a large number of subscribers. It also means that if we were to drop hand grenades into the exchange and blow up a certain proportion of the selectors, the service would get worse for *all* subscribers, in that they would be more likely to find all the lines engaged. This kind of *plasticity*—the ability of one area to take over the function of another that is out of action—is often found at higher levels of the brain (as for example in recovery from stroke), and makes deductions from lesions more difficult still.

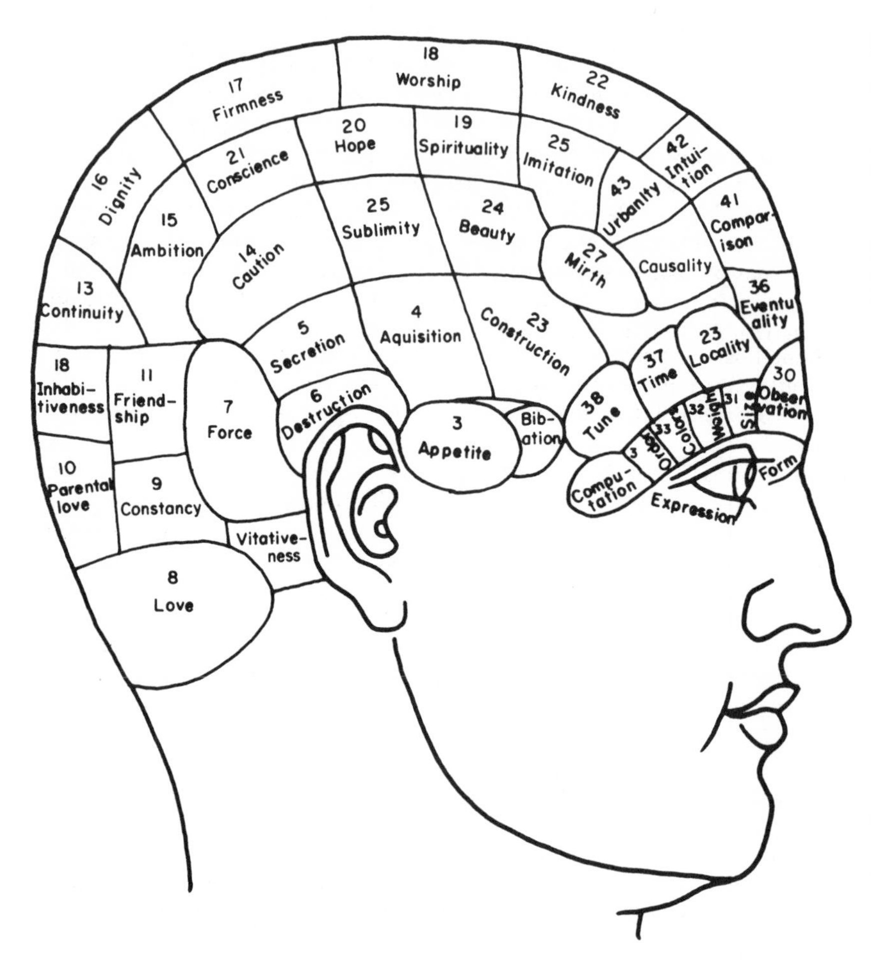

Fig. 1.13 Phrenological head: an early attempt at cerebral localization (Fowler).

Thus there are profound intellectual and technical difficulties with all the methods of investigating the brain currently in use. It is as if we tried to find out how our pocket calculator works by sticking pins in it and measuring the voltages on them, by passing large currents through it, or by noting its reactions to having bits knocked off it with a hammer. It is really astonishing that we know anything at all about the brain.

2

Nerves

Although it had been known since the time of Galen that it was the nerves that were responsible for fast communication between the body and the brain, the question of how they did it was settled only some 30 years ago; nervous conduction is now one of the best-understood processes in the whole field of physiology, and its elucidation has been a major triumph for that branch of the subject known as *biophysics*—the application of purely physical methods to biology.

That there was some link between electricity and nervous and muscular action had been sensed as far back as the end of the eighteenth century; but Galvani's celebrated observation of the twitching of frogs' legs when in contact with certain combinations of metals led to the discovery of electrical currents rather than to an understanding of nerves. This had to wait for the development of sensitive galvanometers that could register the passage of very small electrical currents; by the end of the nineteenth century it was clear not only that nerves and muscle could be activated by electrical stimulation, but conversely that their normal activity was always accompanied by changes in electrical potential. It did not follow, of course, that these electrical changes were the *cause* of neural communication, since activity of many other kinds of tissue also gave rise to electrical effects. It was not until 1939 that it was finally established that electrical currents were not only generated as a side-effect of nervous conduction, but were also *necessary* for conduction to take place at all.

If we lay one end of a nerve across a pair of electrodes, and stimulate the other end electrically (Fig. 2.1), after a short *latency* due to the time taken for excitation to travel from the stimulating to the recording electrodes we see a characteristic transient change in the voltage across the recording electrodes, the biphasic *action potential*. If now we lay a series of platinum strips across the nerve, the latency remains unchanged. But on making electrical contact between the strips, for example by immersing their ends in a bath of mercury, it is found that the latency is immediately reduced, but returns to its previous value as soon as the mercury is removed. Clearly the presence of the mercury increases the apparent conduction velocity of the nerve; this can only be because it allows currents to pass more easily from one part of the nerve's surface to another. In other words, the currents generated by an active nerve are not just an accidental by-product of transmission, like the noise from a car: they are an essential determinant of the entire process.

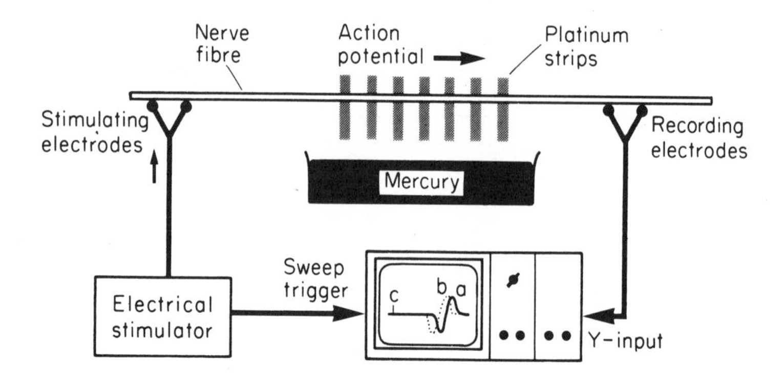

Fig. 2.1 Dependence of action potentials on electrical currents. A nerve axone is stimulated electrically (left) at the same time as a trigger signal is sent to an oscilloscope to start the sweep: (c) is the stimulus artefact that results from direct electrical conduction down the fibre. After a latent period, recording electrodes (right) pick up a diphasic action potential (a). On raising the trough of mercury, so that the platinum strips are in electrical continuity, the latent period is reduced (b). (After Hodgkin, 1939)

The flow of electrical current along nerves

Now nerve fibres, with their conductive central core of axoplasm surrounded by an insulated membrane often reinforced with extra non-conductive layers of myelin, are clearly very like ordinary insulated wires. Could action potentials simply be transmitted by passive conduction, in the same way that signals pass along a telephone wire? For most nerves, the answer is a clear no: action potentials show a number of properties that do not fit such a simple model, and in fact it turns out that they are actually conducted very much better than they would if the nerve fibres were merely acting like simple electric wires. To get an idea of how bad a passive conductor a nerve fibre is, we need to consider what are called its *cable properties*. These depend both on how good its insulation is—the resistance of the axon membrane—and also on how much resistance is offered to currents flowing longitudinally through the axoplasm. If we consider a unit length of axone (Fig. 2.2), we can call the associated transverse resistance of the membrane R_M, and the logitudinal resistance of the axoplasm R_L: the whole axon can be thought of as simply consisting of a large number of these units joined end-to-end. If we now assume that the external medium offers only a negligible resistance to current flow (the justification for this assumption will become apparent later), then we can treat all the outer ends of the individual R_Ms as if they were short-circuited together, producing the ladder-like network of resistors shown in Fig. 2.2 that is called the *equivalent circuit* of the nerve fibre.

Now imagine a potential V_0 applied to the end of this ladder. In the first compartment, the current it generates has a choice: it can either flow out through R_M, or continue through R_L to enter the next compartment. What actually happens is that some fixed fraction of the total current takes the former route and the rest takes the latter: the bigger the resistance of the

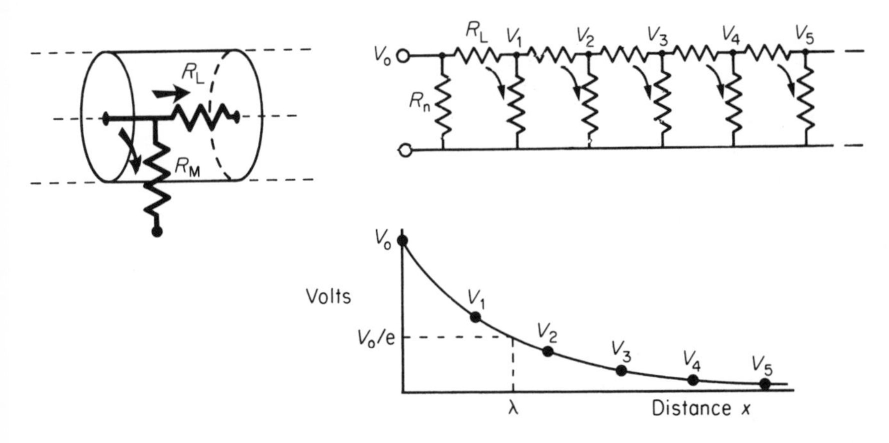

Fig. 2.2 Passive spread of current along an axon. Left, each unit length of the fibre can be thought of as having a longitudinal resistance R_L and a transverse or membrane resistance R_M. Thus the whole axon may be represented by the ladder-like equivalent circuit shown at top right. If a voltage V_o is applied at one end, the voltage measured at different distances x along the axon will fall off exponentially as shown at the bottom, because of leakage through the membrane. The space constant λ is the distance for which the voltage has fallen by a factor e

membrane, R_M , and the smaller that of the axoplasm, R_L, the greater will be the tendency of the current to carry straight on rather than leak away through the membrane. Unless the insulation is perfect, or the axoplasm infinitely conductive (in either case R_L/R_M is zero), the current entering the second cell will be smaller than that entering the first by some fixed ratio. In the same way, the current entering the third cell will be smaller than that entering the second, and so on all the way down the chain; the result is that the *potential* seen at each compartment will fall in a fixed ratio as one goes from compartment to compartment down the line, resulting in an exponential decline in voltage as a function of distance along the axon, at a rate that will depend on the ratio R_M/R_L (Fig. 2.2). Thus V is given by $V_0 e^{-x/\lambda}$, where e is the well known constant whose value is about 2.718, and λ is a parameter called the *space constant* that describes how quickly the voltage declines as a function of the distance x. λ is in fact the distance you have to go before the voltage has dropped to 1/e (about 37 per cent) of its original value V_0. It turns out in fact that λ is actually equal to $\sqrt{R_M/R_L}$; as would be expected, the more leaky the axon, the smaller R_M is, and the shorter the space constant.

What does all this mean in practical terms? Consider an ordinary myelinated frog nerve fibre some 14 μm in diameter, and assume for the moment that the myelin is uninterrupted along its length, without nodes of Ranvier. With 1 mm as our unit of length, it turns out that although the axoplasm is intrinsically a vastly better conductor of electricity than myelin—their specific resistances being of the order of 100 ohm cm and 600 megohm cm respectively—because the cross-sectional area of the axoplasm is so small, the ratio of R_M to R_L is not very great: R_M comes out as about 250 megohm and R_L about 14 megohm, giving a space constant of some 4 mm or so. In other words, a potential generated at one end of such an axon will have dropped to less than half at a distance of 4 mm, to about a tenth of its original value after a centimetre, and by two centimetres will only be some 1 per cent of the original stimulus, and probably undetectable in the general background electrical noise. In other words, axons are quite incapable of acting as reliable passive conductors of electricity over distances of more than a centimetre or two at most.

This is not really because nerve fibres are made of unsuitable materials—we have already seen that axoplasm is more than five million times better at conducting electricity than myelin—but rather because of their small size. Suppose we were to increase the diameter of our axon by a factor D; R_M would then be reduced by the same factor (since the area of membrane per unit length would be increased by D), while R_L would be reduced not by a factor D but by D^2, since the longitudinal resistance depends on the cross-sectional *area* (Fig. 2.3). Consequently R_M/R_L would be multiplied by D, and

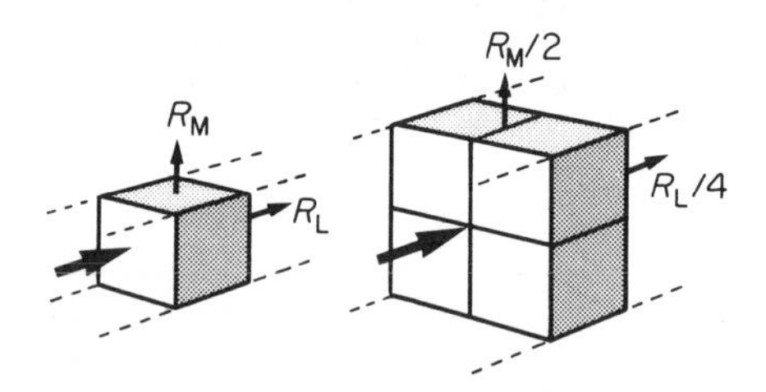

Fig. 2.3 Doubling the diameter of an axon reduces R_M by a factor of 2, but R_L by a factor of 4. (The fact that real axons are not usually square does not affect this result!)

the space constant would be increased by a factor $\sqrt{D}$. Now the longest nerves in the human body are about a metre in length; so to conduct reliably over this distance we would need to increase λ from 4 to about 400 mm; because of the square root relationship, this would mean having a fibre of some 10 000 times its previous diameter, in other words 140 mm instead of 14μm! (This analysis is not *quite* fair because we have assumed a constant thickness of myelin: but even if we allow this thickness to increase in proportion with the axone itself, the fibre will still need to be a hundred times bigger. Bearing in mind the fact that many important fibre tracts in the body contain millions of fibres, one must still conclude that passive conduction is not a practical possibility over long distances.)

The cyclical regeneration of action potentials

In any case, nerve action potentials show a number of properties that demonstrate quite clearly that they are not passive, but actively *regenerated* as they pass along the fibres: amplified in such a way as to overcome the enormous losses that they experience through membrane leakage. If one records the action potential from a single electrical stimulus at different points along a nerve fibre, one finds that its amplitude does not in fact decrease at all as a function of distance, but stays at a constant value. Even more strikingly, the size of this action potential is not even a function of the size or nature of the stimulus that initiated it in the first place; so long as the strength of the stimulus is above a certain *threshold* value (below which no action potential is seen at all), neither the amplitude nor the shape or speed of the action potential is in any way influenced by the nature of the original stimulus, a property known as the *all-or-nothing* law. These two features, the all-or-nothing law and the existence of a threshold, are never shown by voltages transmitted through passive conductors. A nerve is very like a burning cigarette: once lit, the temperature and rate of advance of the burning region is not a function of the temperature of the flame which originally ignited it, so long as this was sufficient to light it at all. Here the combustion is continually regenerative: the heat of the burning tip raises the temperature of the next region to the point where it too catches fire, and so on all along its length. In other words, there is a continuous cyclic process in which heat triggers combustion and combustion generates heat, this heat coming of course from the stored chemical energy of the tobacco.

It turns out that this is a surprisingly close analogy to the mechanism of propagation of the action potential. What happens is that the original stimulus to the fibre causes local currents to flow passively through the membrane, causing a spread of potential rather as in Fig. 2.2. This voltage is in some way sensed by neighbouring regions of the fibre and triggers a mechanism in the membrane that generates a voltage many times larger (thus introducing an amplification of the original signal) which in turn sets up local currents that cause a potential change still further down the axon . . . and so on, until the potential change has been transmitted from the point of stimulation to the end of the axon (Fig. 2.4). This whole cyclical process is known as the *local circuit* mechanism of action potential propagation. Each cycle consists of three distinct stages: first, there is the mechanism by which a potential at one point results in a passive flow of current and thus in depolarization of regions further down the axon; secondly, the mechanism by which this depolarization

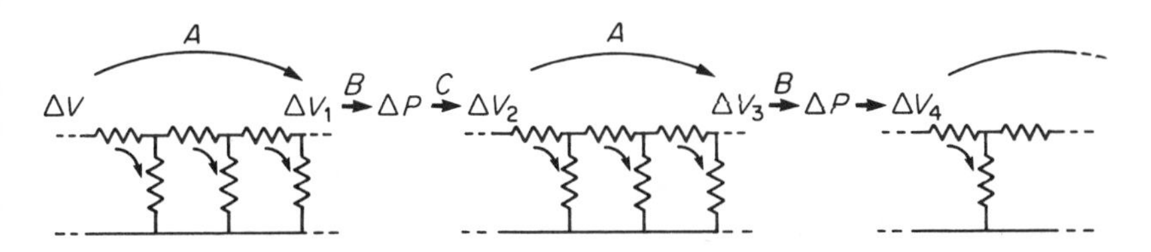

Fig. 2.4 The three components of action potential propagation. A depolarization ΔV at one point on the fibre results, through local current flow (*A*) in a smaller depolarization ΔV_1 some way down the axon. This triggers off (*B*) a permeability change ΔP in the membrane, which in turn (*C*) produces a voltage ΔV_2 which is larger than ΔV_1. The whole sequence is repeated indefinitely (ΔV_3, ΔV_4, etc). The processes *A* and *C* are common to all cells; but *B*, the conversion of changes in voltage to change in permeability, is found only in nerve and muscle

triggers off some change in the membrane; and thirdly, the mechanism by which this change produces a new potential that is much larger than what triggered it off. The first of these processes is essentially what has already been described, and can be readily understood in physical terms by means of equivalent circuits like that of Fig. 2.2; the second two obviously require identification of this mysterious change in the membrane that is supposed to result in amplification of the voltage that triggers it. It turns out that this change consists in a change in the *permeability* of the membrane to certain ions.

Now there are physical mechanisms common to all cells by which changes in ionic permeabilities give rise to changes in potential; what is unique about nerve and muscle cells in that they *also* possess special mechanisms by which such changes are in turn triggered off by changes in potential, thus completing the cycle of three links by which the action potential is propagated over the membrane surface (Fig. 2.4). Thus to understand how nerves work we need to be able to answer two questions: *How do ionic permeabilities affect membrane potential?* and *How do membrane potentials affect ionic permeabilities?*

The dependence of membrane potential on ionic permeabilities

Imagine that we have a system of two compartments, A and B (Fig. 2.5), and that initially A contains a strong solution of KCl, and B a weak one; and suppose that the membrane separating them, initially impermeable, suddenly becomes permeable to potassium ions. Clearly there will now be a tendency for K^+ to diffuse through the membrane down the concentration gradient between A and B. Since the ions carry a positive charge, compartment B will become more and more positive with respect to A as they migrate in this way, setting up an electrical gradient that will tend to oppose the entry of further ions from A. Eventually there will come a point where the *concentration gradient* from A to B will be exactly equal and opposite to the *electrical gradient* from B to A, and the system will be in equilibrium, since there will be no net flow of ions across the membrane. The resultant electrical potential between A and B is then called the *equilibrium potential* for potassium, E_K. To work out how big this

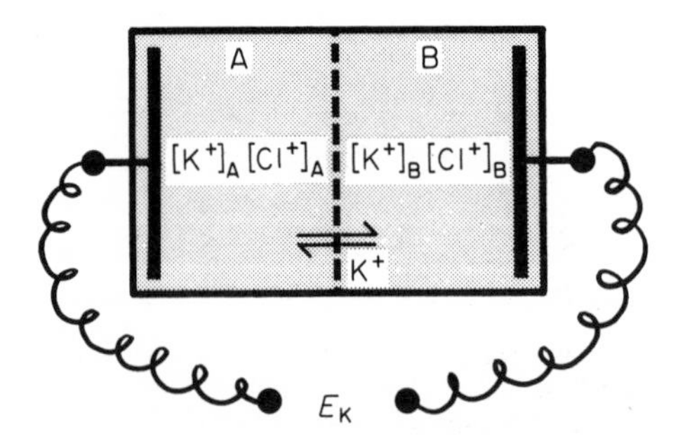

Fig. 2.5 Two compartments each containing potassium chloride, at different concentrations, separated by a barrier permeable only to potassium ions; at equilibrium there is a potential difference between them of E_K volts

potential will be, consider the energy involved in moving one potassium ion from A to B. The work done in moving it against the electrical gradient will be given by its charge e multiplied by the potential difference E_K; since the system is in equilibrium, this work must be exactly equal to the energy gained in moving down the concentration gradient, which can be shown to be

$$\frac{RT}{N} \ln \frac{[K]_A}{[K]_B}$$

where T is the absolute temperature, R the gas constant and N is Avogadro's number. So we can write:

$$eE_K = \frac{RT}{N} \ln \frac{[K]_A}{[K]_B}$$

$$\text{or} \quad E_K = \frac{RT}{F} \ln \frac{[K]_A}{[K]_B} \quad \simeq 58 \log_{10} \frac{[K]_A}{[K]_B} \text{ mV at } 20°C$$

where F is Faraday's constant, equal to Ne. This relationship (the *Nernst Equation*) is true for any ion in equilibrium across a membrane to which it is freely (and solely) permeable, with the proviso that if the charge on the ion is not + 1 (as in the case of Cl^- or Ca^{2+}) we need to include this ionic charge z as well:

$$E_x = \frac{RT}{zF} \ln \frac{[X_A]}{[X_B]}$$

Only a tiny percentage of the ions need to cross the membrane to set up such an equilibrium, so that the concentrations of the ions on each side remain effectively unchanged, and the equilibrium potential is set up virtually instantaneously after a sudden permeability change of this kind. Thus from the purely electrical point of view, suddenly making the membrane permeable to potassium—perhaps by opening up little channels in it that allow K^+ through but nothing else—is rather like connecting a battery of voltage E_K across our equivalent circuit in Fig. 2.2.

Now suppose we extend the model by imagining a membrane with both potassium and sodium channels, which can be independently opened and closed, and that A and B now contain solutions of NaCl and KCl in different proportions on each side (Fig. 2.6). The potential will then be given by $E_K = 58 \log_{10} \frac{[K]_A}{[K]_B}$ when the potassium channels are open, and $E_{Na} = 58 \log_{10} \frac{[Na]_A}{[Na]_B}$ when the sodium ones are open. So by opening one or other set of channels we can switch the potential from one value to another. What about if *both* are open? The answer is that the voltage will depend on the relative permeabilities of the two ions, the ratio of the total numbers of each channel that are open. The more freely K^+ is able to move through the membrane, the nearer the potential will approach E_K; and the greater the permeability to Na^+, the closer it will approach E_{Na}. The final, compromise, potential is in fact given by an

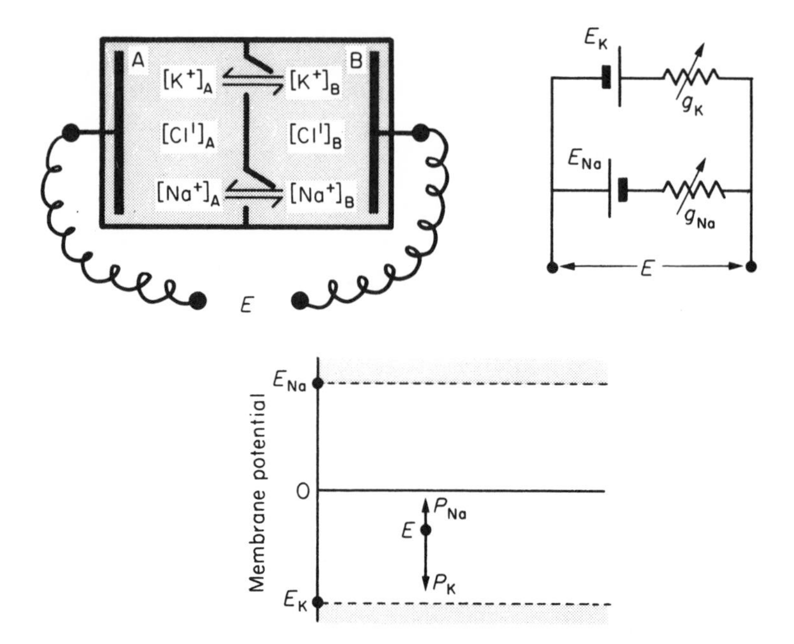

Fig. 2.6 Above left: equilibrium potential, E, between two compartments containing sodium and potassium chloride and separated by a barrier having a permeability P_K to potassium and P_{Na} to sodium; right, its equivalent circuit: g_{Na} and g_K are the electrical conductivities determined by P_K and P_{Na}. Below, the membrane potential E can be thought of graphically as being an equilibrium between the pull of P_{Na} towards E_{Na} and of P_K towards E_K

expression called the *constant field equation*, a generalization of the Nernst equation to include cases where more than one ion is present:

$$E = \frac{RT}{F} \ln \frac{P_K[K]_A + P_{Na}[Na]_A}{P_K[K]_B + P_{Na}[Na]_B}$$

Here P is a measure of the permeability of the membrane to a particular ion. It can be seen at once that the Nernst equation is simply a special case of the constant field equation, for if P_K is very large in comparison to P_{Na}, so that the membrane is in effect only permeable to potassium, E simply becomes E_K. The membrane potential will in fact always lie somewhere between the two extremes of E_K and E_{Na}; since the concentrations of sodium and potassium on each side are effectively fixed—the net flow through the channels is in general exceedingly small—this voltage is only a function of the two permeabilities, in fact of their ratio P_K/P_{Na}. In other words, *changes in permeability cause changes in potential.*

Now nerve cells are ideally suited for converting small permeability changes into large potential changes, since the distribution of Na^+ and K^+ across their membranes is a very lopsided one (Table 2.1). Like all cells in the body, they have a powerful sodium pump in their membranes that transfers Na^+ from inside to outside, exchanging them with K^+ partly by direct reciprocal

Table 2.1 Electrolyte concentrations (in mmol/litre) inside and outside two types of active cell. (Data from Katz, 1966)

		Na^+	K^+	Cl^-
Frog muscle (E = −85 mV)	inside	9.2	140	3–4
	outside	120	2.5	120
Squid axon (E = −60 mV)	inside	50	400	40–100
	outside	460	10	540

transport and partly through electrical coupling. In the case of frog muscle fibres, which happen to have been particularly thoroughly investigated, the two equilibrium potentials may be calculated as about − 100 mV inside for potassium, and about + 60 mV for sodium. If we actually put a microelectrode into the muscle fibre and record its *resting potential*, it is found to be about − 85 mV, implying that in the resting state it is much more permeable to potassium than to sodium. We can also try the experiment of bathing the fibre in solutions with different concentrations of potassium: if the membrane were *only* permeable to K^+, it ought to obey the Nernst equation, and we should find that its resting potential is simply proportional to the logarithm of the external potassium concentration. At high external potassium concentrations this turns out to be true, but at lower values the potential is always smaller than what would be expected, because of the added contribution of the sodium ions. The best fit to the experimental points is obtained by using the constant field equation, with a resting sodium permeability about 1 per cent of that of the potassium ions (Fig. 2.7). Similar

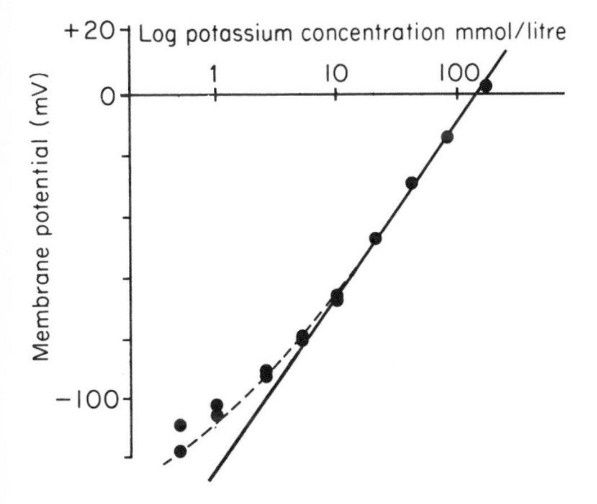

Fig. 2.7 Measurement of membrane potential of frog muscle fibres(●) in response to different external potassium concentrations. The solid line shows the expected relation from the Nernst equation if the membrane were only permeabie to K^+, whereas the dashed line shows the expectation if it is about 1 per cent as permeable to Na^+ as it is to K^+. (After Hodgkin and Horowicz, 1959)

experiments have been done on the giant axons of squids, with the added advantage that because they are so big—often as much as a millimetre across—it is possible to squeeze their axoplasm out with a kind of miniature garden roller, and replace it with fluids of different composition, thus altering the potassium concentration inside as well as outside.

Now if we put a microelectrode inside a muscle fibre or squid axon, and stimulate it to get an action potential, we find that the potential of the inside relative to the outside suddenly reverses from being (in the squid) about − 50 mV in the resting state to a peak of some + 40 mV, and then rapidly declines back to the resting potential again (Fig. 2.8). This is the *monophasic*

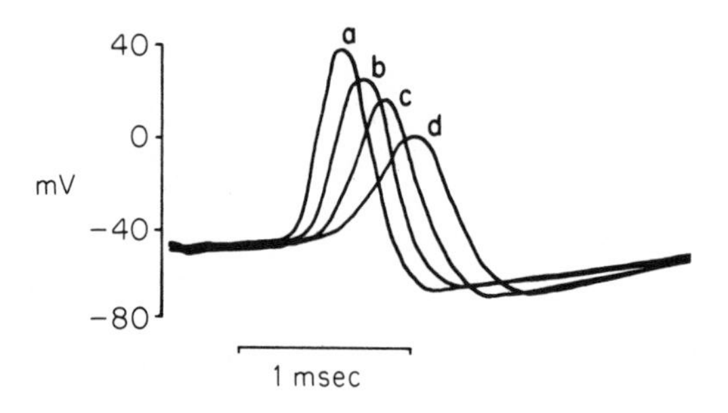

Fig. 2.8 Action potentials in a squid axon, showing the effect of different external sodium concentrations. (**a**): sea water; (**b**): 71 per cent sea water; (**c**): 50 per cent; (**d**): 33 per cent; the sea water was diluted with isotonic dextrose. (After Hodgkin and Katz, 1949)

action potential: the reason for its shape being different from the diphasic action potential recorded with an external electrode (Fig. 2.1) will be explained later. The size of this reversal during the action potential strongly suggests that a sudden increase in Na permeability is occurring, that pulls the membrane potential temporarily towards E_{Na}. By experimenting with various concentrations of sodium ions inside and outside the axon it is possible to demonstrate directly that the peak of the action potential is indeed dependent on E_{Na}: if the ratio of sodium concentration outside and inside the fibre is reduced, this peak declines in amplitude (Fig. 2.8), and eventually the action potential is abolished altogether. To anticipate a little, it turns out in fact that the action potential is produced by a characteristic sequence of changes in ionic permeabilities: there is first a transient increase in sodium permeability, and then—after a short delay—of potassium permeability (Fig. 2.9), and the shape of the action potential can be calculated quite exactly from the time-course of these permeability changes by using the constant field equation. During the initial phase, when P_{Na} is rising but P_K is still at its resting level, the potential moves rapidly past zero towards E_{Na}: it never gets there, however, because P_K has meanwhile started to rise, and P_{Na} to fall, and the potential starts to drop back again to the resting level. In many cases it actually undershoots the resting potential, because there is a period at the end of the whole cycle in which P_K is still elevated, whereas P_{Na} is back to normal: this pulls the membrane even further towards E_K than it is in the resting state.

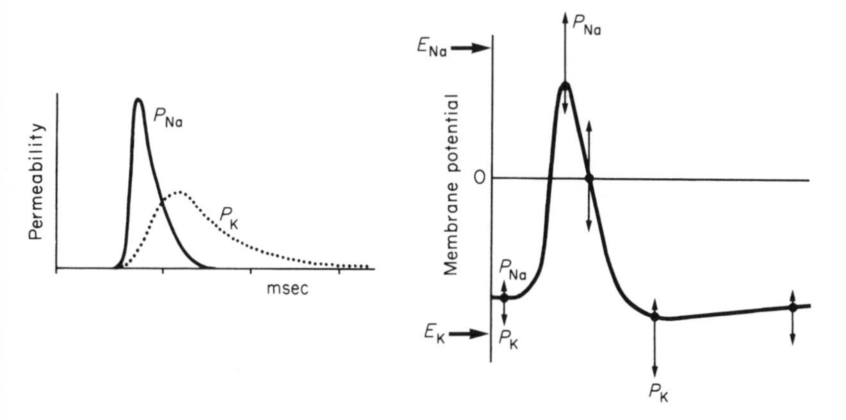

Fig. 2.9 Above, changes in potassium and sodium permeabilities associated with the action potential. Below, how these changes result in the form of the action potential itself. The arrows above and below the line indicate roughly by their length the relative size of P_{Na} and P_K, pulling the potential respectively toward E_{Na} and E_K

Thus the voltage changes that occur during the action potential can be explained entirely in terms of the corresponding changes in permeability; so the only question that remains is how these changes in permeability are themselves created.

The dependence of ionic permeability on membrane potential

One might think that it would be a straightforward matter to determine the way in which ionic permeabilities vary with potential: all one needs to do is to put a microelectrode into a fibre, pass a current through it in order to set up a particular voltage, and then see what the resultant potassium and sodium permeabilities are. In reality, things are not so simple. In the first place, the movement of ions that results from a change in permeability is exceedingly small, and it is not at all easy to disentangle the effects of sodium and potassium ions from one another, if both permeabilities alter at once. More fundamentally, any changes in permeability that may occur in response to a potential change that we impose will themselves tend to alter that same potential. Thus the first experimental requirement is to devise some way of holding the membrane at a particular potential despite whatever permeability changes may be taking place: and this is what the *voltage clamp* technique provides. In essence, this consists of a feedback loop in which the actual membrane potential at every moment is measured with an intracellular electrode and compared with the value at which we want to clamp it, and any difference between the two is sensed and used automatically to increase or decrease a current passed into the fibre through another electrode in such a way as to bring the potential back to where it should be (Fig. 2.10). Thus if for example Na permeability increases in response to a depolarizing current, thus

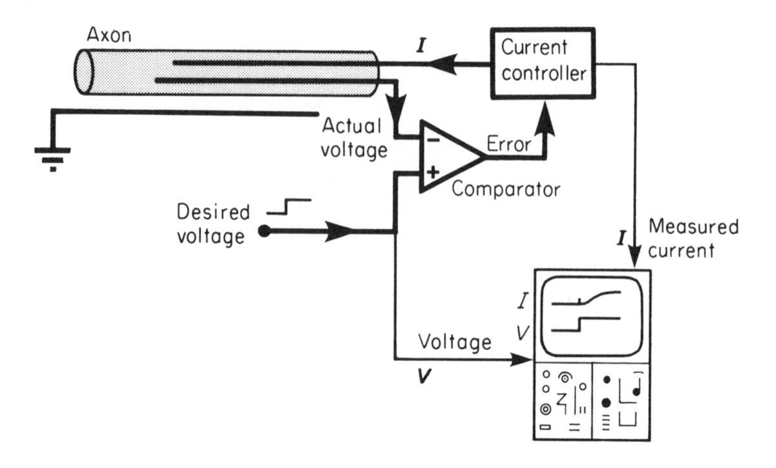

Fig. 2.10 The principle of the voltage clamp. Two electrodes are inserted in the squid axon; the voltage measured by one of them is compared with the 'desired voltage' V, and any difference between the two (error) is used to alter, automatically, the current I passed in to the axon through the second electrode. V and I are simultaneously displayed on an oscilloscope; in this case the current in response to a step change in V is shown (somewhat simplified and schematic)

reducing the membrane potential, the resultant mismatch between the actual value and the desired value of the potential causes the control circuitry to alter the stimulating current until the two are again equal. By designing the control circuits properly, it is possible to make the membrane potential follow the desired value specified by the experimenter as accurately and rapidly as required. Now the current I needed to clamp the membrane at a particular value V at any moment is simply equal to the sum of the currents I_{Na} and I_K carried by the movements of sodium and potassium ions respectively, though their contributions will be of opposite sign, since the sodium ions enter the fibre, whereas the potassium ions leave. For any particular voltage, these two currents will be proportional to the conductivities g_{Na} and g_K of the two ions. Thus if we measure the time-course of the current in response to a step-change in voltage, we obtain a curve that is proportional to the combined conductivities of sodium and potassium, though we cannot immediately disentangle one from the other. Such a curve is shown in (**b**), Fig. 2.11. After a brief spike of current that is due to charging up the membrane capacitance—to be discussed later—it can be seen that there is first an inward flow of current that rises to a peak and then reverses direction to become an outward flow that lasts as long as the voltage is held: on restoring the original resting potential, this outward current declines back to zero over the course of a few milliseconds. The direction of these two phases of current flow suggests that the first, inward, phase is due to Na^+, and the second, outward, one to K^+. To separate these two components it is necessary to devise some way of preventing one or other of them from taking place. One way to do this is to replace most of the sodium ions in the external medium with some other cation

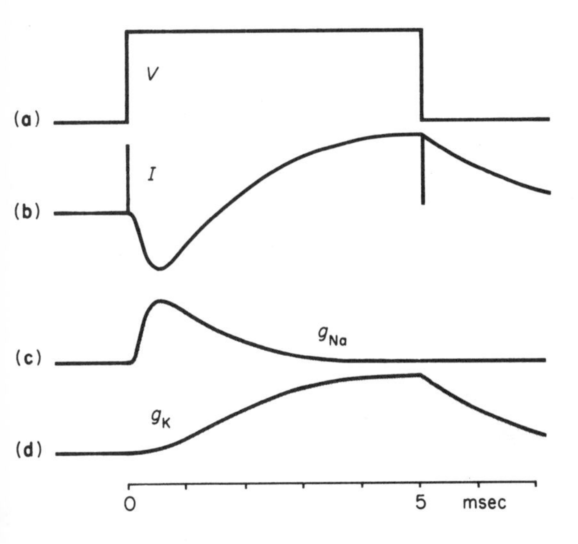

Fig. 2.11 Time course of current I in a voltage clamp experiment on a squid axon (**b**) in response to a 56 mV depolarization V (**a**). Below, (**c**) and (**d**) show the derived time-courses of the changes in sodium and potassium conductivities that give rise to the observed current I. (Simplified, after Hodgkin and Huxley, 1952*a*; Hodgkin, 1958)

such as choline that is too large to get through the sodium channels. Under these circumstances the initial phase disappears entirely, leaving only the outward component; by subtracting this from the original record, one can work out what the time-course of both sodium and potassium permeability must have been (Fig. 2.11). These results can be confirmed by other methods, such as blocking the ionic channels with pharmacological agents such as tetrodotoxin and tetraethylammonium that specifically affect sodium and potassium permeability respectively.

Thus the effect of a sudden step of depolarization is to produce a transient increase in g_{Na}, followed by a delayed but sustained increase of g_K. By systematically measuring these permeability changes in response to steps of different size and starting from different initial voltages, Hodgkin and Huxley (1952*a, b*) were able to derive equations expressing g_K and g_{Na} in terms of membrane potential, that summarized their data and made it possible to predict, in general, how the permeabilities would vary in response to any given pattern of depolarization of the membrane. For once one knows how a system behaves in response to a small step input, then by breaking up any given voltage pattern into a series of small steps and adding the results together one can calculate the response to the whole thing. In particular, starting with the time-course of the intracellular action potential of the squid axon, one can work out in this way what permeability changes would result from it, ending up with the curves already shown in Fig. 2.9. We have now come full circle, for these permeability changes are of course precisely those that result, through application of the constant field equation, in the original action potential that we started with: we now have a complete description of the way in which the action potential is able to regenerate itself. More precisely, the action potential represents the solution of the set of differential equations that embody the results of the voltage clamp experiment, the electrical properties of the membrane, and the constant field equation: the fact that this solution is so nearly identical to the shape of the actual action potential (Fig. 2.12) testifies to the completeness of Hodgkin and Huxley's description of the way in which membrane permeability depends on voltage.

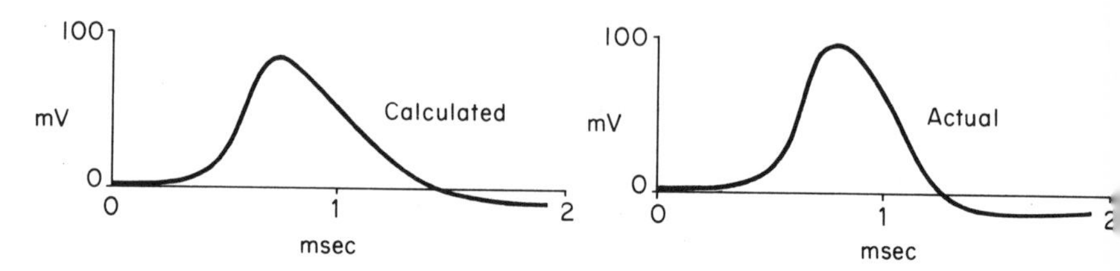

Fig. 2.12 Left, theoretical solution of the differential equations embodying the electrical properties of squid axon, the constant field equation, and the results of voltage clamp experiments. Right, actual action potential in squid axon at 18.5°C. (After Hodgkin and Huxley, 1952*b*)

Thus to summarize what is known of the electrical propagation of the action potential: a local depolarization of a section of nerve gives rise, at first, to an increase in P_{Na} that in turn causes the membrane to become more depolarized as the potential moves towards E_{Na}. Meanwhile, however, P_K starts to rise, and the sodium permeability to fall, causing the potential to start to drop back towards the resting value. This in turn tends to shut off both the sodium and potassium channels; but because of the delayed response of potassium

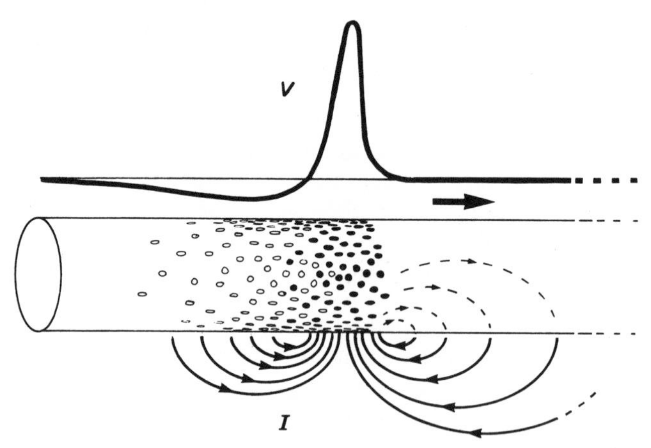

Fig. 2.13 'Snapshot' of a nerve axon with an action potential travelling from left to right. Above, the distribution of potential along its length; below, the associated flow of current. On the axon itself, the black holes indicate the approximate density of open sodium channels, the white ones of potassium channels

permeability, there is a period during which P_K is greater than in the resting state, and the membrane is hyperpolarized: and eventually the resting potential is regained. Meanwhile, the currents generated by this process have spread to neighbouring regions of the fibre, causing them to depolarize and thus initiating, at a distance, the same sequence of changes all over again. In this way the whole pattern of potential and permeability changes is propagated down the fibre (Fig. 2.13).

Conduction velocity

So far, very little has been said about the speed at which all these processes occur. We have traced the sequence of events by which one active region of nerve can trigger off a similar pattern of activity in another one at a distance from it by means of local currents: conduction velocity is simply a matter of how *far* and how *soon* these currents are sufficient to reach the fibre's threshold. Thus the factors that will influence this velocity are: how large the currents are, how high the threshold, how far they spread, and how long it takes them to depolarize the membrane.

In Fig. 2.2 an equivalent circuit of the nerve membrane was introduced, in order to explain the spread of current from one part of the fibre to another. What was omitted from this circuit was that nerve fibres show not only resistance, but also another passive electrical property, *capacitance*: the ability to store charge. Any two conductors separated by a layer of insulation act as a capacitor; the larger the opposed areas of the conductors, and the thinner the insulating layer between them, the larger the capacitance will be. In the case of nerve fibres, the membrane, being both a good insulator and extremely thin, acts as rather a good capacitor, having a capacitance, C_M, of about $1 \mu F/cm^2$, so that our equivalent circuit should really be redrawn in the form shown in Fig. 2.14. Now the effect of having capacitance in a circuit of this sort is to

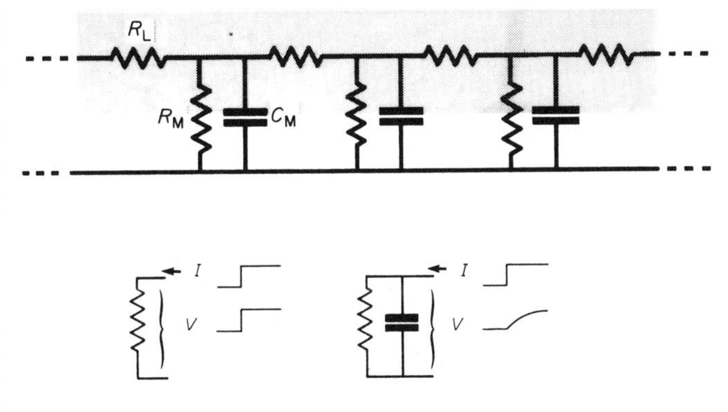

Fig. 2.14 Above, modification of the equivalent circuit of Fig. 2.2 to include membrane capacitance C_M as well as resistance R_M. Below, voltage response of a resistor (left) and resistor and capacitor in parallel (right) to an applied step of current, showing the slow, exponential, rise of voltage in the second case.

make it more sluggish in its responses. If we suddenly pass a current I through a resistor R_M on its own, the voltage across it immediately reaches the value $V = IR_M$; but with a capacitor as well it now takes *time* for the voltage to reach this value, because part of the current must be used to charge up the capacitor to the new level. What is observed is that on applying a step of current of this kind, the voltage rises only slowly to its final value of IR_M, with a time-course that is exponential and given by $V = IR_M (1 - e^{-t/\tau})$, where τ is the *time constant* of the circuit (the time taken for the discrepancy ($IR_M - V$) to fall by a factor of e), equal in this case to $R_M C_M$. For many nerve fibres, this time constant is of the order of a few milliseconds, setting a limit on the rapidity with which the membrane can generate voltages in response to local currents.

The question of how *far* the local currents spread was considered earlier in this chapter. We saw that a voltage generated at a particular point on the membrane declines exponentially as a function of distance, with a space constant λ. The space constant and time constant together give a measure of the speed with which an electrical disturbance is propagated passively along the axon: this speed is in fact proportional to λ/τ, which has the dimensions of a velocity; other things being equal, the velocity of conduction of action potentials ought also to be proportional to λ/τ. Now we saw earlier that λ is determined entirely by the longitudinal and transverse resistance of the axon ($\lambda = \sqrt{R_M/R_L}$), and thus varies with the square root of the fibre's diameter. The time constant $R_M C_M$, on the other hand, does not vary as a function of the diameter: if this is doubled, C_M, being proportional to surface area, is also doubled, but R_M is halved. It follows therefore that the conduction velocity of a fibre ought also to vary with the square root of the diameter, and for unmyelinated fibres this is indeed found to be the case.

However, animals rarely have unmyelinated fibres larger than about 1μm in diameter: the reason for this is that there is a far better way of increasing the conduction velocity of large fibres than simply increasing their size, and this is *myelination*. As we saw in the previous chapter, the effect of myelination is enormously to thicken the layer of insulation round the fibre (except at the nodes of Ranvier); this has the desirable consequence of greatly increasing R_M and reducing C_M. Although the time constant of the membrane is not influenced very much by myelination, because the effects on R_M and C_M largely cancel out, the space constant is greatly increased: the external local currents are forced to travel further before they can gain access to the axoplasm through the nodes. Since in myelinated fibres the active, voltage-sensitive, sodium and potassium channels are virtually confined to the nodes, the action potential in effect jumps from node to node down the fibre, giving rise to the term *saltatory* conduction. The physics of saltatory conduction is slightly different from that in unmyelinated nerve, and one may calculate that its velocity ought to be proportional directly to fibre diameter rather than to its square root. A consequence of this is that if we plot the velocity–diameter relationship for both types of fibre on the same graph (Fig. 2.15), the two curves cross over, at a diameter near 1μm. Consequently, there is no point in having myelinated fibres smaller than 1μm in diameter, or unmyelinated ones larger than this.

The other factors that govern conduction velocity are those concerned with

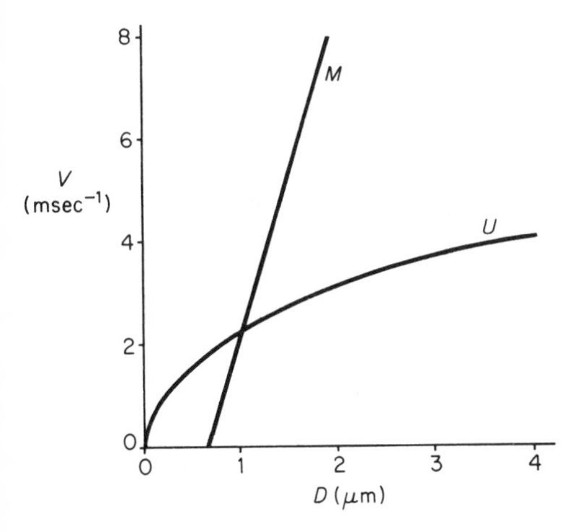

Fig. 2.15 Theoretical dependence of conduction velocity V on axon diameter D, for unmyelinated (U) and myelinated (M) axons. M is extrapolated from observations, U is scaled to fit observations on fast C-fibres. (After Rushton, 1951)

the mechanisms that generate the local currents in the first place. Other things being equal, if the currents generated by the sodium channels are increased in size, or set up more rapidly in response to voltage changes, then the local currents will cause the membrane to reach threshold either further away or sooner than would otherwise be the case. Thus *temperature* has a marked effect on conduction velocity, not because it alters the speed of local currents, but because it speeds up the process of amplification by which potential differences cause the ionic channels to open and shut: as these processes are high-order ones, with a marked temperature sensitivity, conduction velocity itself is sharply reduced at low temperatures. The relationship to temperature is an irregular one, because high temperatures appear to affect the potassium mechanism more than the sodium one, so that the action potential actually gets smaller with increasing temperature; in the squid, conduction ceases altogether if the temperature exceeds some 38°C. The size of the local currents also depends on the ionic concentrations inside and outside the fibre—low external sodium for instance reduces the velocity of conduction because it makes the sodium current smaller—and is influenced by local anaesthetics and other pharmacological agents acting on the permeability mechanisms. It is also a function of the density of sodium channels in the membrane; the nodes of Ranvier have a very much higher density of sodium channels than do ordinary unmyelinated fibres, another factor contributing to the increased conduction velocity of myelinated nerves.

The compound action potential

Now most peripheral nerves contain a variety of fibres of different diameters, some myelinated and some not, so that if we use a pair of external electrodes to record the response to a single shock delivered to the whole nerve, we usually obtain a very complex waveform with a number of peaks and troughs, called the *compound action potential*. Some of this complexity is due to the existence of fibres of different velocity, so that action potentials in different fibres arrive at the recording electrodes at different instants. But some is also due to the fact that we are recording with extracellular electrodes rather than an intracellular one, and it is convenient to begin by considering why this makes a difference.

Consider first just a single axon from the nerve. If we were to insert a microelectrode in it we would record the classical *monophasic* action potential, rising from a resting potential of some – 60 mV to become slightly positive, and then returning over the course of a millisecond or two back to rest, as the action potential passes the point of insertion of the electrode. But if instead we record the same response with a pair of external electrodes, some way apart, we shall in effect register the action potential twice over, as it passes each electrode in turn. As it passes the first electrode, it will make it more negative with respect to the other; and as it passes over the second, the situation will be

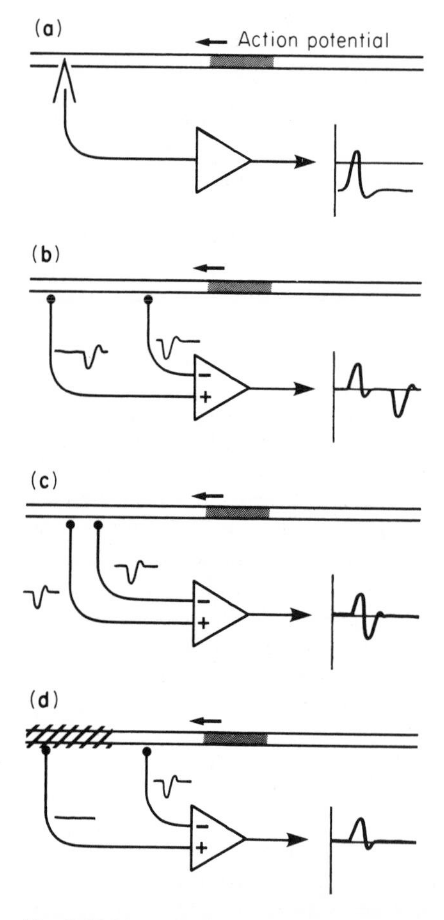

Fig. 2.16 Monophasic and biphasic action potentials. (**a**) Recording a monophasic response with a single microelectrode penetrating the axon. (**b**) A pair of external electrodes produces a biphasic potential. (**c**) as (**b**), but with the electrodes closer together. (**d**) as (**b**), but with the nerve crushed so that the action potential never reaches the second electrode: the result is a monophasic recording

reversed, and the first electrode will now be more *positive* than the second (Fig. 2.16). Thus although there is of course only one action potential, we will now record a *biphasic* response, with a swing of potential first in one direction and then in the other, about a resting value of zero. If we now bring the two electrodes close together, the two individual responses will begin to overlap, producing a smaller response because of the partial cancellation of the two components. Neither of the two swings of potential will have the same time-course as the true intracellular action potential, so that the shape of the whole thing is rather uninformative. One way to get round this problem is to arrange things so that the action potential never actually gets as far as the second electrode, by crushing the nerve between the two recording points. Under these circumstances a monophasic response is obtained that is similar to the intracellular action potential, though a good deal smaller.

Consider now what would be seen if we had not one axon but several. If each of them had the same conduction velocity, and were in good working order, then the situation would not be very different from that of a single axon. But in a real preparation, it is rare for all the fibres in a bundle to be conducting all along their length, and at least some of them are likely to fail to conduct somewhere between the two electrodes. Consequently the biphasic

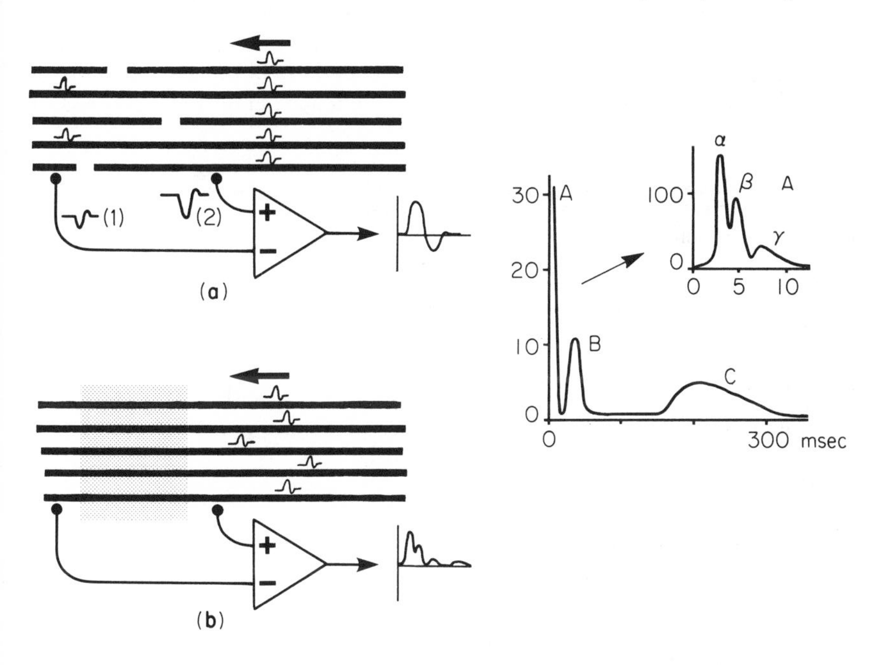

Fig. 2.17 The compound action potential. (**a**) Biphasic recording from whole nerve: the fact that some fibres fail to conduct between the electrodes results in an asymmetric action potential. (**b**) As (**a**), but with the nerve crushed to give monophasic action potentials. Differences between the conduction velocities of individual fibres give rise to a dispersed compound action potential. Right, actual compound action potential from frog sciatic nerve, with the A group shown on an expanded time scale in the inset. (Data from Erlanger and Gasser, 1938)

potential from the whole nerve is likely to be a lopsided one, with some of the fibres contributing biphasically and others only monophasically (Fig. 2.17). If in addition they are all conducting at different speeds, then the whole thing will be drawn out into a series of overlapping responses, the peaks of some corresponding to the troughs of others, in a messy and uninformative way. It is then essential to ensure that *all* the fibres are producing monophasic potentials, by damaging the nerve between the electrodes as before. Under these circumstances, the pattern of peaks in the compound action potential gives a sort of spectrum of the conduction velocities of the fibres in the nerve, though not a very quantitative one, since large peaks may simply be due to large fibres rather than to a large *number* of fibres of a particular velocity. In such a spectrum from a large peripheral nerve, it is often possible to distinguish groups of fibres having roughly similar conduction velocities. Unfortunately, two different systems of classification of these groups have grown up side by side. The first, due mainly to Erlanger and used primarily for motor nerves, divides fibres into groups A, B and C in order of decreasing conduction velocity (C being unmyelinated), with group A further divided into subgroups (Table 2.2). The other system, due to Lloyd and used mostly for sensory fibres, uses a four-fold classification (I, II, III, IV) the last again being unmyelinated. It is clear from the table that there is a good deal of overlap between the classes; in practice, most motor nerve fibres fall into either group Aα ('alpha fibres') or group Aγ ('gamma fibres'); the functional differences between the groups in sensory cutaneous nerves are discussed in Chapter 4.

Table 2.2 Classification of sensory nerve fibres (Lloyd) and motor fibres (Erlanger). (Group B has fallen into disuse in describing motor fibres: it corresponds to group Aδ)

Sensory	Diameter (μm)	Velocity (m/sec)
I	12–20	70–120
II	4–12	24–70
III	1–4	3–24
IV (unmyelinated)	< 1	< 2
Motor		
A α	8–20	50–120
A β	5–12	30–70
A γ	2–8	10–50
A δ	1–5	3–30
(B)	(1–3)	(3–15)
C (unmyelinated)	< 1	< 2

Threshold properties

Once we understand the mutual relationship between membrane potential on the one hand, and ionic permeabilities on the other, it is not difficult to explain many of the basic properties of nerve that make it behave so differently from a simple passive conductor of electricity, in particular the phenomena of *threshold* and of the *all-or-none law*. We have already seen that the relation between potential and permeability is that of a closed cycle or feedback loop, in which potential changes cause changes in permeability, and these in turn cause changes in potential (Fig. 2.18). But if we compare the two cases of sodium and potassium, the functional behaviour of these feedback loops is very different. In the latter case, a depolarization causes an increase in P_K,

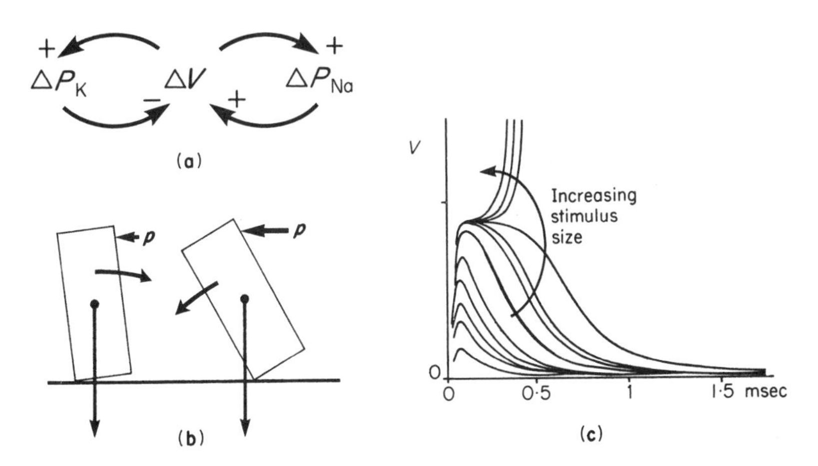

Fig. 2.18 (**a**) The relation between small depolarizations ΔV and changes in permeability to potassium, ΔP_K and sodium, ΔP_{Na}, illustrating the existence of negative feedback in the former case, and positive in the latter. Whether the system as a *whole* shows negative or positive feedback depends on the way P_K and P_{Na} are affected by ΔV. (**b**) A domino on edge: for a small push p it shows negative feedback and is stable; for a larger push P it shows positive feedback and topples over. (**c**) Extracellular voltage response to stimulating currents of increasing size applied near the recording electrode in crab nerve, showing stability for small stimuli and instability for larger ones: close to the threshold the voltage teeters on the brink. (Partly after Hodgkin, 1938)

which then tends to oppose the depolarization by bringing the membrane potential nearer to E_K: a good example of a *negative feedback* system that tends to stabilize the membrane near its resting potential. The case of sodium is the exact opposite: here, depolarization again causes an increase in permeability, but this tends to depolarize the membrane still further. Here we have not negative but *positive* feedback. Now positive feedback is a property usually associated with explosive systems: if we light a firework, the initial heat is sufficient to decompose part of the gunpowder, in turn releasing more heat which sets off yet more of the gunpowder, and so on: the positive feedback accelerates the reaction until the whole of the stored chemical energy has been released and converted into a big bang. In the same way, depolarization of the nerve membrane causes an increase in P_{Na}, which depolarizes the membrane still further, causing yet more of an increase in sodium permeability, and so on: if this were all that were happening, the potential would move smartly off to E_{Na} and stay there. (The reason it doesn't is of course both that the increase in P_{Na} is only a temporary one, and that meanwhile the potassium permeability has also started to rise.) So whether the membrane as a *whole* is stable or unstable depends crucially on the relative contribution of the potassium and sodium feedback loops: if the former predominates, the membrane will tend to resist any potential changes applied to it; if the latter, it will tend to respond with an explosive increase in P_{Na} and a large swing of voltage. In practice it turns out that the relative contribution of the two mechanisms depends in a non-linear way on the size of the stimulus applied: small stimuli seem to produce more of a potassium response than a sodium one, so that the membrane remains stable, whereas larger stimuli tip the balance in favour of sodium and result in instability—in other words, the regenerative action potential. The situation is in fact very like that of the domino of Fig. 2.18, standing on its edge. At rest, it is in a state of negative feedback: a small push tends to raise its centre of gravity, resulting in a turning couple that restores the status quo. A large push, however, that brings the vertical projection of the centre of gravity past the corner on which it rests, converts the domino's stable equilibrium into an unstable one: the further it is pushed, the greater the couple acting to topple it over. Like nerve, it thus exhibits a *threshold*; there is a certain size of stimulus that converts its normal condition of negative feedback into one of positive feedback, and it then falls over and releases all its stored potential energy. If one records close to the point of stimulation of a single axon, while stimulating with finely graded potentials in the region of the threshold, one can often see the membrane potential teetering on the brink between stability and instability (Fig. 2.18).

In general, therefore, any factor that favours the potassium mechanism rather than the sodium one will tend to raise the membrane threshold. Two important instances of this occur in the *refractory period* and in *accommodation*.

Refractory period

If we try to stimulate a nerve with a pair of shocks, gradually reducing the interval of time between them, we find that there comes a point when the threshold for the second shock begins to rise relative to that for the first; eventually, as we go on decreasing the interval between the stimuli, we find that we cannot activate the nerve a second time at all, no matter how large the current we use (Fig. 2.19). This period, during which it is impossible to stimulate the nerve for a second time, is known as the *absolute refractory period*; the period during which it can be stimulated, but only by using a larger current than usual, is called the *relative* refractory period. The latter corresponds quite well with the period just after the peak of the action potential during which P_K is still raised relative to its resting level, thus tending to stabilize the membrane potential. The absolute refractory period seems to be due mostly to a property of the sodium channels. We saw earlier that in the voltage clamp experiments the sodium permeability rose quickly in response to a step of depolarization, and then declined spontaneously, leaving the channels in an inactivated condition which lasts as long as the voltage is

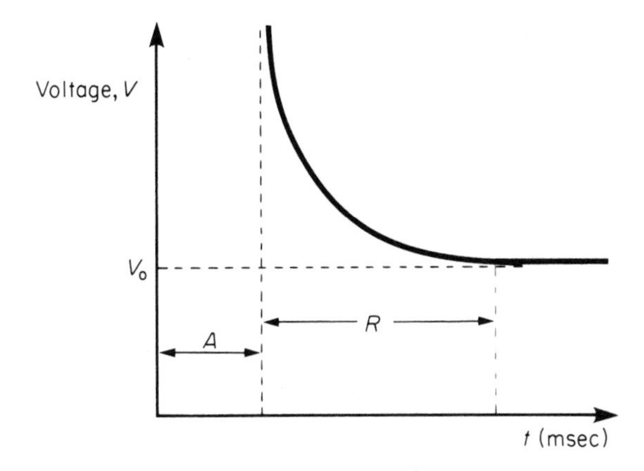

Fig. 2.19 Refractoriness of nerve: the voltage V required to stimulate an axon at different times t after a previous superthreshold stimulus, showing the absolute refractory period A and the relative refractory period R. V_0 is the threshold when a single stimulus is used.

maintained. It turns out in fact that even when the voltage is returned to its original value, it takes a certain period of time for the sodium channels to revert from their inactivated state to one in which they can once again respond to changes of voltage. Thus after the peak of the action potential has passed, there is a period of recovery during which the sodium mechanism is unresponsive, making the membrane absolutely stable to stimuli of any size. The existence of the refractory period is of considerable functional importance, since this is what prevents the action potential from being conducted in both directions at once: because the local currents flow almost equally both ahead of the action potential and behind it (Fig. 2.13) it is essential that the region over which it has just passed should not be reactivated all over again; its refractoriness prevents this happening.

Accommodation

One normally measures a nerve's threshold by using a small voltage step or pulse in which there is a sudden change of potential. If instead we try to depolarize the nerve more slowly, we find that we need to depolarize it further before it will respond with an action potential; indeed, if the rate of depolarization is sufficiently slow we may find that the nerve never responds at all, however far we depolarize it (Fig. 2.20): the nerve has *accommodated* itself to the changing potential. This phenomenon can be readily explained if we think in terms of the balance between the sodium and potassium mechanisms. In the voltage clamp experiments, we saw that a sustained step of depolarization gave rise to an immediate but transient increase in sodium permeability, and a delayed but sustained increase in that of potassium. Thus there is only a short period during which the sodium mechanism dominates: time is on the side of

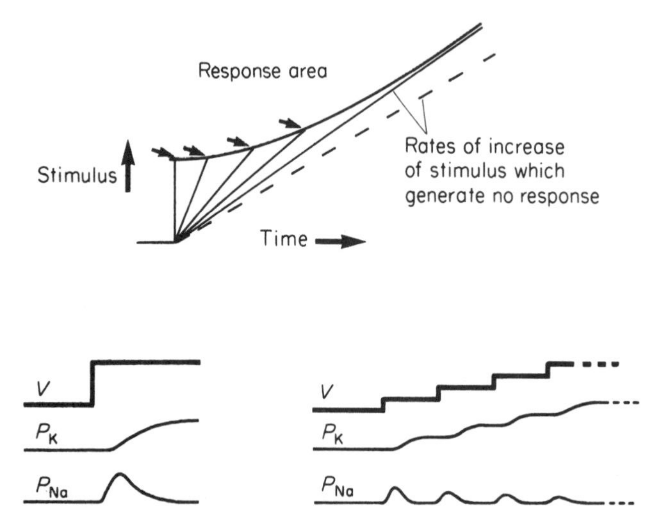

Fig. 2.20 Accommodation. Above, the threshold for generating an action potential (arrows) depends on the rate of depolarization of the axon: if this is too slow the fibre may never fire at all however much it is depolarized (After Fabre, 1927). Below: left, the changes in P_K and P_{Na} in response to a clamped step of voltage; right, in response to a series of small steps, simulating a slowly increasing depolarization: potassium permeability increases steadily, while sodium permeability declines through inactivation

stability. Suppose for example we were to stimulate a nerve not with one large step of depolarization, but with a staircase-like sequence of little ones (Fig. 2.20). It is clear that whereas P_K increases cumulatively with each new step, P_{Na} does not, since it is only transient; furthermore, the transient increase in P_{Na} will steadily decline with increasing depolarization, because of the steadily increasing degree of sodium inactivation. Thus the more gradually we depolarize a nerve fibre, the more we push the sodium/potassium balance in favour of potassium, and the further we need to depolarize it in order to reach the threshold; and if we depolarize it slowly enough, there will come a point where P_{Na} is *never* great enough relative to P_K for the nerve to fire at all, and the membrane will therefore completely accommodate. We shall see later that the mechanism of accommodation can sometimes be an important determinant of the way in which sensory receptors respond to slowly changing stimuli.

The all-or-none law

Any system with positive feedback will tend to behave in a manner approximating to all-or-none behaviour. The violence with which a barrel of gunpowder explodes depends very little on the temperature of the flame used to light it, so long as it is big enough to set it off at all. The domino of Fig. 2.18 ultimately hits the ground with much the same force, whether the original push was large or small. Much the same, but not *exactly* the same: clearly, if the energy of the push is appreciable in comparison with the domino's stored potential energy, the force with which it strikes the ground will be increased. More exactly, the energy released on falling over will be $P + E$, where E is the stored potential energy, and P the energy imparted by the original push. In the case of nerves, the all-or-none law is not found to be strictly obeyed if one records close to the point of stimulation—within a space constant or two—since the stimulus energy then contributes in part to what is recorded. But as the action potential is propagated further and further away from its origin, this contribution becomes increasingly negligible, and it eventually settles down to its standard form. What we have, in effect, is not just one domino but a whole line of them (Fig. 2.21): when one falls, it imparts a fraction of its energy to the next, sufficient to knock it over, and so on in turn all the way down the line. Imagine for the sake of argument that one-tenth of a falling domino's energy is used in knocking over the next. Then the first domino imparts an energy $(P + E)/10$ to the second, which in turn imparts $((P + E)/10 + E)/10$ to the third, and so on: it is clear that the contribution of the original push, P, to the energy with which the nth domino hits the ground will get vanishingly small as n gets larger, and that this force will in fact settle down at a constant level: the system as a whole *will* then obey the all-or-none law exactly.

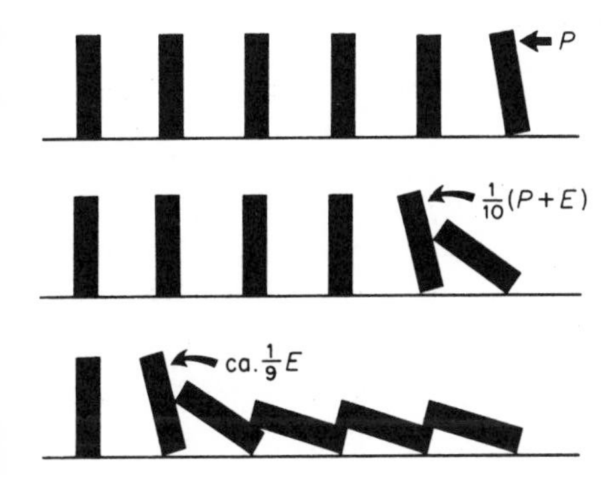

Fig. 2.21 All-or-none behaviour of a row of falling dominos

This law is of fundamental significance in the nervous system, and it is worth reflecting on its functional implications. Why should it have evolved? It clearly imposes very severe limitations on the kinds of messages that nerves can convey, prohibiting direct transmission of graded quantitative information (of the kind conveyed, for example, by the varying concentration of a hormone in the blood), the only messages permitted being of the binary 'yes/no' variety. The answer certainly lies in the problems of trying to send messages along cables that are so leaky that currents cannot be conveyed passively more than a matter of millimetres. An engineer faced with such a problem—as found for example (on a somewhat larger scale!) in transatlantic submarine cables—would probably deal with it by introducing a series of booster amplifiers at intervals along the cable, to restore the losses caused by leakage. In the case of nerve axons, we have already seen that the length over which they are required to conduct is so vastly greater than the space constant, that many thousands of such stages of amplification would be required. What are the characteristics of a chain of amplifiers of this sort? Imagine for example a thousand hi-fi amplifiers connected end to end, so that the output of one forms the input of the next. Now all amplifiers, however good their quality, suffer from two defects: they introduce *noise*, and they create *distortion*. Noise includes both the hiss that arises inevitably in any electrical system—including neurones—from the random movements of the electrons or ions in its conductors, and also disturbances picked up from external sources of interference; if we connect up a number of noisy amplifiers in series, the noise generated by each one of them is amplified all the way down the line and added to those of the others, making the final output very much noisier than if there were only one amplifier. Distortion arises through inaccuracies in the linearity of the amplification. This too becomes exaggerated if a number of amplifiers are connected in series: if for example the gain of each amplifier is 1 per cent greater than it should be, then the gain of the whole set of a thousand will be too large by a factor of some 2000; and if 1 per cent smaller than it should be, then the overall gain will be 1/2000 of the correct value. Thus accurate transmission of quantitative information becomes almost impossible: the system almost automatically becomes all-or-none in character, since signals either vanish or become saturatingly huge. The only solution is to be less ambitious about *what* one is trying to signal. If for example one limits oneself to only two possible signals—'yes' or 'no'—then distortion no longer matters: the signal is either there or not there, and no regard need be paid to how large it is. If we also arrange for each amplifier to have a threshold that is higher than the normal noise level, but allows through the signal 'yes', then we can get rid of noise as well. In other words, the only kind of system that is capable of transmitting messages reliably over distances that are much bigger than the space constant is precisely what we have found in the nerve axon itself: a series of regenerative amplifiers (the voltage-sensitive sodium channels) exhibiting a threshold that prevents the fibre from producing spurious signals in response to its own noise. There is no advantage in such a system for conduction over shorter distances, and in practice it is found that short neurones (as for example the bipolar cells of the retina) do not normally use action potentials at all, but rely on the much simpler and more informative method of passively propagated electronic potentials.

Neural codes

So if it is not open to us to use the *size* of an action potential to convey quantitative information, what possibilities remain? One way is to use the *number* of action potentials. It is very common in the nervous system to find a large number of fibres running in parallel from one location to another, all apparently carrying the same kind of information. In such cases, the magnitude of a stimulus can be coded by *how many* of the fibres are active at any one moment. A good example of this, as we shall see, is in the nerve from the vestibular apparatus; here the fibres are all found to have different stimulus thresholds, so that increasing stimulation leads to more and more of them firing at once, a phenomenon known as *recruitment*. In the same way, muscular contractions are often controlled by varying the number of active fibres innervating them. One may also use the number of action potentials per unit of time, the *firing frequency*, to convey quantitative information. As we shall see in the next chapter, the normal way in which a neurone responds to stimulation, whether external as in the case of receptors, or in the case of a central neurone, to the synaptic effects of other neurones connected to it, is by firing repetitively at a frequency that depends in quite a simple way on the magnitude of the stimulus. Conversely, the degree of tetanic contraction of a muscle depends on the frequency with which it is stimulated. There is an analogy here with the use of FM (frequency modulation) rather than AM (amplitude modulation) in radio transmission. In amplitude modulation, the amplitude of the radio-frequency carrier wave is a direct copy of the sound wave being transmitted (Fig. 2.22); the radio receiver decodes this signal by converting the envelope of the radio wave back into a sound wave. The disadvantage of such a system is that any variations in the amplitude of the wave caused by transmission itself—fading, or noise generated by radio interference—get incorporated in the sound reproduced by the receiver. In FM transmission this is no longer the case: here it is the frequency of the radio wave rather than its amplitude that conveys the sound information, and disturbances that affect its amplitude no longer matter, since it is only the frequency of the received signal that is decoded by the receiver, producing reliable and relatively noise-free transmission.

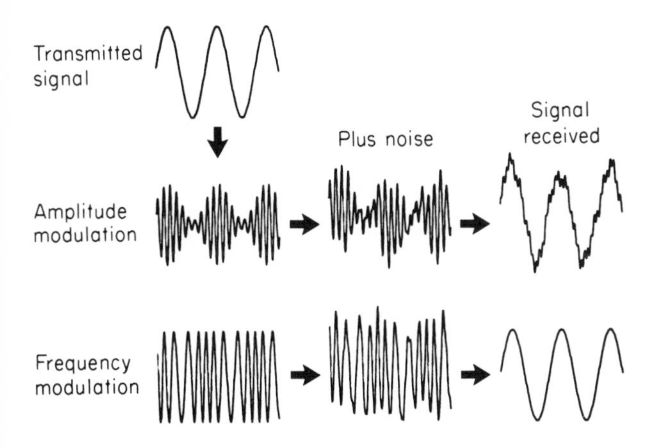

Fig. 2.22 Amplitude and frequency modulation, showing how noise added to the modulated radio wave results in much more interference in the decoded audio signal in the case of amplitude modulation (AM) than for frequency modulation (FM)

Frequency modulation is by no means the only way in which quantitative information may be conveyed down a single all-or-none or binary channel. A familiar example of a different kind of binary code is the Morse code, where information is carried in the temporal pattern of the only two possible signals—dot and dash. However, coding of such sophistication has never been observed in actual neurones, and we shall see in the next chapter that the mechanism by which neurones are caused to fire repetitively makes it unlikely that information could actually be carried by the nervous system in this form. In computers, quantitative information may be conveyed by a single pulse by varying the time at which it occurs relative to some kind of internal clock within the computer. Again, it is unlikely that this form of coding is used by the brain, both because the slowness of conduction would make it difficult to maintain accurate timing between events, and also because of the apparent absence of anything equivalent to an internal reference clock. Thus in practice we are limited to frequency modulation, and to methods such as recruitment that make use of the spatial patterning of activity across a set of fibres. In the next chapter we examine the mechanisms by which these patterns are set up.

3

Receptors and synapses

By now the reader should have a clear idea of how information may be carried from one part of an excitable cell to another: by passive electrical conduction when the distances are short enough to permit it, and otherwise by means of action potentials. We need now to consider how this information is generated in the first place, not only by stimuli in the outside world, but by the action of other neurones.

Sensory receptors

The general term for a process by which energy of one form is converted into energy of another is *transduction*. The energy incident on a receptor cell, whether electromagnetic, mechanical or chemical, must be turned into electrical energy in the form of potentials across the cell membrane, since this is the only language that interneurones understand. In general the effect of the stimulus energy is to alter the *permeability* of the cell membrane to certain ions, resulting in a flow of current and a movement of the potential towards some new equilibrium value. The specificity of a receptor for a particular type of stimulus comes about because not all the kinds of energy impinging on a cell are able to cause these permeability changes; transduction is in effect preceded by a specialized *energy filter* that allows certain types of energy through to cause electrical effects, and not others (Fig. 3.1). This filtering is seldom absolute: the receptors of the eye, for example, though exquisitely sensitive to light, will also respond to mechanical deformation if it is severe enough. If in the dark you shut your eye and press on the side of it with your finger, you will see a faint blue patch of light called a phosphene: the receptors respond to mechanical stimulation by sending the brain exactly the same message that they would have sent if a real blue light had been present. Specificity of receptors is in fact a relative matter, and many receptors are in fact surprisingly unspecific in what they respond to: thus fine discriminations between one type of stimulus and another are to a large extent the work of the central nervous system rather than of the receptors themselves.

Transduction in the Pacinian corpuscle

One receptor that because of its peculiar anatomical structure happens to be very specialized indeed has been studied in great detail, and that is the *Pacinian*

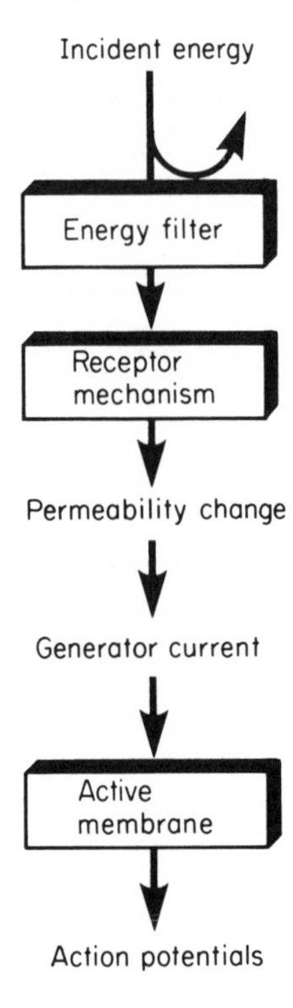

Fig. 3.1 Schematic representation of the stages by which a receptor generates action potentials in response to incident energy

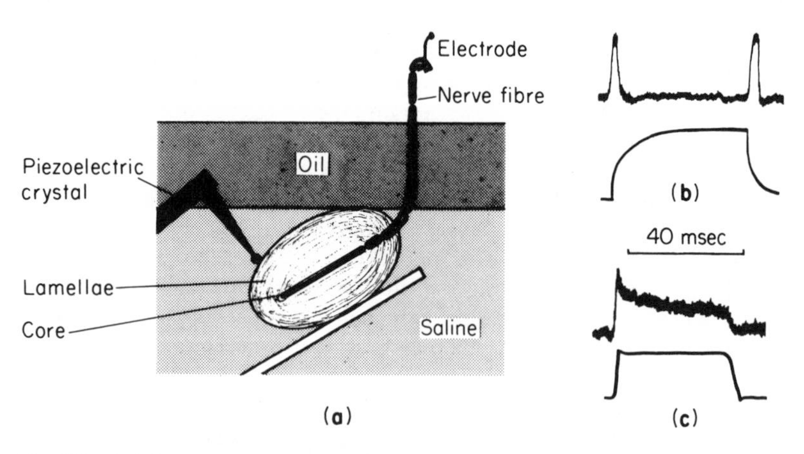

Fig. 3.2 (**a**) A Pacinian corpuscle dissected out, and arranged for mechanical stimulation by means of a piezoelectric crystal, and electrical recording. Right, generator potentials recorded from a Pacinian corpuscle before (**b**) and after (**c**) removal of the outer lamellae, in response to a brief maintained deformation (shown below in each case). (After Loewenstein and Mendelsohn, 1965)

corpuscle (Fig. 3.2). Here the naked tip of an otherwise myelinated axon is sheathed in concentric onion-like layers called *lamellae* that shield it from virtually every type of stimulus except that of mechanical deformation. It is a convenient preparation for studying the transduction process in general, since individual corpuscles may be isolated, and their electrical responses recorded whilst precise mechanical stimuli are applied to the capsule's surface. A convenient way of stimulating it is to hold against it a probe that is mounted on a small piezoelectric crystal: when a voltage is applied across the crystal, it changes its shape slightly and causes a controlled deformation of the capsule, and a change in membrane permeability. Now a complicating factor when trying to measure the relation between a stimulus and the permeability changes that result from it is that if the permeability changes are big enough they will trigger off action potentials that will in turn interfere with the very permeabilities one is trying to measure. Consequently it is helpful to disable the active properties of the axon by poisoning it with a substance like tetrodotoxin that blocks the voltage-dependent sodium channels. If this is done, we find that mechanical stimulation of the ending results in a depolarization of the axon—the *generator potential*—whose magnitude is a roughly linear function of the size of the stimulus. To show that this change in potential is indeed due to a change in permeability of the receptor membrane, we need to see what happens if we artificially depolarize it to different levels by passing a current into its axon, and then stimulate it mechanically. If the effect of the pressure is to make the membrane more permeable to certain ions, then this should alter the equilibrium potential of the membrane to some new value, E_S. So whatever the artificial level of potential we start at, we should find that stimulation results in a movement towards E_S. If we pass more and more current into the cell, forcing its resting potential nearer to E_S, the generator potential should get smaller and smaller; when E_S is reached, nothing should happen at all on stimulation; and if the resting potential is forced beyond E_S, we should find that the response is reversed, giving a hyperpolarization rather than a depolarization. Actual experiments show that this does indeed happen, and that this *reversal potential* is very close to zero. The simplest explanation is that when we deform the cell membrane we open up ionic channels in it which are permeable to all ions and so act as a sort of short-circuit: perhaps distorting the membrane simply increases its leakiness.

Adaptation

A striking feature of the response that is obvious in Fig. 3.2**b** is that its time-course is very different from that of the stimulus itself. If we apply a prolonged but constant deformation to the surface, we find that the generator potential rises quite rapidly to a peak, and then spontaneously falls back again to the resting potential; when the stimulus is removed a second peak is generated. The cell seems, in fact, to respond to *changes* in the degree of stimulation rather than to its steady level, a very common type of receptor response that is called *adaptation*. It turns out that much of this adaptation is due to the mechanical properties of the lamellae that surround the ending, that act as an energy filter. If we strip them off, and apply the stimulating probe directly to the surface of the axone, we find that although the generator potential still falls off after its initial rise, it remains depolarized so long as the stimulus is maintained, with no hint of a second peak of depolarization when the stimulus is removed (Fig. 3.2**c**). This kind of response is known as *incomplete* adaptation, as opposed to the *complete* adaptation seen when the lamellae are intact, and the response falls to zero.

There is a simple mechanical model that explains how this filtering out of steady levels of stimulation may come about (Fig. 3.3). We can think of the end of the axon itself as behaving in a simple elastic manner, so that any force applied to it results in a corresponding deformation, and hence in a change in permeability. The lamellae on the other hand behave very differently because they are separated from one another by layers of viscous fluid: when a steady pressure is applied to a particular part of the capsule, the lamellae in that region slowly collapse as the fluid between them oozes sideways to neighbouring regions. The lamellae thus act very like the oil-filled cylinders or

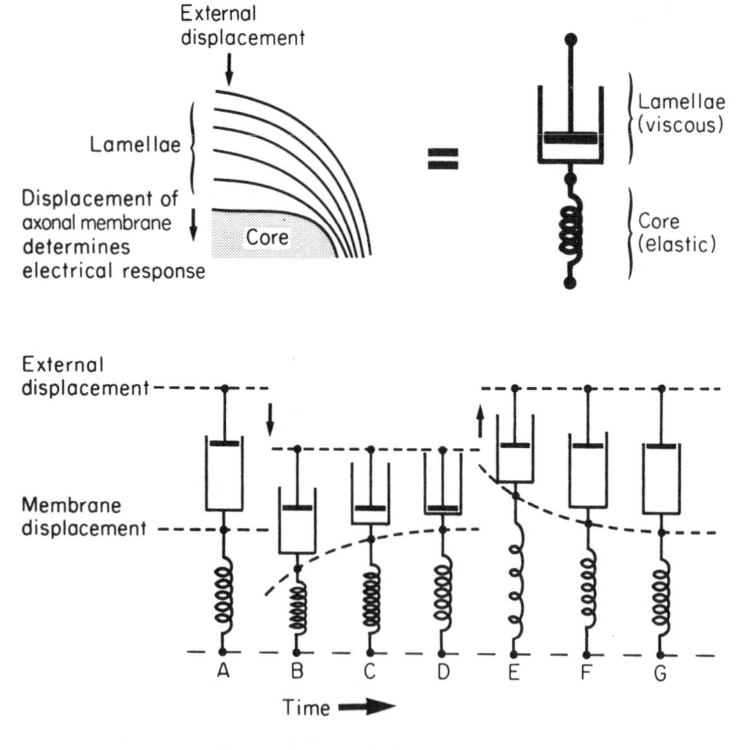

Fig. 3.3 Above, highly schematic representation of the mechanical elements of the corpuscle, and their mechanical 'circuit' in the form of a dashpot and spring. Below, response of such a model to a steady displacement applied just before B and removed just before E, showing adaptation of the degree of distortion of the elastic element, and hence of the electrical response of the receptor

dashpots that make up part of the shock absorbers fitted to car suspensions or to the tops of swing doors, whose function is to resist sudden movement; they are what are known to engineers as *viscous elements* (represented in the figure by the conventional symbol of a dashpot), while the axon tip acts as a purely *elastic element* (the spring symbol). In the case of the Pacinian corpuscle, both these elements are connected in series: if we apply a sudden steady displacement to the outer end of the viscous element, at first there is no time for it to collapse, and the displacement is taken up by the elastic element, which is thus compressed. But in being compressed (B, Fig. 3.3), it exerts a force on the viscous element which then tends to collapse (C,D), through the sideways oozing of fluid described above. The elastic element, the nerve itself, therefore gradually resumes its original shape, and the potential returns to its resting value. But if the stimulus is removed (E), the whole process is reversed: at first the elastic element must stretch to take up the new displacement, but in doing so it pulls on the viscous element, expanding it again (F,G), so that in the end all is as it was originally. It is clear that deformation of the axonal ending only occurs in association with *change* in the stimulus that is applied, and that in the case of a steady stimulus it will be deformed both at the onset and cessation of the stimulus. If we suppose that inward and outward deformations of the cell membrane are equally good at causing changes in permeability, then we can see how the biphasic generator potential of Fig. 3.2 comes about. This kind of mechanical filtering is a common one in the body, and we shall meet it again in discussing the stretch receptors that are found in muscles. It is sometimes called a *high-pass filter* because it passes high-frequency vibratory stimuli much better than low-frequency vibrations of the same amplitude, since the rate of change of deformation is very much smaller in the latter case.

Adaptation in the Pacinian corpuscle cannot be entirely due to the lamellae, since we have seen that there is still a decline in response at the beginning of a period of constant stimulation even when they have been stripped away. This component of the adaptation appears to be an electrical property of the membrane itself, found not only in other types of receptor but in other neurones as well, and may be called *membrane adaptation*; it is sometimes described as accommodation, though somewhat misleadingly since it is not in fact due to the same mechanism as that underlying the true accommodation described in Chapter 2. But before considering membrane adaptation, we need first to understand how the changes in permeability that have just been described normally lead to the generation of action potentials.

The initiation of impulses

Consider a neurone—not necessarily a receptor—whose resting potential is E_r with a set of channels which when open tend to short-circuit the membrane and lead to an equilibrium potential around zero. Somewhere between E_r and zero there will be a threshold potential θ for triggering an impulse (though we must bear in mind that the value of θ will depend in general on the rate of depolarization, because of accommodation). If we suddenly open these channels and keep them open, the potential will move towards zero, and may at some point cross the threshold, setting off an action potential. The usual stereotyped sequence will then ensue, terminating in a recovery phase in which P_K will be elevated and the neurone relatively hyperpolarized as its potential is pulled towards E_K. As P_K declines to normal after the impulse, the potential will rise again, not just to the original resting potential but past it (since we suppose that the channels are still open) towards zero. What happens next will depend on the rate at which this depolarization occurs. If it is sufficiently fast (and θ correspondingly low), the threshold will be crossed once more, and a second action potential will be generated; then a third, a fourth, and so on: impulses will continue to be generated so long as the channels remain open (Fig. 3.4). The greater the short-circuiting current, the faster the

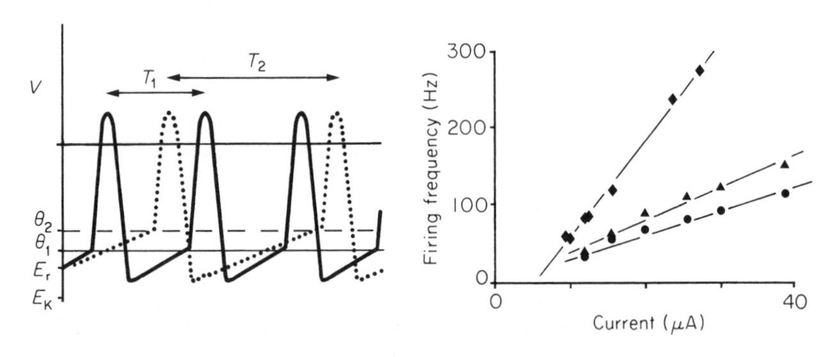

Fig. 3.4 Top, mechanism by which a steady current may initiate repetitive firing. The solid and dotted lines represent the schematic responses to a greater and lesser current respectively: θ_1 and θ_2 are the threshold levels corresponding to the associated rates of depolarizations. The frequency in each case is the reciprocal of the interval T between spikes. Right, experimental relation between injected current and resultant steady firing frequency for three motor neurones (data from Granit *et al.*, 1963)

rate of depolarization will be after each impulse, and so the sooner the nerve will fire off again. Thus the frequency of the repetitive firing will depend on the degree of short-circuiting that we have produced: the more channels are open, the higher the frequency. But if the rate of depolarization after the first action potential is too slow, θ may rise so much because of accommodation that the membrane potential never reaches it, and the neurone will fail to fire for a second time. This is precisely what does happen in the Pacinian corpuscle, where a steady deformation normally produces only one action potential when the stimulus is applied, and another when it is removed: the accommodation of the ending is too great for the rate of repolarization after the action potential. But this is not at all typical of sensory receptors, which generally respond to a steady stimulus with a continuous train of impulses. Under these conditions, adaptation will manifest itself as a steady decline in firing frequency after application of the stimulus, as may be seen in the responses from various receptors shown in Fig. 3.5. If the receptor is of the completely adapting sort, it will eventually stop firing altogether; otherwise

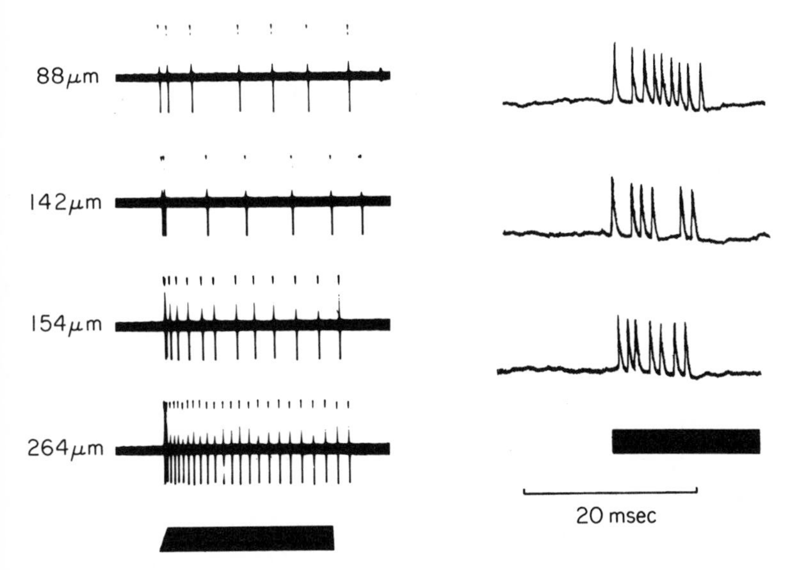

Fig. 3.5 Examples of adaptation in different receptors. Left, tactile receptor in the skin of a cat's leg, with the duration and four different degrees of indentation shown (after Mountcastle, 1966). Right, hair receptor, cat skin, showing complete adaptation after a few impulses (Hunt and McIntyre, 1960)

the frequency will decline to some steady level.

There are many receptors in which adaptation is at least partly due to energy filtering of a kind analogous to what is done by the lamellae of the Pacinian corpuscle. The muscle spindle stretch receptor for example has a mechanical high-pass filter that behaves in a very similar way; in the eye, incident light gradually bleaches away the photopigment responsible for the transduction process and thus lowers the receptor's sensitivity during steady illumination (the full mechanism is actually rather more complex than this and will be discussed in Chapter 7). But in most cases at least some of the adaptation is also the result of membrane adaptation. We can demonstrate this by by-passing the transduction stage altogether and sending currents directly into the cell through a microelectrode. Under suitable conditions, a steady current will imitate the effect of a short-circuiting permeability change, and will elicit repetitive firing. If the frequency of impulses declines throughout such a stimulus, even though the current is constant, we can be sure that membrane adaptation rather than energy filtering is responsible. Very similar effects, of the same timescale, can be seen when neurones are stimulated by steady currents (Fig. 3.6), suggesting that membrane adaptation is a general property of neurones and not merely confined to receptors. Its mechanism is uncertain, though it has been suggested that it might be due to slow changes in P_K, similar to the voltage-dependent increase in P_K that produces accommodation, but on a timescale that is some hundred times longer.

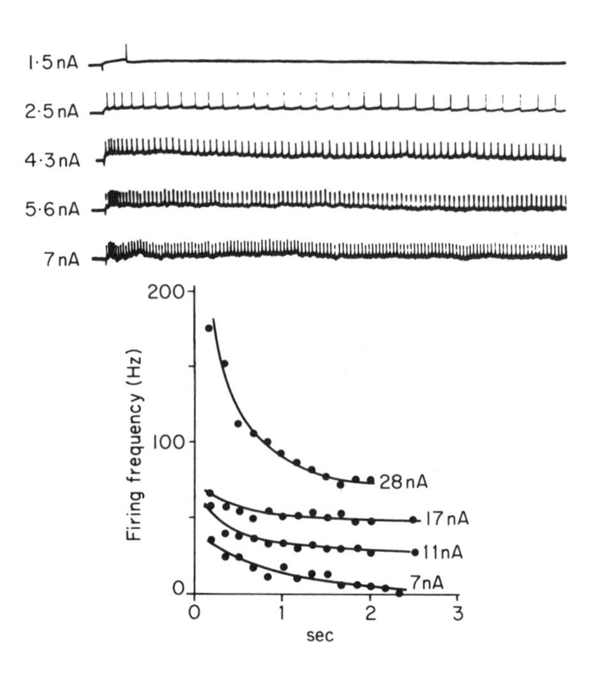

Fig. 3.6 Membrane adaptation in motor neurones. Above, spike responses to steady injected currents of the strengths indicated (Oshima, 1969). Below, Decline in frequency with time in such an experiment (data from Granit *et al.*, 1963)

Figure 3.7 may help to summarize the transduction processes that have been described so far. The energy impinging on a receptor is first filtered, possibly by virtue of surrounding structures as in the Pacinian corpuscle, or in the case of some birds the little coloured oil droplets that contribute to the colour selectivity of the visual receptors, and it then brings about—by mechanisms that are yet unknown—a change in the permeability of the neuronal membrane. This in turn produces a generator current which tends to depolarize the axon, and if large enough will fire off an action potential. If conditions are right, this may be followed by repetitive firing at a frequency dependent on the degree of stimulation, although this frequency will in general decline because of membrane adaptation. Otherwise, if the degree of accommodation is too great in relation to the rate of depolarization after the first impulse, only a single action potential will occur. A further source of adaptation may be the energy filter itself, if its properties are such that it tends to let through changes in incident energy rather than steady levels. This filter is sometimes also under the control of the central nervous system, as in the well known case of the iris of the eye, or the efferent innervation of the muscle stretch receptors. Finally, it is worth pointing out that the receptor and the nerve axon that joins it to the central nervous system need not be one and the same cell: there are many cases—for example the hair cell receptors of the ear—where the receptor is quite separate from the axon, though the region of

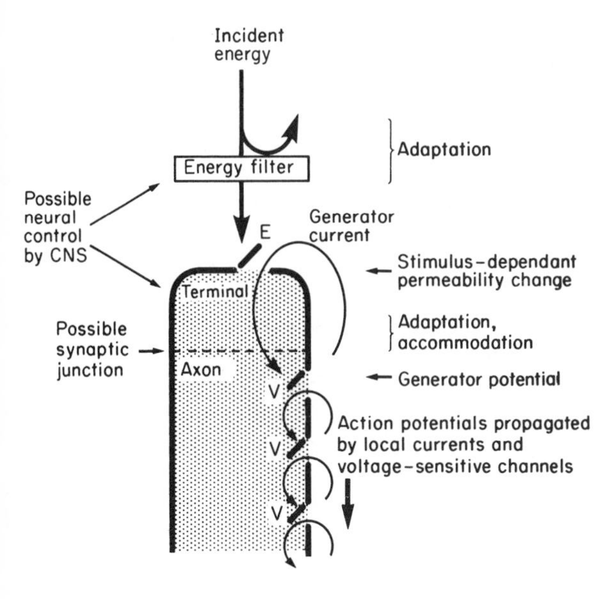

Fig. 3.7 The sequence of events in a typical receptor, showing the sites at which adaptation and control by the CNS may occur. In some cases, axon and terminal may be two separate cells, with a synaptic junction at the level of the dashed line. Energy- and voltage-sensitive channels are labelled E and V respectively.

contact between the two has a low resistance that may allow the generator current from the receptor to initiate firing in the axon. Even where receptor and axon are part of the same cell, it is usual to find that the area of membrane that is specialized for transduction cannot actually carry action potentials because it lacks voltage-sensitive channels. Thus in the Pacinian corpuscle, the currents generated by the deformation-sensitive ending only cause action potentials because they flow to the portion of axon where the myelination begins, and depolarize it; and it is likely that impulses are in fact normally generated at the first node of Ranvier. We shall see later that exclusive specialization of neuronal membrane into regions that are active, where permeability is a function of voltage, and those that are devoted to generating permeability changes in response to *other* kinds of stimuli, appears to be the universal rule.

Functions of adaptation

It may perhaps seem strange that the pattern of impulses generated by what is supposed to be a pressure receptor should be so very different from the time-course of the pressure that it actually experiences (Fig. 3.2); it looks as though the receptor is throwing away useful information. Is there any advantage in signalling changes rather than steady levels?

In the first place, it is not quite true to say that information has been thrown away. So long as the brain 'knows' how the Pacinian corpuscle responds to different patterns of pressure, it can in principle reconstruct the time-course of the original stimulus from the coded signals that it receives from the receptors, so that no information is really lost. One advantage of adaptation may thus be that it makes for *economy* of nervous impulses. If a stimulus is such that it tends to remain constant for long periods of time, with only occasional shifts to some new value—for example the pressure sensed by Pacinian corpuscles in one's buttocks during a long lecture—then there is little point in sending a stream of information to the brain which only tells it, in effect, that nothing has happened. Since the same information could have been carried by many fewer impulses, by sending a message only when something new happens that might call for a response, one general function of adaptation could be said to be to get rid of unnecessary action potentials—it reduces the *redundancy* of the messages that are conveyed. The same argument applies equally to interneurones at higher levels in sensory systems, where adaptation to particular *patterns* of stimulation is called *habituation*. One is perhaps less horrified the fourth time a head of state is assassinated in a year than the first; but this is not because one's sensory receptors have adapted. Again, the fact that one is not continually aware of the somatic sensations produced by one's own clothes is often attributed to adaptation of touch receptors, but is really due to habituation.

A second possible reason for the widespread existence of adaptation in sensory systems is that it may improve sensitivity by increasing the *signal-to-noise ratio* of the receptor. This is a concept that is fundamental to understanding the coding of sensory information, and is not a difficult one to appreciate. Any signal, whether it consists of frequencies of action potentials or simply of varying voltages as in a telephone wire, is inevitably subject to a certain degree of uncertainty on account of the all-pervasive random *noise* that is an inescapable feature of the physical world. If we measure the frequency of firing of a sensory fibre under conditions that are as near constant as we can achieve, we shall find nevertheless that the frequency we observe is not fixed, but undergoes continual random perturbations (Fig. 3.8). This noise may be due to small changes in the temperature or chemical environment of the receptor, to slowly acting properties such as fatigue that are not under our control, or, ultimately, to the fact that the ions whose movement generates the potentials we measure are themselves in continuous random thermal motion, so that the currents they carry must equally be subject to a certain degree of unpredictability. Thus we can never say that a nerve is firing exactly 70 times a second: the best we can do is to estimate with more or less confidence that its frequency lies somewhere between 69 and 71 Hz. This in turn puts a limit on the amount of information that a fibre can carry. First of all, there is an upper limit to its firing frequency that is set by the refractory period, and is typically of the order of 500 Hz. One might think that if a nerve can fire at any of the infinite number of frequencies between this limit and zero, then each of these possible frequencies could convey a separate message to the brain, and so the number of different possible messages at any moment would be limitless. But this is not so, for the existence of the noise that has just been described means that the brain may not be able to *discriminate* between frequencies that lie close

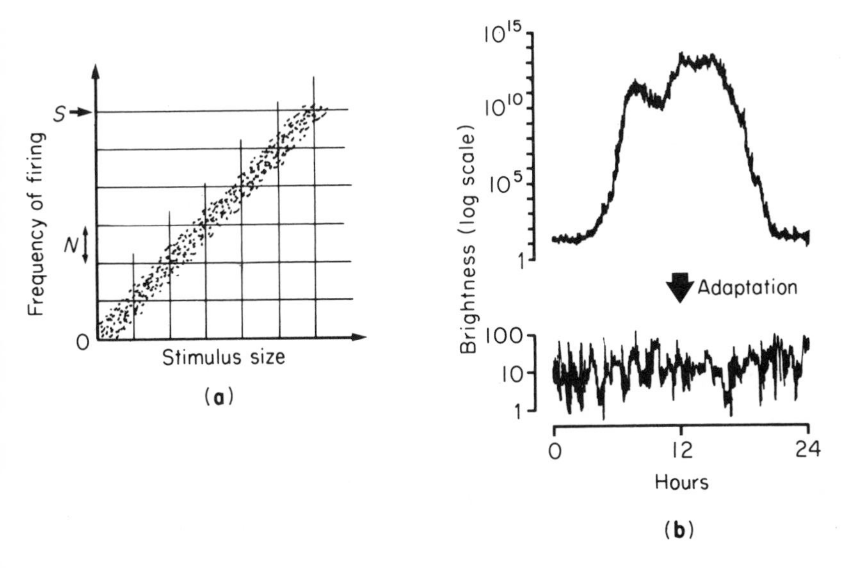

Fig. 3.8(a) Signal-to-noise ratio. Schematic relationship between firing frequency and stimulus size for a hypothetical sensory fibre, measured on a large number of occasions (single data points), showing the band of frequency scatter, of width N, associated with any particular value of the stimulus. As a result, frequencies must be separated by N before they can reliably be discriminated from one another. Thus the number of significantly different frequencies, and hence of discriminable stimuli, is given by S/N, where S is the maximum firing frequency of the fibre. In this case the signal-to-noise ratio is only about 6. **(b)** Intensity of light that might fall on a typical retinal receptor throughout one day, showing that the rapid fluctuations that convey visual information are centred on a level that changes more slowly. A receptor capable of responding to the entire range of steady intensities would necessarily be relatively insensitive to the smaller fluctuations. Below, adaptation has the effect of filtering out the slow changes of intensity level: the receptor now need only cope with some two log units of intensity, and can thus respond better to small fluctuations within this range

together, because of the impossibility of determining exactly what the frequency actually *is* at any moment. More specifically, if we call the size of the largest signal that a nerve fibre can carry S, and the amplitude of the ever-present noise N, then the number of different frequencies that can be discriminated reliably from each other is only of the order of S/N (Fig. 3.8). For example, if the noise in a fibre leads to an uncertainty of about 1 Hz in determining its frequency, the ratio S/N—the signal-to-noise ratio—will be 500: in other words, at any moment the nerve can only convey one of 500 distinguishably different messages. Now if one thinks of this as being equivalent to an accuracy of one part in 500, or 0.2 per cent, this may not seem too bad a performance. But the problem is that the *dynamic range*—the ratio of the largest stimulus normally encountered to the smallest—over which most receptors have to operate is exceedingly large. In the case of the eye, for example, the dynamic range corresponds to the ratio between the brightness of the sun and the visual threshold in the dark, and is of the order of 10^{15}. If there were no adaptation in the eye, and each receptor coded a particular level of light intensity directly as a particular steady frequency of firing, its 500 possible output levels would have to be spread—pretty thinly—over the entire 10^{15} range of possible inputs. Clearly, a just-discriminable difference in receptor firing would correspond in general to a very large difference in light intensity, and our power of perceiving small differences in intensity would be very much worse than it actually is. But in practice the whole of this 10^{15} range is never present in our field of view at the same time, and—for reasons that are explained in Chapter 7—the ratio between the darkest and lightest parts of our field of view at any particular instant is typically only about 1:100. It is true that in the course of the day the *absolute* level on which this range of brightnesses is centred may fluctuate very widely indeed, as the sun rises and sets, and night follows day. But these shifts of absolute level, as well as being of little interest to us, are comparatively slow. Adaptation, acting as a high-pass filter, will tend to get rid of them, leaving behind the significant and relatively rapid changes of intensity that are generated by our eye-movements as we look around, and which lie in a dynamic range that is not so very different from that of the nerve fibres themselves. Adaptation, in other words, provides a kind of automatic sliding scale by which the limited signal-to-noise ratio of our neurones can be shifted to match the range of inputs we are interested in, ignoring slower and larger changes of the baseline that could only be accommodated by sacrificing overall sensitivity. It thus acts like the automatic level control sometimes fitted to tape recorders, which automatically adjusts the amplification to compensate for the average level of sound that is being recorded, but lets through the more rapid fluctuations that constitute the sound itself. In the case of the Pacinian corpuscle, for example, it means that we can be made aware of very small changes even when superimposed on large steady background pressures—for example in the soles of the feet when standing—because adaptation has again shifted the scale of the receptor to allow for the steady background. Finally, for proprioceptors such as joint receptors and muscle spindles that essentially give information about position of the limbs, the fact that adaptation implies response to *rate of change* rather than to steady levels means that such receptors, if completely adapting, will essentially signal *velocity* of the limbs rather than position, which may be more relevant in the control of certain kinds of movement such as throwing.

Synaptic transmission

We now need to consider the initiation of activity in neurones other than receptors, driven not by the outside world but by the activity of other neurones that make contact with them at specialized regions, the *synapses*. At a typical synapse, a branch of the afferent axon forms a swelling, the terminal *bouton*, the further side of which forms an enlarged area of intimate contact with the post-synaptic cell: in the case of the neuromuscular synapse, the endplate, this area is much increased by the presence of invaginating folds (Fig. 3.9). In most cases there is a clear *synaptic cleft* between pre- and post-synaptic membranes, typically of the order of 20 nm wide: but sometimes there may be

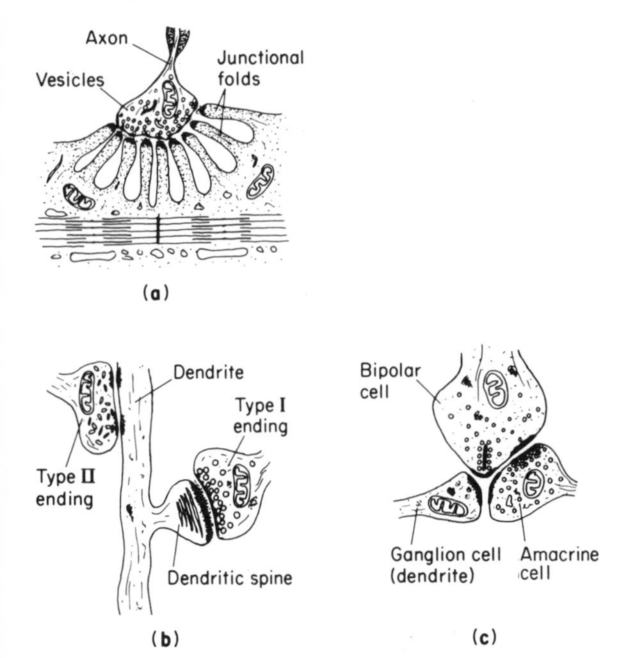

Fig. 3.9 Somewhat stylized representations of synaptic types. (**a**) Neuromuscular junction. (**b**) Two types of presynaptic axonal endings, synapsing with a dendrite; the synapse on the right is with a dendritic spine. (**c**) A three-way synapse from the retina; the junction between bipolar and amacrine cell probably permits the flow of information in both directions

a much closer apposition of the two membranes, with areas of fusion, forming a tight junction or a gap junction. At all events, it is clear that it is across this synaptic area of contact that information is passed from one cell to the other. How may this come about?

Electrical transmission

One might imagine that action potentials could be propagated across the synapse by the same mechanism of local current flow that sends the impulse along the axon itself: and there are indeed a number of instances where synaptic transmission is of this simple electrical kind. The best known example is perhaps that of conduction between the muscle fibres of the heart, and in the brains of both mammals and other species one may find occasional examples of electrical synapses; but rather stringent structural conditions have to be met before this mode of synaptic transmission will work. Figure 3.10 shows an idealized electrical synapse and its equivalent circuit. It is clear that the currents generated by the presynaptic bouton have two alternative routes: they can either cross the gap and enter the post-synaptic cell, or they can

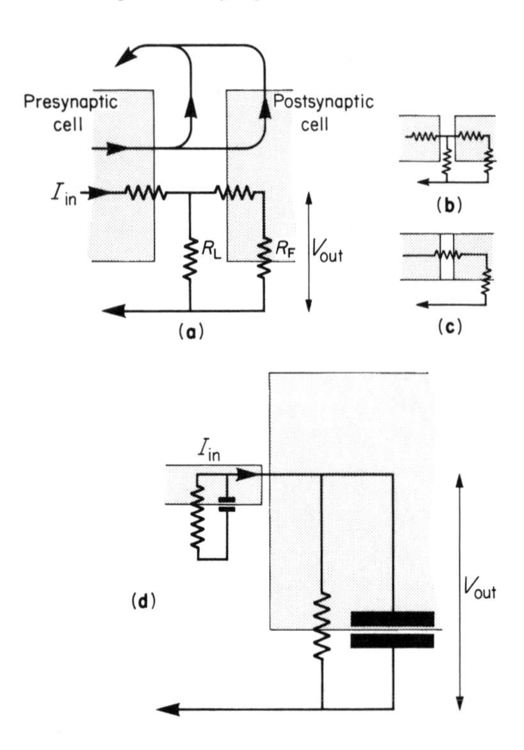

Fig. 3.10 The requirements for electrical transmission. (**a**) If a presynaptic current I_{in} is to create a sufficiently large depolarization V_{out} of the post-synaptic cell, sideways leakage through R_L must be small: thus the forward resistance R_F needs to be much less than R_L, as is achieved if the gap is reduced (**b**) or in a tight junction (**c**). (**d**) Even if R_L is negligible, if the post-synaptic cell is large in comparison with the presynaptic ending, its large capacitance and small membrane resistance result in a low impedance, and I_{in} may still be insufficient to cause a threshold change in V_{out}. This would be the case for an ordinary neuromuscular junction

simply leak out sideways through the synaptic cleft. The greater the fraction of current that takes the former route, the greater will be the degree of electrical coupling between the two neurones, since by entering the post-synaptic cell the currents will cause potential changes that may, if large enough, trigger a new action potential. The degree to which the current chooses one route rather than the other will in turn depend rather critically on the width of the synaptic gap. The smaller it is, the lower will be the impedance to current passing into the post-synaptic cell, and the higher the impedance to current escaping sideways. Those synapses that are known to operate electrically are invariably found to be of the gap or tight junctional form, reducing or preventing this sideways leakage of current. In the case of the cardiac muscle fibres, for example, tight junctions between cells provide for low-resistance pathways that allow the free passage of action potentials from cell to cell just as if no barriers existed—the whole mass of cardiac muscle acts as if it were a single cell. But even if no leakage at all occurs, and the transmembrane impedance at the junction is reduced to zero, there is *still* no guarantee that an action potential will be able to pass successfully from one cell to another. The size of the local currents that flow during the passage of an action potential along an axon is strongly dependent on the size of the axon itself. The larger it is, the greater the number of sodium channels per unit length and so the larger the active currents that can be generated. But equally, larger axones have greater capacitance and smaller transverse resistance per unit length; consequently the currents have to be that much larger to achieve a particular threshold level of depolarization. In other words, the currents automatically keep up with the increased requirements as the fibre's diameter is increased. But if we imagine a small axon whose diameter suddenly gets bigger at a particular point, or—as comes to the same thing—a small axon with a low-resistance electrical synapse joining it to a cell body of larger size, this is obviously no longer the case (Fig. 3.10). To trigger an action potential, the larger cell requires larger currents which the small axon may well be unable to provide; if a burning thread is attached to a rope, the heat the thread generates may be insufficient to ignite the rope. One can calculate, for example, that electrical transmission across the ordinary neuromuscular junction is *in principle* impossible even if the junction were a low-resistance one (which it is not): the impedance ratio on the two sides is much too large for the axonal currents to make any significant impression on the potential of the muscle cell. It is clear, therefore, that in such cases an *extra* source of amplification in addition to that provided by the action potential mechanism is needed. It turns out that this additional amplification is chemical in nature: the knot between thread and rope is soaked in petrol.

Chemical transmission at the neuromuscular junction

In a typical synapse in the central nervous system, we can see that the presynaptic ending contains a large number of *vesicles* (Fig. 3.9), each about 50 nm across and containing a transmitter substance that is released from the ending when it is invaded by an action potential. The mechanism by which this release occurs, and how the transmitter then excites the post-synaptic cell, have been most clearly demonstrated at the neuromuscular junction, and it is helpful first to consider this type of synapse in detail before going on to consider others. The transmitter at the neuromuscular junction is acetylcholine; each vesicle contains some 10^5 molecules of ACh, and when an action potential arrives at the endplate it triggers the release of the contents of some 200–300 vesicles: each vesicle appears to obey a kind of all-or-none law in that it either empties completely into the synaptic cleft, or not at all. The transmitter thus released must diffuse across the synaptic cleft—a process that takes a millisecond at most—before it can act on the muscle cell. To see what it does there when it arrives, it is best to work backwards from the electrical changes that are observed in the muscle cell when the axon is stimulated.

If we put a microelectrode close to the endplate, and poison the muscle with tetrodotoxin so that our observations of the primary electrical events are not obscured by any subsequent action potentials that may be generated, we see that a single action potential in the afferent nerve is associated with a characteristic electrical response in the muscle cell, called the *endplate potential* (Fig. 3.11). Although the original action potential only lasts a millisecond or so, the endplate potential or EPP is relatively prolonged. Most of this prolongation is due to the capacitance of the membrane, and knowing the value of the appropriate time constant, we can estimate the duration of the current flow that must have produced the potential change. It turns out to have a time-course not very different from that of the original action potential, though delayed in time by the millisecond or so of synaptic delay that is the result of diffusion across the synaptic cleft. This brief current discharges the membrane capacitance, which subsequently must recharge relatively slowly through the resting membrane resistance (Fig. 3.11). The fact that the current

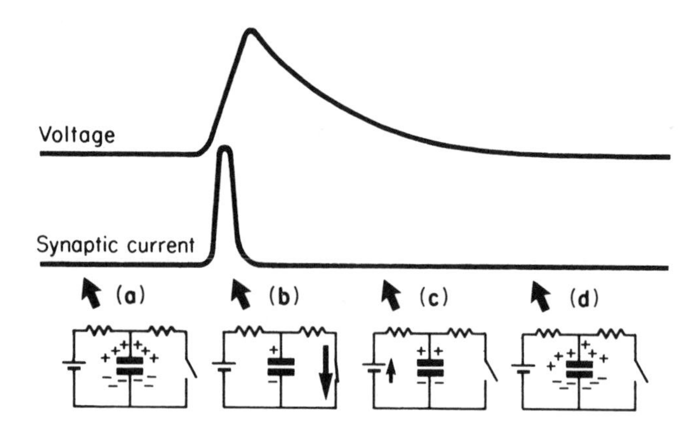

Fig. 3.11 Above, relation between endplate potential and synaptic current (somewhat idealized); below is a schematic equivalent circuit of the post-synaptic membrane showing: (**a**) membrane capacitance charged to resting potential E_r; (**b**) rapid discharge through opening of unselective channels; (**c**) channels closed, capacitance recharges relatively slowly; until (**d**) equilibrium is restored when the capacitor is fully charged. (Arrows at (**b**) and (**c**) represent current flow; (**a**),(**b**),(**c**),(**d**) are each about 10 msec apart)

hardly lasts longer than the afferent impulse is largely due to the presence at the ending of high concentrations of the enzyme *cholinesterase* that mops up the acetylcholine almost as soon as it arrives. If this enzyme is blocked by an anticholinesterase such as eserine, one finds that the current flow, and hence the endplate potential, is enormously prolonged, leading to a depolarization block of the muscle fibre.

The first clue as to the source of the current produced by the arrival of acetylcholine comes from measuring the reversal potential of the response, in exactly the same way as was described in the case of the Pacinian corpuscle. A steady current is passed into the muscle cell in order to set the resting potential at a new artificial level, and the size of the EPP is then observed. Just as with the corpuscle, it is found that the EPP gets smaller and smaller as the resting potential is reduced to near zero, and that if the membrane is hyperpolarized, the EPP is reversed. The conclusion is therefore that the effect of acetylcholine is to open channels that allow sodium as well as potassium to pass through the post-synaptic membrane, and thus produce something like a short-circuit. Since the number of ACh molecules released by each impulse, and hence the number of channels opened, is very large, it is clear that this is the mechanism whereby the relatively small currents in the axon can trigger off the relatively enormous currents needed to initiate an action potential in the muscle cell: the source of these currents is the muscle cell itself.

If one records from the endplate with very high sensitivity, one finds that even when the afferent fibre is not stimulated there are continual spontaneous potential changes taking the form of a random succession of *miniature endplate potentials* having roughly the same shape as a normal evoked EPP, but about 0.2 – 0.3 per cent of its size. These miniature potentials have been shown to be due to the fact that the presynaptic ending, even at rest, releases individual vesicles randomly at a very low rate. What is significant about this observation is that this rate of spontaneous release is strongly dependent on the resting potential across the presynaptic terminal, and if this is artificially reduced—for example by changing the external potassium concentration—the average rate of vesicle release increases sharply. By extrapolation, one can show that the size of a normal EPP is about what would be expected if the action potential simply had the effect of temporarily increasing the rate of spontaneous release of vesicles. This rate is also strongly influenced by the concentration of calcium ions at the ending: if it is increased, one again finds that the rate of vesicle release goes up. It appears in fact that what happens when the action potential invades the terminal is that channels in the presynaptic ending are opened that permit the entry of calcium (rather than sodium), and that this in turn stimulates the emptying of the vesicles into the synaptic cleft. How this comes about is not fully understood: possibly the calcium ions promote some kind of attachment between vesicle and presynaptic membrane that results in the voiding of its contents. Certainly with the electron microscope one can see at some synapses a regular array of structures on the inside of the presynaptic membrane that may well be the sites of this kind of attachment. At all events, we have now completed the causal chain that leads from an impulse in the afferent nerve to excitation of the muscle cell. An action potential arrives at the presynaptic terminal, allowing entry of calcium through special voltage-sensitive channels; the calcium encourages the vesicles to release their acetylcholine, which diffuses across the synaptic cleft and causes an unselected increase in the permeability of the post-synaptic cell; this then generates a current which may depolarize the surrounding membrane sufficiently to trigger off an action potential (Fig. 3.12). As in the case of the Pacinian corpuscle, it is found that the subsynaptic membrane, being specialized in having short-circuiting channels that are opened by acetylcholine, does not have enough of the voltage-sensitive sodium channels to be able to propagate an action potential, and is therefore electrically inexcitable: it initiates impulses only by generating local currents which pass through the surrounding region that *is* excitable. In some muscle fibres—for example, the slow fibres of the frog—no action potential is generated at all: here there is not

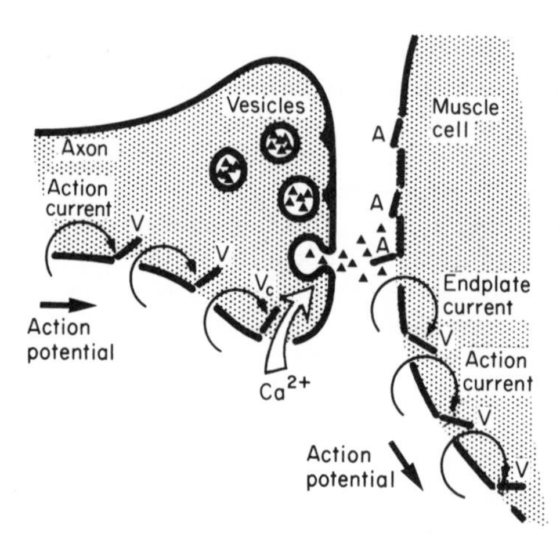

Fig. 3.12 Schematic representation of the sequence by which nerve action potentials lead to muscle action potentials at the neuromuscular junction. Channels sensitive to voltage are labelled V; V_C is a voltage-sensitive channel that permits calcium entry; channels sensitive to acetylcholine (which is represented by the black triangles) are labelled A

just one endplate on the cell, but a large number of synapses distributed all over its surface; activation of the afferent fibres thus causes a widespread passively summated EPP over the whole cell, leading to a relatively slow contraction of the muscle fibre.

Central excitatory synapses

Once the principles of operation of the neuromuscular junction are understood, those of central excitatory synapses present little extra difficulty. The most frequently studied synapse of this kind is one forming part of the *monosynaptic reflex arc* in the spinal cord that generates the tendon jerk response. This reflex pathway will be discussed more fully in Chapter 10; at the moment, it is sufficient to note that it consists of exactly two neurones: a 'primary' (Ia) afferent fibre carrying impulses from stretch receptors in a muscle synapses excitatorily with a motor neurone in the ventral horn of the spinal cord, whose axon retuns to innervate the same muscle group from which the afferent fibre came (Fig. 3.13). Tapping the tendon of the muscle causes a brief stretch of the sensory ending, firing the Ia fibre, which then excites the motor neurone and causes a reflex twitch of the muscle—the familiar knee-jerk response, if we use the patellar tendon. The advantage of this reflex pathway from the experimenter's point of view is that the afferent fibres are readily accessible in the dorsal root for controlled stimulation, while the post-synaptic cell bodies are large enough to be punctured by a microelectrode with little difficulty. If we apply a single brief shock to the Ia fibres whilst recording from the motor neurone, we find that the post-synaptic response consists of a small depolarization rather similar in shape to the EPP, called the *excitatory post-synaptic potential* or EPSP; if it is large enough, it may trigger off an impulse in the motor neurone. In exactly the same way as in the case of the neuromuscular junction, it is possible to show that this potential is the result of a transient increase in permeability to sodium and potassium ions with a duration comparable to that of the action potential: the EPSP itself is relatively prolonged because of the long time constant of the cell membrane (Fig. 3.13). One may therefore assume that the arrival of an impulse releases some transmitter substance from the vesicles visible in the presynaptic endings and that this substance diffuses across the synaptic cleft and causes the opening of short-circuiting channels. Could this transmitter also be acetylcholine?

It would be nice if we could simply extract the vesicles from the ending and see what was in them: but this is seldom a practical procedure. However, it is often possible to treat nervous tissue with histological stains that are selective for particular transmitters or their metabolic precursors, or for enzymes that are associated with them. In the case of acetylcholine, which is widely found as

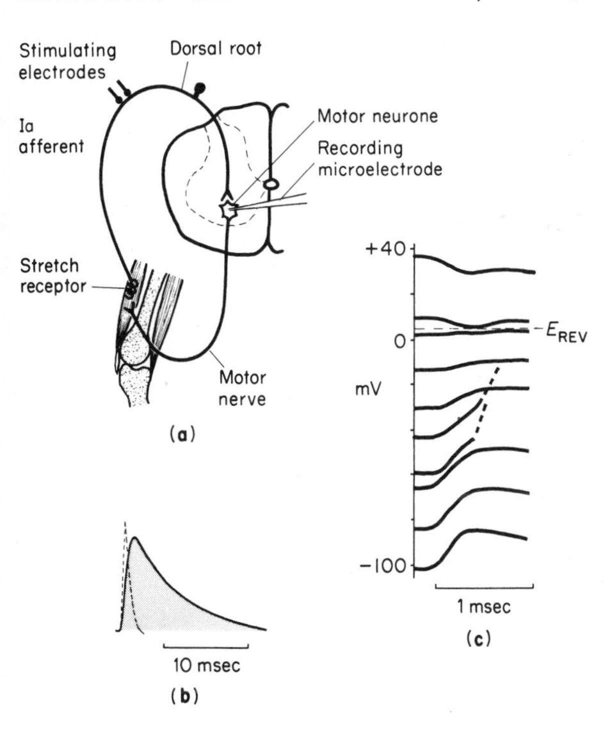

Fig. 3.13 Excitatory synaptic action in the monosynaptic stretch reflex. (**a**) The neural circuit: Ia afferents from stretch receptors in a muscle enter the dorsal root and then synapse excitatorily with motor neurones in the ventral horn which innervate the same muscle. (**b**) EPSP (shaded) recorded with a microelectrode in a motor neurone after a single stimulating shock to the afferent fibres. The dotted line shows the time-course of the synaptic current. (**c**) EPSPs recorded at different initial resting potentials produced by passing current steadily in or out of the neurone by means of a double-barrelled microelectrode. The response reverses at E_{rev}, not far from zero potential. (Partly after Curtis and Eccles, 1959 and Coombs *et al.*, 1955)

a transmitter within the brain as well as at the neuromuscular junction, one may stain for the enzyme cholinesterase, whose presence suggests strongly the use of acetylcholine itself. But the mere presence of a possible transmitter in the presynaptic endings is not sufficient evidence by itself that it is actually being *used* as a transmitter. There are a number of further criteria that have to be met. We need to confirm, for example, that application of the supposed transmitter actually causes the *same effects* as the real transmitter. It is not enough simply to note that both are excitatory: they must both open the same channels, and so have the same reversal potential. Ideally it should also do so in plausibly small concentrations, but this is a difficult criterion to meet because the post-synaptic membrane is very much less accessible from outside than it is to transmitter released in the proper way from the terminal. We must also demonstrate that the real and supposed transmitter have the *same pharmacology*: that they are blocked by the same pharmacological agents, and that substances that inhibit the inactivating enzyme for one, and hence prolong its action, do so for the other. Ideally, one should also be able to show that afferent action potentials really do release the supposed transmitter, but this also is often technically extremely difficult to establish adequately. In fact it is only at a relatively small proportion of the synapses in the central nervous system that we are certain of the identity of the transmitter; fortunately, this is not a matter of great importance since knowing what the transmitter is at a particular synapse does not really help us to understand what the synapse *does*. In the case of the afferent terminals of the monosynaptic reflex arc, we have no certain knowledge of the nature of the transmitter, except that it is certainly *not* acetylcholine. Substances which have been shown to be excitatory transmitters at other sites include (apart from acetylcholine) adrenaline, noradrenaline, dopamine, gamma-aminobutyric acid, serotonin, glycine and various small peptides; acidic amino acids such as aspartate and glutamate, though they often cause depolarizations that are superficially similar to EPSPs, have not been shown conclusively to be transmitters in the sense of meeting the criteria described above. But whatever the transmitter released by the Ia fibres, it is clear that it opens unselective channels in the motor neurone, which in turn cause depolarizing currents to flow. How does this generate post-synaptic impulses?

Synaptic integration

The situation is not quite the same as at the neuromuscular junction: there, in striated fibres at least, each muscle fibre receives only one endplate and it is often the case that a single afferent action potential will trigger off a single impulse in the muscle. But things are very different at a typical central neurone: here there is not one afferent synapse, but an enormous number—typically some 10 000. It is *not* generally true that a single action potential arriving at any one of them will fire off the whole cell, and indeed Ia afferent terminals release only a few vesicles for every impulse they receive. In fact the whole point of having such widespread convergence of interactions between cells in the central nervous system is that post-synaptic activity is a function of the *integrated* discharge of all its afferent terminals. Neurones thus exhibit what is called *spatial summation*: what matters in determining whether or not a cell fires is whether some point on its surface is sufficiently depolarized to exceed the threshold for producing an action potential. This in turn will be a function of the currents generated by all the synapses that happen to be active at that moment, wherever they may lie on the cell body or dendrites (though naturally the further away an excitatory channel is, the more of its current will be lost through leakage). This summation is not in general linear unless the synapses are sufficiently separated from one another for no interaction to occur *between* them. If a particular point on the cell surface is short-circuited by an excitatory synapse, short-circuiting another point very close to the first will have little additional effect: because the membrane is already depolarized, the second synapse will contribute less additional current than it would do if it were acting on its own (Fig. 3.14). Thus although sometimes the effect of stimulating two separate afferents to a motor neurone is the sum of the effect of stimulating each separately, more often it is substantially less, a phenomenon known as occlusion. Neurones also show *temporal summation*: because the potential produced by a brief synaptic current falls off relatively

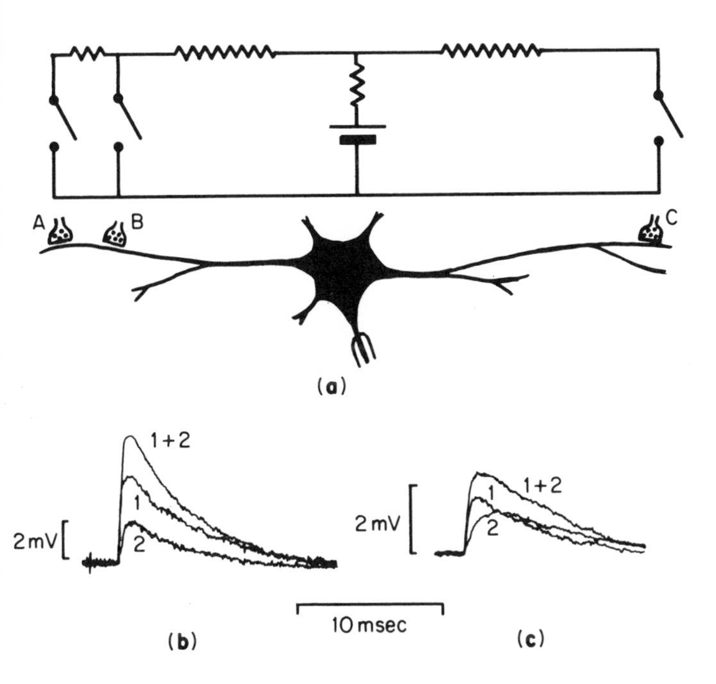

Fig. 3.14 Linear and non-linear addition of EPSPs by spatial summation. (**a**) Schematic neurone with its electrical analogue, showing three excitatory synapses. If A is active, activation of B as well will have little additional effect; but the excitation produced by C will add linearly to either A or B. Below, actual records of linear (**b**; equivalent to A + C) and non-linear (**c**; equivalent to A + B) addition of EPSPs obtained by stimulating different afferent nerve bundles. (Partly from Burke, 1967)

slowly, it is possible to get summation of the effects of repeated stimulation of a single ending if the frequency of firing of the afferent fibre is high enough. In fact, because of both this smoothing effect of the membrane time constant, and the very large number of endings synapsing with each neurone, most of which will probably be tonically active at any moment, it is probably more helpful to forget about the quantized nature of the action potentials at afferent synapses. It is rather all of them together provide a combined inward current that is continually varying as the result of changing patterns of afferent discharge, and which in turn makes the cell fire repetitively in exactly the same way as was described in the case of the Pacinian corpuscle, at a frequency that is a function of the total current. It turns out—again, rather as in the Pacinian corpuscle—that since so much of the cell body and dendrites of a spinal cord neurone is taken up with post-synaptic receptor sites that are inexcitable because they lack the voltage-sensitive sodium channels, the cell body as a whole has a higher threshold than the part that lies immediately adjacent to the axon and is free of synapses, called the *axon hillock.* It is possible to show that whenever a motor neurone is excited to fire by its afferent connections, the action potential actually starts not in the region near the excitatory synapses themselves but rather at the axon hillock, from which it spreads both forwards down the axon, and also backwards over the cell body and also possibly the dendrites. The axon hillock thus acts as a sort of detonator for the rest of the cell, and what determines the frequency of firing at any moment is simply the total current density in this region. Other things being equal, the nearer an excitatory ending is to the axon hillock the greater will be its influence on the firing of the cell; synapses on distant dendrites will be relatively less effective.

Inhibitory synapses

Now a nervous system in which the only connections were excitatory would not be a very useful one: clearly there are occasions on which the proper response to a stimulus is inhibition rather than excitation, withdrawal rather than attack, relaxation rather than activation. In particular, the way in which our muscles are generally arranged in pairs that oppose one another implies that excitation of one muscle is usually associated with inhibition of the other, by a process of *reciprocal innervation*. In the tendon jerk, for example, the reflex contraction of the muscle that is stretched is accompanied by a relaxation of its antagonist. In this case, the inhibition of the corresponding motor neurones is brought about by branches of the afferent fibres from the stretch receptors, that after entering the dorsal cord send excitatory branches to interneurones which in turn form *inhibitory synapses* with the motor neurones in the ventral horn (Fig. 3.15). One might wonder why a seemingly unnecessary interneurone is interpolated in this pathway. The reason may lie in a general rule that seems to be true of the transmitters used by cells in the central nervous system, namely that a neurone always releases the same transmitter, acting in the same way, at all its terminals (Dale's hypothesis). Since the stretch receptor fibres are excitatory to the motor neurones of the agonist

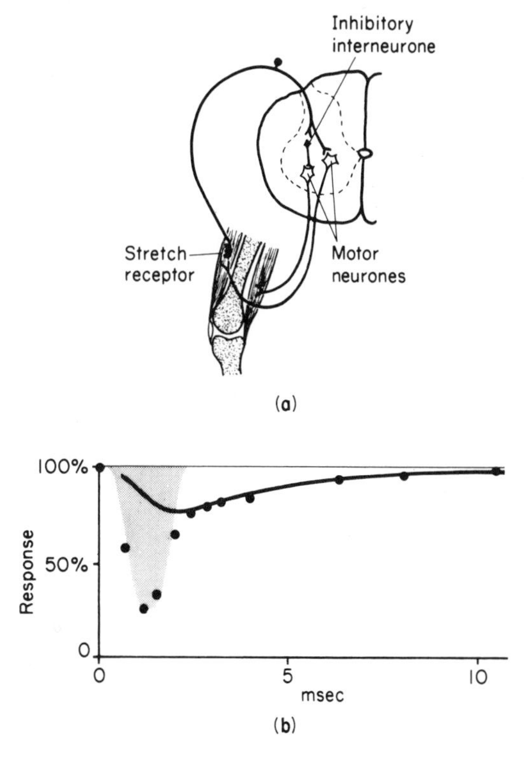

Fig. 3.15(a) Schematic neural circuit for reciprocal innervation of flexor and extensor muscles by stretch receptor afferents. The inhibitory interneurone is shown in black, excitatory cells in white. **(b)** Inhibition of the monosynaptic reflex, and its electrical correlates. Data points show the size of reflex evoked at different times after a shock that stimulates inhibitory afferents. The line shows, on the same time scale, the time-course of the IPSP associated with the inhibition, and the shaded area is the approximate time-course of the inhibitory synaptic current. The degree of inhibition in this case appears to be related both to current and potential. (Data from Araki *et al.*, 1960)

muscle, they cannot also be inhibitory to the motor neurones of the antagonist: thus they must first excite an interneurone, by opening the same kind of short-circuiting channels that they open in the motor neurones, and this interneurone must then inhibit the antagonist motor neurones.

The existence of reciprocal inhibition in the tendon jerk reflex is convenient from the experimenter's point of view, since it is possible first to insert an electrode in a motor neurone, and then stimulate various dorsal root fibres until some can be found that produce either excitation or inhibition of the motor neurone. If we then give a single shock to the inhibitory fibres, and follow it with an excitatory stimulus delivered after different time delays, we can measure the time course of the inhibitory effect by measuring the size of the subsequent response to the excitatory stimulus: some results of this kind are shown in Fig. 3.15. Here, an inhibitory shock clearly leads to a depression of the excitatory response that lasts for many milliseconds. One may also of course simply see what happens to the motor neurone's potential when the inhibitory shock is delivered in the absence of any excitation. One then finds that the stimulus is followed by a potential change in the neurone, of rather similar time-course to an EPSP, but of opposite polarity: this hyperpolarization is called the *inhibitory post-synaptic potential* or IPSP. To find the origin of the IPSP, we can follow our usual procedure of passing various steady currents in or out of the cell by means of one half of a double-barrelled electrode, and using the other half to measure the resultant size of the IPSP, and hence determine its reversal potential. What we find then is that the IPSP, unlike the EPSP, gets larger and larger instead of smaller as the resting potential is reduced to zero; but that if we artificially hyperpolarize the membrane, the potential is reduced in size and eventually reverses at about – 80 mV, the reversal potential for the IPSP (Fig. 3.16). This voltage lies somewhere between the equilibrium potentials for potassium and chloride ions, so our guess might be that the action of the inhibitory transmitter here is to increase the permeability not unselectively as in the case of the EPP, EPSP and generator potential, but specifically only to chloride and potassium. One way of finding out whether this is true, and in general exactly what ions will or will not pass through these inhibitory channels, is to introduce different ions into the cell through a double-barrelled electrode, thus altering their

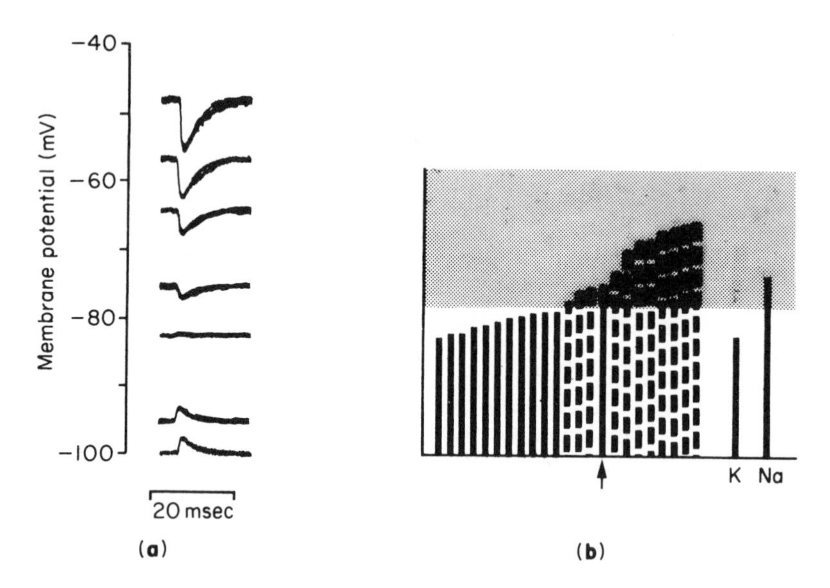

Fig. 3.16 Ionic mechanism of the IPSP. (**a**) IPSPs recorded at different initial levels of depolarization, as in Fig. 3.13, showing a reversal potential of about –80mV (after Eccles *et al.*, 1964). (**b**) The correlation between hydrated ion size and effect on reversal potential. The bars indicate by their height the mean hydrated diameter of a number of different ions: the black bars are those that do, and the grey those that do not, influence the reversal potential for IPSPs when injected iontophoretically into spinal neurones. Apart from the bar marked with an arrow (HCO_2), for this effect there is clearly a critical diameter that lies between those of sodium and potassium. (Data from Ito *et al.*, 1962)

concentration ratios across the membrane, and see what effect, if any, this has on the size or polarity of a subsequent IPSP. Clearly if the ion we introduce cannot pass through the inhibitory channels, it will have no effect on the IPSP; if it can, it will alter its reversal potential. A systematic study of this kind has shown that there is a clear distinction between those ions that do alter the IPSP and those that do not (Fig. 3.16), a distinction essentially based on their hydrated size. Small ions such as chloride and potassium appear to be able to pass through the inhibitory channels, while those such as sodium that are too large do not. The apparent exception to this rule, shown in Fig. 3.16, actually confirms the notion of inhibitory channels of a definite and limited aperture: HCO_2^- gets through because it is somewhat sausage-shaped! Of course in real life it is only chloride and potassium ions that are present in sufficient numbers to contribute significantly to the IPSP, and at other sites in the CNS, the IPSP may be due to the action of channels solely permeable to potassium or to chloride ions.

As in the case of excitation from Ia fibres, the transmitter here is not known for certain but elsewhere in the central nervous system both GABA (gamma-aminobutyric acid) and glycine have been confirmed as inhibitory transmitters. In the spinal cord, however, GABA fails to meet the pharmacological criterion mentioned earlier, for although the inhibition here is blocked by the convulsant poison strychnine, as is the action of glycine, that of GABA is not. Where GABA *is* known to be an inhibitory transmitter it is blocked by another convulsant, picrotoxin, which is ineffective at this site in the spinal cord. Incidentally, the convulsant effect of these inhibitory blockers illustrates another general function for inhibition in the brain: if we have a large network of cells that are connected together in highly convergent and divergent pathways that are entirely excitatory, we have a situation that is potentially explosive. Stimulation of any one cell is likely to lead to a chain reaction involving the progressive spread of activity over a large area, and this is precisely what is observed with convulsants like strychnine and picrotoxin. At least as much inhibition as excitation is required if this sort of explosive response is to be avoided, and we shall see later, in Chapter 14, that special systems exist in the brain to regulate the general level of neural activity through diffuse inhibition—very like the damping rods in a nuclear reactor—and thus prevent fits of this kind from occurring.

There is another site in the spinal cord where inhibition is relatively easy to study. The axons that leave the motor neurones of the ventral horn on their way to the muscles also send off branches that turn back into the cord and innervate—excitatorily—small interneurones called *Renshaw cells* (Fig. 3.17): from Dale's hypothesis we would expect the transmitter at this synapse to be acetylcholine, and this is found to be the case; it acts in the usual way, by causing an unselective increase in membrane permeability. But these Renshaw cells themselves send off short axons that in turn synapse with the pool of motor neurones by which they are stimulated, and their synapses are inhibitory. They are relatively easy to study because one can activate the Renshaw cells by stimulating the motor neurones antidromically in the ventral root. This kind of feedback inhibition is a common one in sensory systems as well, as we shall see, and really embodies yet another kind of adaptation.

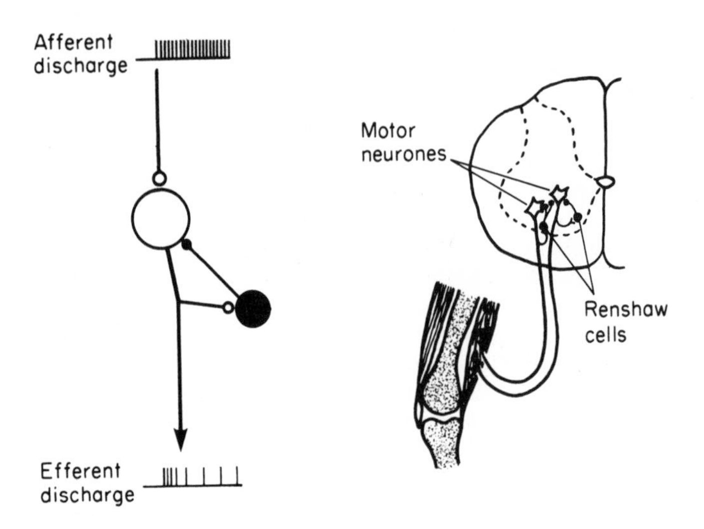

Fig. 3.17 Feedback inhibition. Left, schematic representation, showing how a sudden maintained burst of afferent activity is converted into an adapting efferent response. Right, Renshaw cells in the spinal cord, showing feedback inhibition of motor neurones (inhibitory cells and endings shown in black)

During a period of constant afferent stimulation, the efferent discharge will be large at first, but then reduced as the inhibitory pathway comes into play, producing a transient response that is essentially the same as incomplete adaptation (Fig. 3.17).

Voltage and current inhibition

We have now completed the chain of events by which stimulation of inhibitory afferents leads to release of a transmitter that then opens up channels in the post-synaptic membrane large enough to allow potassium and chloride through, but not sodium, leading in turn to a hyperpolarization as the membrane potential moves towards the reversal potential for the IPSP. But how does this result in actual inhibition? The answer is not quite as simple as might appear at first sight. In some cases, the time-course of the inhibition mirrors quite accurately the time-course of the IPSP itself. Other things being equal, a hyperpolarization means that the potential has to be driven further than would otherwise be the case in order to reach threshold, and so one might well expect a close degree of correlation between the degree of hyperpolarization at any moment and the degree of inhibition. But in many instances, as can be seen in Fig. 3.15, there is an extra peak of inhibition at the beginning that cannot be explained in this way, and in some cases one may find this short-term component even in the absence of the IPSP-like slow component. It turns out that the shape of this peak is very similar to the time-course of the burst of current that generates the IPSP: this current is shorter in duration than the IPSP, being roughly the same as that of the action potential, because as usual the decline of potential back to its resting level is prolonged by the membrane capacitance. Thus there appear to be two separate components of the inhibition that may be observed: one that is closely related to the *potential* at any moment (voltage inhibition), and is relatively easily explained, and one that seems to be associated with the *current* (current inhibition). To see how current inhibition arises, consider two synaptic endings, one excitatory and one inhibitory, lying close to one another on the post-synaptic cell membrane. It is clear that if both happen to be active simultaneously, they will to some extent cancel each other out, since one is a current source and the other a current sink: the currents that would otherwise be generated by the excitatory ending, and might eventually initiate an impulse at the axon hillock, are mopped up before they have got any distance at all. But this inhibitory effect would clearly only operate while the inhibitory channels are actually open: as soon as they close the excitatory current would be free to exert its effects at a distance as before. This kind of inhibition is quite distinct from the voltage effect. Imagine for example a hypothetical channel whose associated reversal potential happened to be exactly the same as the resting potential. Clearly such a mechanism could not generate an IPSP; but it would still be *inhibitory* because while it was open it would tend to clamp the membrane potential firmly at the resting level, by draining away current generated by any nearby excitatory synapses that happened to be active. In a sense, the increase in P_{Cl} during the IPSP does just this: because E_{Cl} is normally close to the cell's resting potential, this increase contributes nothing to the hyperpolarization: in fact it actually makes the amplitude of the IPSP less than it would be if there were only an increase in P_K. The importance of Cl^- lies in its current effect, in clamping the membrane potential close to its resting level.

We can now perhaps see why it is that some inhibitory effects seem to be of the current type and some of the voltage type, or a mixture of the two. If the excitatory and inhibitory synapses involved happen to be close to one another, the inhibition will be predominantly of the current type, and of short duration. If they are separated, for example on different dendrites, they will not interact directly with one another, but their effects will simply add at the site of initiation of the action potential, producing the voltage effect. Spatial summation of excitation and inhibition is thus rather complex, and—as in the case of summation of EPSPs—not just a matter of simple linear addition. Again, inhibitory endings that are very close to the axon hillock region will be particularly good at preventing the cell from firing, because they will ambush the excitatory currents just before they reach the detonator region. Indeed it is frequently found in the central nervous system that powerful inhibitory synapses are found clustering near the axon hillock and acting as a sort of guard ring around it.

Presynaptic inhibition

A phenomenon that puzzled early investigators was that sometimes one could find clear evidence of inhibition of a cell—in the sense that stimulation of certain fibres resulted in a reduction of the usual response to stimulating other afferents—*without* any corresponding change in its potential or permeability. In some cases this could be explained as 'remote inhibition'; in other words, an interaction between neighbouring inhibitory and excitatory endings of the type just described, on a dendrite so far from the recording site that although the inhibitory synapse was capable of cancelling the effect of the excitatory one through current inhibition, it could not generate currents large enough to be measurable to the cell body. But when histologists first noticed that not all the synaptic endings that could be seen in the cord were between axone and cell body or dendrite, but that on the contrary there were many occasions when an ending appeared to terminate against *another* ending that in turn synapsed in the conventional way (Fig. 3.18**a**), it became clear that another possible explanation of inhibition in the absence of any detectable change in the post-synaptic cell was possible: that of *presynaptic inhibition*. The notion here is that the ending on the second terminal may somehow hinder the latter's excitatory action, and so cause inhibition of its effects without influencing the post-synaptic cell in any way. Further evidence of such a mechanism came to light in the phenomenon known as *primary afferent depolarization*, or PAD. Selective stimulation of the small fibres of the dorsal root tends to cause a somewhat long-lasting inhibition of the monosynaptic response to stimulation of the Ia afferents. If one records from Ia fibres close to the cord, one finds that stimulation of these smaller fibres also leads to depolarization of similar time-

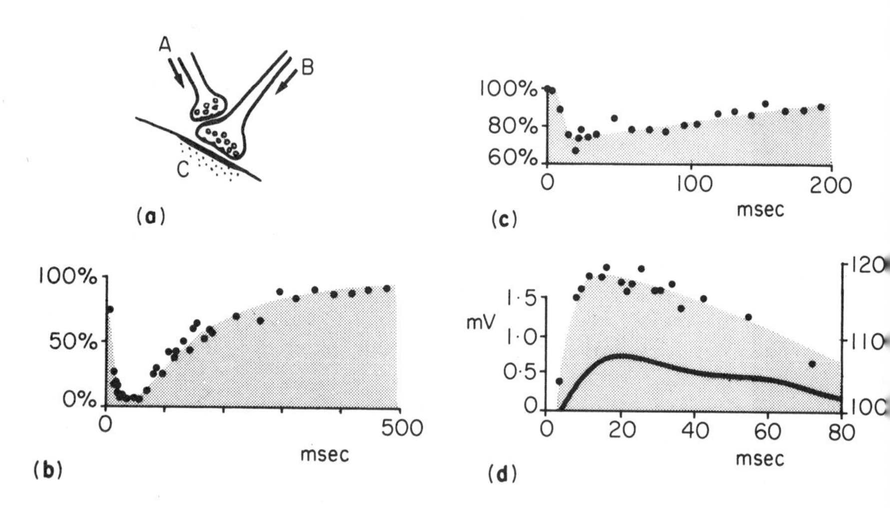

Fig. 3.18 Presynaptic inhibition. (**a**) Schematic representation of a terminal A that presynaptically inhibits the excitation of C by B, by synapsing with B's terminal. (**b**) Time-course of depression of monosynaptic reflex after brief stimulation of afferents producing presynaptic inhibition, and (**c**) the time-course of the EPSP size in the same experiment. (**d**) Time-course of primary afferent depolarization: the solid line shows the potential measured across a Ia afferent fibre after a stimulus generating presynaptic inhibition; the shaded area and data points show the corresponding increase in excitability due to the depolarization. (Data from Eccles, 1963; Eccles *et al.*, 1961, 1962*b*)

course (Fig. 3.18**d**). Since one can also observe that the primary afferents are among those that receive synapses on their terminal boutons from interneurones in the dorsal horn, it seemed likely that the PAD was simply the result of the action of these axo-axonal contacts on the terminals, and a side-effect of the mechanism by which they were inhibited from releasing as much transmitter as they normally would. This has now been almost entirely confirmed: the presynaptic transmitter (which may possibly be GABA) causes an unselective increase in the permeability of the excitatory ending, leading to depolarization. This action is blocked by the convulsant picrotoxin, but prolonged by such CNS depressants as chloralose and the barbiturates.

One might well wonder *why* this should lead to inhibition: surely an increase in the excitability of the ending ought to cause an increase in the amount of transmitter released by each impulse. The answer seems to lie in the relationship that is thought to hold between terminal depolarization and rate of release of transmitter. For those synapses where it is possible to study this relationship quantitatively, it is found that there is a very sharply rising increase in the rate of release of transmitter as a function of the potential across the ending: in other words, during the entry of the action potential it is really only the *peak* of the impulse that contributes significantly to the number of vesicles that are released. Now if we open up short-circuiting channels in the terminal itself, although admittedly the consequent depolarization of the resting potential should increase the steady rate of transmitter release, equally it will reduce the peak potential of any afferent impulses, by pulling the membrane potential towards zero (Fig. 3.19). Since the peak is what counts,

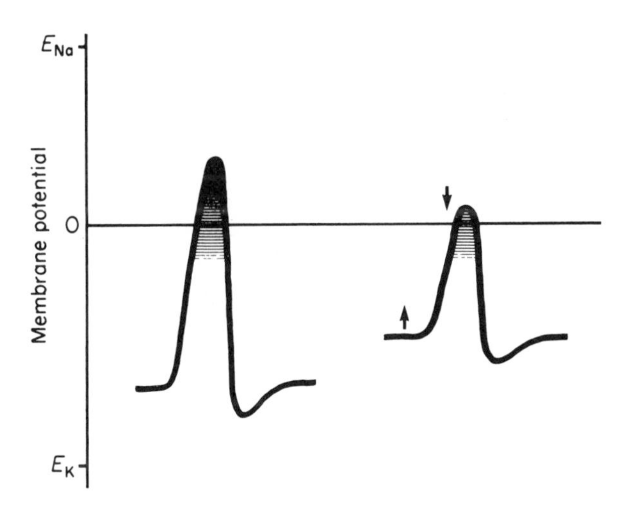

Fig. 3.19 Effect of depolarization of terminal on size of peak of action potential (schematic). Left, a normal action potential. Because of the steep dependence of rate of transmitter release on depolarization, only the peak of the action potential (shaded) contributes significantly to the amount of transmitter released. Right, partial short-circuiting causes a tonic depolarization, but also reduces the peak of the action potential, since the effect is to pull the voltage at every moment towards zero (arrows). Consequently there is a relatively large reduction in the amount of transmitter released

this latter effect will more than compensate for the resting depolarization, and the consequence will be a *reduction* in the amount of transmitter released by the terminal, and hence in the size of the ensuing EPSP. However, it must be said that not all investigators believe that such a mechanism will explain every kind of presynaptic inhibition. In particular, the extremely long time-course of presynaptic inhibition is puzzling (though one may postulate some kind of reverberatory neural circuit that makes the inhibitory interneurones fire for a longer time than the duration of the stimulus), and other explanations for presynaptic inhibition are possible. For example, in considering the way in which the neural elements of the central nervous system are densely packed together with apparently rather little extracellular space, we ought not to neglect the possibility that ionic concentration changes, of a kind that we can usually neglect when considering transmission in peripheral axons, may cause long-term changes in permeability and excitability. In particular, the role of the ubiquitous *glial cells*, that occupy most of the space not taken up by the neurones themselves, is far from clear. Although they do not carry impulses, microelectrode recording shows that they act very like 'potassium electrodes', and show potential changes that reflect the local concentration of potassium in their environment. Since this in turn depends on the average degree of activity in the neurones around them, they are well placed to mediate some kind of long-term regulatory action, that could be spatially integrated through the gap junctions that they make with one another. For all we know, glial cells *may* play some more active role than merely acting as a kind of ionic buffering system for central neurones, conceivably being responsible for certain types of pre-synaptic inhibition, or even possibly the longer-term changes associated with memory (Chapter 14), and this will certainly be an area of interest in the future.

Functionally, presynaptic inhibition has the advantage of being rather more precise and specific in its actions than post-synaptic inhibition. In the latter case, an inhibitory ending acts on the post-synaptic cell as a whole, regardless of what the source of excitation may be (although it will affect some excitatory afferents more than others because of the spatial effects outlined earlier). But pre-synaptic inhibition provides a mechanism whereby certain inputs may be disabled while others are left unhindered: a *gating* function, which implies control over *which* of the many types of input to a cell may or may not be allowed to influence it. In the case of a motor neurone, we shall see that many different neuronal pathways converge to form synapses on it: tendon jerk reflex afferents, inhibitory afferents for reciprocal inhibition, descending fibres from the higher levels of the brain, afferents controlled by pain and other receptors in the skin, Renshaw cells, and many others. It would clearly be an advantage to be able to alter the strength of some of these inputs independently of the others, and presynaptic inhibition provides a mechanism for doing this.

Long-term changes in excitability

All the phenomena described so far in this chapter are really rather brief in duration, and fall into the category of what has been described as 'millisecond physiology'; but some other synaptic phenomena are of a rather longer time scale. The first of these concern the effects of repetitive stimulation, effects most easily seen at the neuromuscular junction or at autonomic ganglia. If an afferent fibre is repetitively stimulated at a high frequency, one usually finds that the evoked post-synaptic potentials gradually decline in size (Fig. 3.20); if the stimulation is stopped, and the fibre allowed to recover, the size of the potential in response to a single test shock gradually regains its former value:

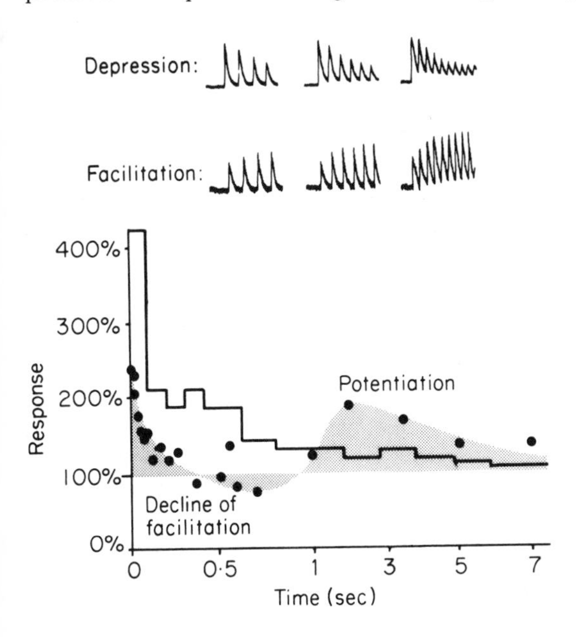

Fig. 3.20 Responses to repetitive stimulation. Above, depression and facilitation of endplate potential in curarized rat diaphragm muscle when stimulated at a high rate: in the lower set of responses, showing facilitation, the external calcium ion concentration was reduced by a factor of about 10 (Lundberg and Quilisch, 1953). Below, the size of the EPP in response to a single test shock at different times after a period of tetanic stimulation, in a magnesium-poisoned preparation (data points and shaded area); the line shows the average rate of miniature endplate potentials during the same period (data from Hubbard, 1963)

the whole phenomenon is known as *post-tetanic depression*. But if we partially block the release of transmitter during the stimulating period, by reducing the amount of available calcium, we find that instead of getting smaller and smaller during stimulation, the evoked potentials get bigger—a *facilitation*. At the end of the period of stimulation, the response to a single test stimulus returns to normal within about half a second, but this is then followed by a delayed and prolonged increase that is known as post-tetanic *potentiation*. These observations are readily explained by considering the metabolism of the transmitter that is released, in this case acetylcholine. Clearly, tetanic stimulation of the afferent fibre is likely to cause the loss of a certain proportion of the transmitter contained in its vesicles and immediately available for release. Normally this loss is made up by mobilization of stores of transmitter in the ending, and ultimately by the resynthesis of acetylcholine from its precursors—including choline that is recycled after the cholinesterase has finished with it at the post-synaptic membrane. But it appears that this replenishment is not quite fast enough to keep up with the extreme rate of release caused by high-frequency tetanic stimulation. Consequently the number of vesicles steadily declines, and so in turn does the amount of transmitter released per impulse, resulting in smaller post-synaptic potentials. When release is blocked, this decline no longer occurs. The potentiation that may then be observed may be due to the fact that the mobilization of transmitter, as well as its release, is in some way stimulated by the arrival of action potentials, helping to compensate for the expected depletion as a result of vesicle release. When this release is blocked, the increased production of acetylcholine is revealed by the fact that the amount of transmitter released per impulse is then correspondingly larger. The mechanism by which neural activity stimulates transmitter mobilization as well as release is at present unknown. In this context it is worth remembering that in many neurones the transmitter is synthesized in the cell body rather than in the endings. It is transported there by means of a specialized system that is probably associated with the activity of neurotubules and neurofibrils within the axon, at rates that may be as high as 15 mm/hour and it is quite possible that this transport mechanism may also be stimulated by the electrical activity of the axone. Another way in which the effectiveness of a synapse may undergo slow modification is by gradual inactivation of the calcium channels in the presynaptic ending as a result of its activity, resulting in a smaller calcium entry and hence less transmitter release: in some invertebrate neurones such a mechanism has been shown to be responsible for a kind of habituation.

The development of synaptic connections

Finally, there is the interesting and really rather basic question of how synaptic connections are set up in the first place. The human brain is perhaps the most complex structure known to Man, and if we understood the rules that govern the way in which its innumerable and intricate synaptic connections are specified and formed, we would have come a long way in our understanding of its function. Bearing in mind that there are some 10^{10} neurones in the brain, and that each one receives and gives on average some 10^4 synaptic connections, it is quite inconceivable for these patterns to be specified *in detail* by the instructions for building the brain embodied in our DNA. Though the broader structure of the brain—in terms of tracts that connect nuclei and other subpopulations that are relatively homogeneous within themselves—might be genetically specified, one must conclude that the connections between individual neurones are either essentially random, or more probably that they are in some way governed by our own sensory experience. The latter is an attractive hypothesis, since it implies that the structure of the brain may in a sense be capable of *self-organization*, of adapting itself to the particular tasks and the particular types of sensory stimulation it has to cope with. In this sense the brain may be thought of as rather like a telephone exchange, in which, when first built, only the broad outlines of the connections between its various elements are specified by its designer, and up to a point almost every unit in it is potentially capable of connection with every other. Once in use, the actual pattern of links at any moment is clearly a function of the patterns of impulses that subscribers have sent to it from their telephone dials. It is also clear that something of this sort must be present to explain the modification of synaptic connections as a result of experience that is implied by the existence of memory, a topic that will be pursued further in Chapter 14. This is a very active area of research, and it is becoming clear that in particular cases there do exist mechanisms by which the brain can in effect build its own connections so as to adapt itself to a particular task, and that the instructions given it by the genetic code are essentially rather vague.

Let us consider first a question that may already have occurred to the reader. Clearly a synapse will only function properly if on the post-synaptic membrane there exist receptors that match the transmitter released by the presynaptic terminal. Is there then some mechanism that guides a developing axon towards only those cells that have receptors corresponding to its own transmitter? Or is it rather than when a nerve fibre approaches another neurone, in some way it stimulates the manufacture of the appropriate kind of receptor site? Our clearest information on this point comes from the study of the formation of the neuromuscular junction. If acetylcholine is applied locally to different parts of a muscle cell's surface, it is found that only the region underlying the neuromuscular junction will respond with depolarization: presumably the cholinergic receptors are confined to the post-synaptic membrane. If we now cut the nerve fibre to the muscle we find that progressively more and more of the muscle cell's surface becomes sensitive to the transmitter, a phenomenon called *denervation hypersensitivity*. But if a new axon starts to grow towards the muscle cell, this process is reversed, and once again we find that the response to acetylcholine becomes limited to the region where the new junction is developing. It seems therefore that the presence of the nerve ending, presumably by the release of some substance, either attracts the receptors, or at least encourages their formation while suppressing those that are present elsewhere (a process reminiscent of the trophic affects described in Chapter 1). Furthermore, it appears that the development of this hypersensitivity is itself in turn a stimulus that attracts nearby axons and leads them to form new synaptic junctions. A normal frog muscle fibre has only one neuromuscular junction, and if a severed motor nerve is placed in its vicinity it will not form additional endplates to it. But if the original innervation is cut, it is found that the resulting hypersensitivity is also accompanied by the acceptance of a synaptic junction from a fibre that previously was ignored. In the same way, transplanting an extra limb at an inappropriate site in many amphibia leads to new fibres growing out from the central nervous system to innervate it. It is difficult to escape the conclusion that under these conditions the muscle must be releasing some kind of *nerve growth factor* that attracts potential axons, and indeed a small protein having just these properties has recently been identified and isolated.

Similar work on the regeneration of neural connections in amphibia (regeneration is not observed—at least not over such large distances—in the mammalian central nervous system) has shown that this guidance of neurones on to their targets can sometimes be even more specific; not just on to the correct type of cell, as defined by its receptor properties, but even on to the correct part of an extended mass of such cells. In the frog, there is an orderly projection of the fibres of the optic nerve to the frog's 'visual brain' (the tectum), that preserves the topology of the retinal image. If the optic nerve is cut, it is found that the fibres not only regenerate back to the tectum, but do so in such a way as to retain, at least approximately, their correct spatial arrangement. The mechanism by which this specificity of connection arises is not yet understood. It is perhaps worth emphasizing that in these experiments the guidance is anatomical rather than functional: if, after cutting the optic nerve, the frog's eye is rotated in its orbit through 180°, the pattern of the regenerating fibres is *not* also rotated through 180°. As a result, the animal's subsequent visual behaviour is inverted, with upward movements in response to objects in the lower visual field, and so on. In other words, there is no suggestion in these experiments that the pattern of *activity* in the incoming fibres can influence the pattern of their connections. But other recent experiments in mammals have shown that connections can in certain circumstances be altered by the pattern of neural activity, in such a way that only *useful* connections are formed, or useless ones are lost. For example, the cells of a cat's visual cortex (Chapter 7) are usually found to be driven in almost equal numbers by each eye, and many by both. But if one eye of a kitten about 5 weeks old is kept closed, even if only for a few days, one finds when it has grown up that the number of its cortical cells that are driven by the eye that had been closed is very greatly diminished. There has been no *anatomical* interference here: the only difference between the two eyes was the degree of their neural activity during the period of closure, so it must follow in this case that the synaptic connections have been influenced by the pattern of neural activity experienced. An even clearer example of degeneration of useless connections and the growth of useful ones has been demonstrated in the spinal cord of the kitten. If part of the innervation of unrelated pairs of synergic muscles is cut, and the cut ends crossed and reunited as in Fig. 3.21, regeneration will occur both of the motor nerves and of the sensory fibres coming from the muscle stretch receptors. In the normal animal, the afferent fibres form monosynaptic excitatory connections with the motor neurones both of the parent and its synergist; but after the cross-union, part of this excitatory projection will at first—quite inappropriately—be to motor neurones of the unrelated muscle pair. If, after 6 months or so, one records the electrical responses of the motor neurones to stimulation of the various possible afferent pathways, one finds that the inappropriate synaptic connections have weakened, and that conversely some new and appropriate monosynaptic connections have developed that did not previously exist. As can be seen in the figure, growth of new connections appears to be limited to those motor neurones whose axons were severed in the course of the experiment, suggesting some mechanism akin to denervation sensitivity, by which the injured neurone becomes more receptive to the formation of new synapses. Here again, the pattern of connections seems to be determined by

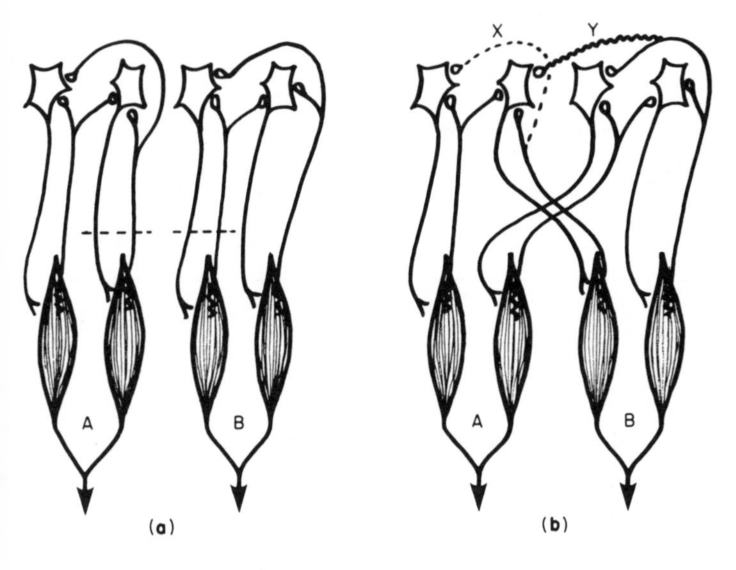

Fig. 3.21 Plasticity of neural circuits. (**a**) The connections existing before the experiment, between stretch receptors in two independent pairs of synergistic muscles, A and B, and their motor neurones. After cutting the nerves at the level indicated by the dashed lines, they were reunited as shown on the right. After recovery, certain inappropriate connections (dashed line *X*) were found to have weakened, while new appropriate ones (wavy line *Y*) were found to have appeared. (After Eccles *et al.*, 1962*a*)

the pattern of activity in the nerve fibres themselves, and synapses are formed and maintained only between afferent and efferent neurones whose activities are mutually correlated because they are connected to synergistic muscles. One can see how such a mechanism might well—in principle—determine the correct wiring of the monosynaptic pathways in the first place, whereby the stretch receptor afferents only form contacts with the properly corresponding motor neurones; and it is not too wild an extrapolation to imagine how similar processes might lead to the development of specific and appropriate neural connections in the central nervous system in general, and thus contribute to the brain's ability to *learn*: but our knowledge of such processes is still in its infancy.

4

Skin sense

This chapter is concerned with the information that comes both from *exteroceptors* in the skin, in many respects similar to the *interoceptors* of the gut and other visceral organs. The whole system is often loosely termed the *somatosensory* system, but this strictly also includes receptors from muscles and joints, which as *proprioceptors* are considered in Chapter 5. Many types of stimulus can produce sensations from the skin; and to a large extent the receptors and pathways of cutaneous sensation are *modality-specific*, responding preferentially to such specialized categories as pressure, cold, warmth and so on. But the concept of 'modality specificity' is not quite as straightforward as might be thought at first sight.

Sensory modalities

The concept of modalities comes about through our natural urge to classify the objects around us; the reason that difficulties arise in its use is that there are many *criteria* by which objects may be classified, and unless one is clear about which type of classification is referred to, misunderstandings become inevitable. If we consider all the kinds of things that may come in contact with the skin, we might group them according to their physical effects (as mechanical, thermal, etc.), or according to the sensations they produce (pain, tickle, softness), or even in terms of the types of peripheral nerve fibres they stimulate. Each of these classifications will in general divide the whole set of stimuli into different patterns of subsets (Fig. 4.1) which may or may not correspond with one another. Now if it happened that in each system of classification the boundaries were identical, as in (**a**) and (**c**) of the figure, then no difficulties would arise, and we could say with certainty that the fibres were modality-specific. For example, if we found a particular type of fibre that responded only to heating of the skin, and that this in turn was also a clear and distinct class of sensation, then one could say that the fibres in question were specific for that particular stimulus or sensory modality. But in practice, things are seldom so simple, and there is no uniquely valid way of classifying either the physical attributes of objects or the sensations they evoke. In particular, there is a danger of introducing a degree of tautology: one may be influenced by one's knowledge of one of three levels of classification when

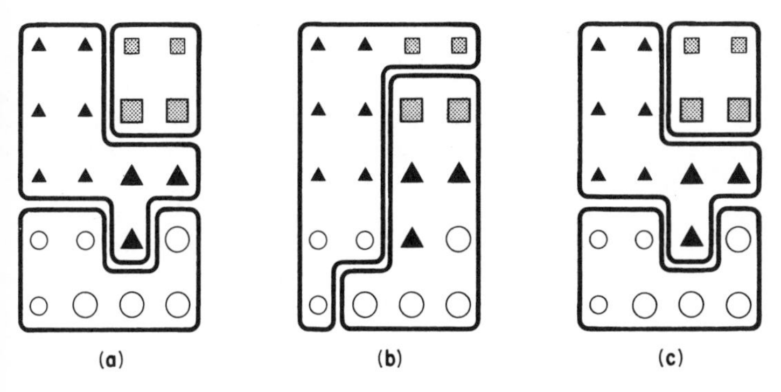

Fig. 4.1 A set of miscellaneous objects classified according to shape (**a**), size (**b**), and colour (**c**)

drawing up the boundaries for the others. If just for a moment you forget all you have been taught and ask yourself what you really *feel* to be the categories of cutaneous sensation, your list is likely to include not just the familiar stereotypes of pain, warmth, pressure and so forth, but also other sensations that are just as immediate and apparently 'primary': tickle, itch, softness, roughness, hardness, stickiness, wetness, sharpness, and many others. It is doubtful whether a man who had never read a physiology book would naturally consider the classical modalities to be more 'primary' than the others. Our classification of the physical classes of stimuli is almost equally biased: for instance, the all-important factor of local curvature of the skin, that gives rise to the sense of sharpness and roughness, is usually wholly ignored. On the other hand, to try and produce some physical quality that could be said to correspond with the obvious sensation of pain, it is customary to invent a special class of physical stimulus, whether mechanical, thermal or even chemical, that causes tissue damage of some kind and may therefore be called 'noxious', and sensed by 'nociceptors': yet many kinds of pain are not associated with tissue damage at all.

In other words, there is a danger of unconsciously falsifying what might be called 'natural' classifications of sensations or physical types of stimulus; and if our modalities are thus *defined* by what is observed in sensory fibres, then it must follow tautologically that the fibres appear to be modality-specific! A discussion about whether a particular system is modality-specific, or whether on the contrary a particular mode of stimulation gives rise to a characteristic *pattern of activity* amongst a set of afferent fibres that is indicative of that class of stimulus (as for example the letter 'A' does to our retinal fibres), amounts in the end simply to an argument about how we happen to name what we perceive. An example may make this clearer. Imagine a simple-minded creature—perhaps some kind of slug—whose cutaneous sensations fell into only three categories: 'wet', 'earth', and 'nice'—the last being the result of contact with a slug of the opposite sex. A slug who studies physiology and investigates the responses of his own somatosensory neurones would find that some fibres—what we would call 'light touch' receptors—fire during both 'earth' and 'nice', while others ('cold') fire during 'wet' and sometimes during 'earth', and so on: he would deduce in fact that his fibres were *not* modality-specific, but that 'earth', 'wet' and 'nice' were coded in the form of particular patterns of activity. A human physiologist would completely disagree: what *he* would report would be highly specific fibres responding to the traditional categories of 'warm', 'cold', 'light touch' and so on: but the argument would clearly be about the *naming* of sensory categories, not about the observations themselves. A further, insidious, bias that may creep into investigations of sensory systems—this applies with equal if not greater force to other special senses such as vision and audition—is that by having preconceived notions as to what the categories of stimulus are, based on categories of primary fibres rather than on what might be important to the organism in controlling its behaviour, one may tend to limit oneself to those categories when trying experimentally to evoke responses from higher levels of the brain. This error may become self-perpetuating: if one explores the neurones of the somatosensory cortex using only stimuli of light touch, warmth, cold, or one of the other traditional modalities, then naturally all the cells that respond at all must fall into one of these categories. If there were cells responding to more useful things like stickiness or wetness, one would never discover them; and so the myth would be perpetuated. So it is very important to bear in mind these reservations about over-simple categorizations of stimuli into modalities when considering the specificity of receptors and of central neurones to cutaneous and other kinds of stimulation.

The receptors and their central connections

The afferent fibres from cutaneous receptors are bipolar cells: their bodies lie in the *dorsal root ganglia* near the spinal cord, and axons run all the way from the sensory endings in the skin to their terminals within the central nervous system (Fig. 4.2). The dorsal roots are connected in an orderly way to different areas of the skin, and one may draw maps of the body surface showing the *dermatomes* or regions projecting to each dorsal root (Fig. 4.3). The demarcation of the different zones is not actually as sharp as such idealized representations suggest, and because of overlap between adjacent dermatomes, each point on the body surface is connected to at least two dorsal roots; overlap is more marked for touch than it is for pain or temperature. Sensory fibres from the *viscera* are found in both the sympathetic and parasympathetic divisions of the autonomic nervous system. The former pass in peripheral sympathetic nerves to the sympathetic chain, and thence via the dorsal root ganglia (where their cell bodies are) to the dorsal root; parasympathetic afferents of the sacral region travel with the corresponding efferents and again have their cell bodies in the dorsal root ganglion, while the cell bodies of afferents in the vagus are in the inferior (nodose) ganglion.

At the peripheral end, the fibres branch and terminate either as naked endings or in terminal *encapsulations*, of which many varieties have been described. One such is the *Pacinian corpuscle*, whose responses to pressure were discussed in the previous chapter. Others include *Meissner's* corpuscle, *Merkel's*

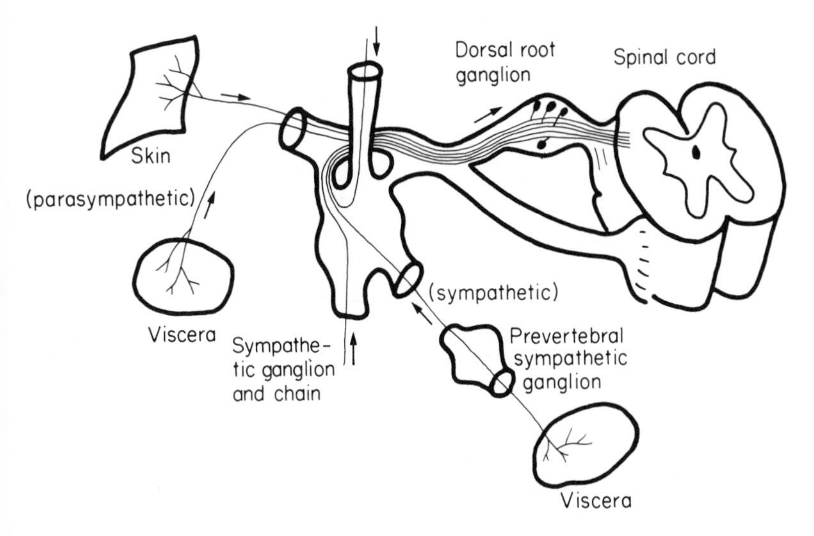

Fig. 4.2 Afferent pathways from the skin and viscera (schematic)

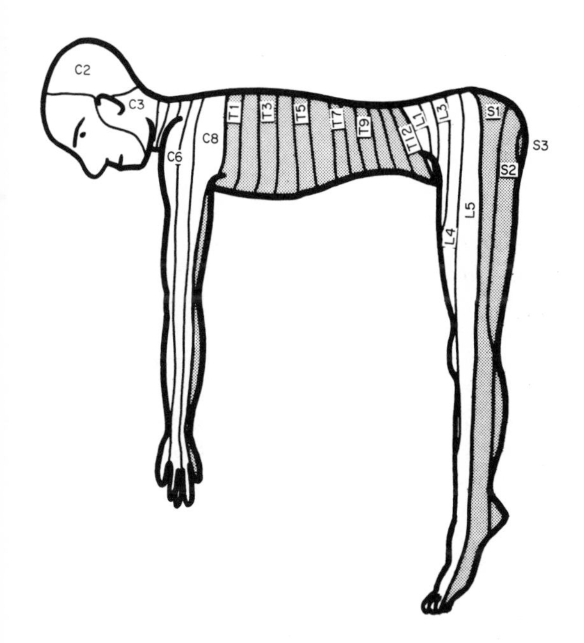

Fig. 4.3 Pattern of dermatomes in Man: C, cervical; T, thoracic; L, lumbar; S, sacral. The boundaries are not actually as sharply defined as this schematic representation implies, and there are discrepancies between maps produced by different authorities (data from Keegan and Garrett, 1948)

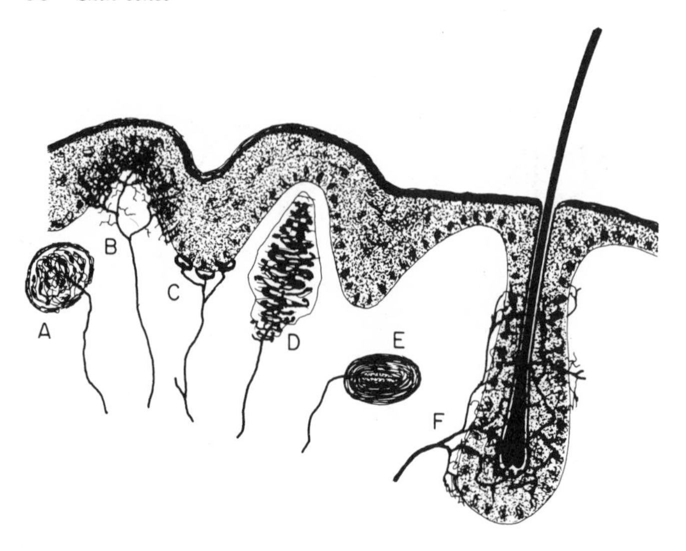

Fig. 4.4 Types of sensory ending in skin (idealized). A, Krause bulb; B, free endings; C, Merkel's discs; D, Meissner's corpuscle; E, Pacinian corpuscle; F, free endings associated with hair bulb

discs, and the end-bulbs of *Krause* and of *Ruffini* (Fig. 4.4). It is by no means clear what functional significance these various morphological differences may have, or whether indeed some of them may not simply be immature versions of others, or even artefacts of histological preparation. Encapsulated endings are found mainly in hairless or *glabrous* skin: the palms of the hands and soles of the feet, the lips, eyelids, mucosal surfaces and parts of the external genitalia. Some, notably the Pacinian corpuscles, are distributed in visceral structures and in joints and ligaments and deep connective tissue. Free or naked endings are abundant in hairy skin, some innervating the hair follicles themselves and sensing hair movement, and are also to be found in both glabrous skin and in deep fascia and visceral organs. Their afferent fibres are small and sometimes unmyelinated, falling into group C and group Aδ (or III and IV, with a few in II): the fibres from encapsulated endings are mainly of group Aβ (or II).

This broad division of afferent fibres into two groups on the basis of their size and type of ending is also reflected in their mode of termination within the central nervous system. The larger fibres, from encapsulated mechanoceptors, turn upwards soon after entering the dorsal horn of the spinal cord, to form a pair of large ascending tracts called the posterior or *dorsal columns* (Fig. 4.5). These continue ipsilaterally up to the level of the medulla and terminate in the *dorsal column nuclei* (*gracile* and *cuneate*); the gracile receives afferents from sacral, lumbar, and lower thoracic segments, and the cuneate from higher regions. From these nuclei, second-order fibres cross to the other side and continue up as the *medial lemniscus* to the *ventral posterolateral* and other posterior nuclei of the thalamus (VPL), and thence relay through the *internal capsule* to a region of cerebral cortex called the *somatosensory cortical area* or *SI*

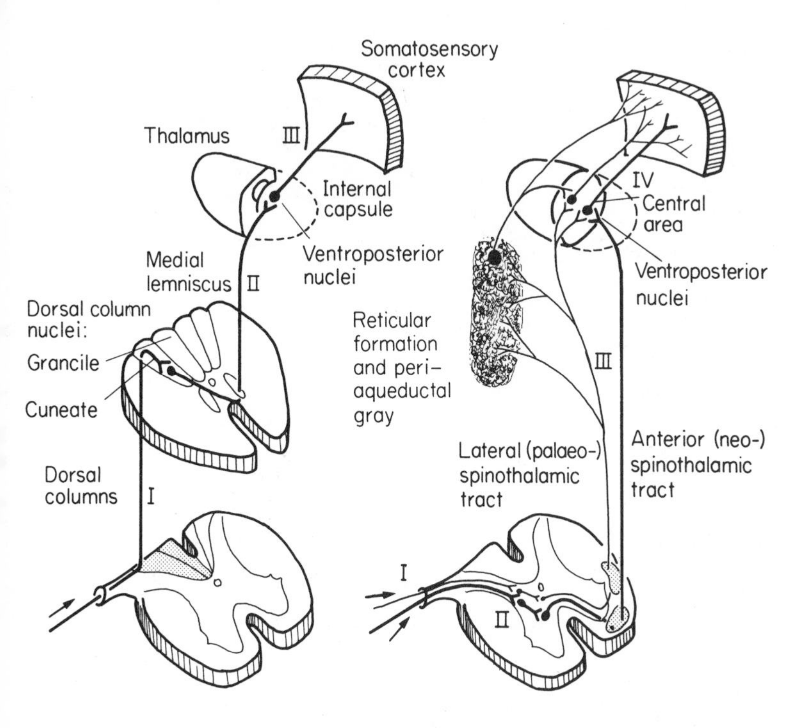

Fig. 4.5 The main ascending somatosensory pathways. Left, the lemniscal system; right, the neo- and palaeo-spinothalamic divisions of the anterolateral system. (The thalamus is shown in section to make the position of the central and ventroposterior nuclei clearer)

(areas 3, 2 and 1 in Brodmann's classification, based on the different histological appearance of different cortical regions). Throughout this system—often called the *lemniscal system*—the general topological relationship between the representations of different areas of the skin is preserved, so that the somatosensory cortex itself embodies a map of the opposite side of the skin surface, the *sensory homunculus* or 'little man' (Fig. 4.6). Although it is topologically correct in the sense that neighbouring parts of the body surface are on the whole represented by neighbouring regions of cortex, it is very much distorted in shape: those areas such as the hands and lips with the greatest cutaneous sensitivity and acuity have a much larger area devoted to them than regions like the trunk and back. A second somatosensory area, *SII*, is found in primates (Fig. 4.6) which differs from SI in receiving somatosensory information from both sides, and to some extent in the modalities to which it responds.

The smaller afferents, derived from free endings and also from some of the encapsulated ones, and concerned with temperature, pain and light touch, do not immediately ascend on entering the cord: instead, they synapse in the central grey matter with interneurones, which in turn excite third-order neurones whose axons cross to the other side and proceed upwards as part of the *spinothalamic* projection (Fig. 4.5). There are two of these spinothalamic

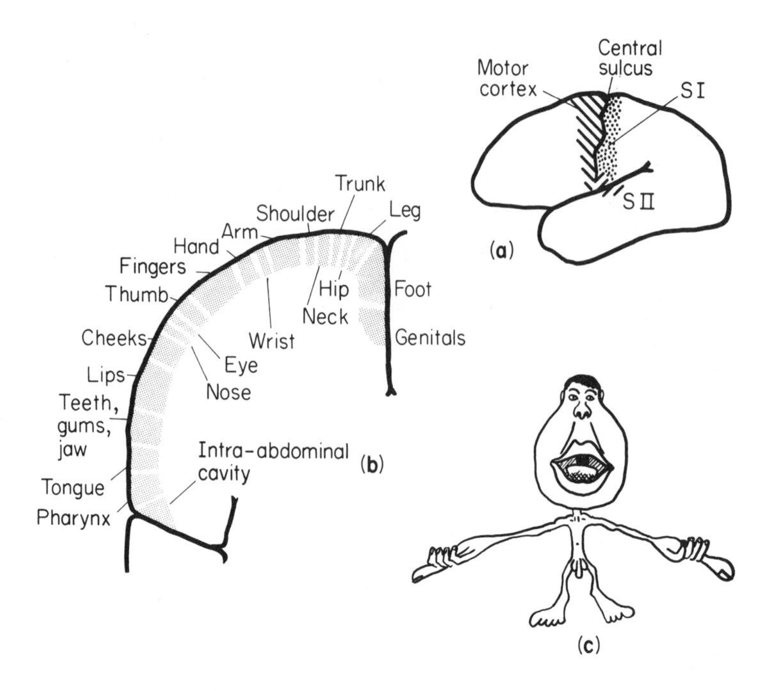

Fig. 4.6 (**a**) Lateral view of human brain, showing approximate positions of somatosensory areas SI and SII. (**b**) Coronal section through the primary somatosensory cortex in Man, showing the approximate areas associated with different parts of the body. (**c**): Sensory homunculus, distorted so as to indicate by the relative size of different parts of the body the relative areas devoted to each in the somatosensory cortex (after Penfield and Rasmussen, 1950)

pathways, the *anterior spinothalamic* tract and the *lateral spinothalamic* (the whole system is called the *anterolateral system*); though they are both evolutionarily older than the more recent lemniscal system, the anterior tract is more highly developed in higher animals than the lateral, and for this reason they are also known as the neo- and palaeo-spinothalamic pathways; however, more recent work suggests that one should not exaggerate the difference between them. The former projects only to the border of VPL, and to nearby regions that are not wholly somatosensory, terminating in large bushy arborizations (in contrast to the more compact lemniscal endings). The palaeo-spinothalamic afferents project to central and intralaminar regions of the thalamus, and also rather diffusely to the reticular formation of the medulla and pons: they are concerned more with pain and temperature than with touch. Thus on the one hand we have the new, fast lemniscal system with its precise and orderly projection of accurate mechanical information directly up to the cortex; and on the other, the older, slower, and more diffuse projection of less precise, but in a sense more immediately important information—often with an affective or emotional quality to it—by the anterolateral system. The differences between the two systems can be of diagnostic value in certain kinds of disorders of the spinal cord. Thus a hemisection of the cord, that interrupts all ascending fibres on one side, will result in a loss of deep pressure and vibration sense below the level of the section on one side, and loss of pain, temperature and light touch on the other. Deliberate anterolateral chordotomy is in fact sometimes performed to deal with otherwise intractable pain of peripheral origin. Finally, some cutaneous fibres ascend to the cerebellum in the posterior spinocerebellar tract (see p. 105).

Neural responses to cutaneous stimulation

The larger, Aβ or group II, fibres from the skin all respond specifically to mechanical stimuli. Some are completely adapting, and thus only respond to changes in the deformation of the skin: they are thought to originate from the Pacinian and Meissner corpuscles and from the endings in hair follicles. Because of their adapting properties, they are particularly sensitive to vibration, and thus are well suited to contribute to the sense of roughness when the hand is passed over a textured surface. They may well help one to sense when an object being lifted between the fingers begins to slip, and thus assist in regulating the pressure with which such an object is grasped. (Robot hands designed for grasping and lifting objects are sometimes provided with a similar sense, in the form of microphones built into the gripping surfaces, whose output is used in a feedback loop to increase the pressure when the object is slipping.) Other fibres show only incomplete adaptation and can therefore signal static deformation as well: they probably come from Merkel's discs, and possibly Ruffini endings.

In order to be effective in stimulating a fibre, a stimulus must lie within a particular area of the skin called the *receptive field*. The size of this field is partly a consequence of the unavoidable spread of the stimulus itself—any indentation of the skin, however localized, will cause deformation of the layers of the skin over a much wider area—but also the result of the branching of the afferent fibre and consequent distribution of its endings over an extended region. In the case of the fibres innervating hair follicles, for example, one finds that each fibre may innervate as many as a hundred follicles, and each follicle in turn receives branches from several fibres. Thus the receptive fields of the individual fibres are quite large, and also show a considerable degree of *overlap* (Fig. 4.7). Such a pattern of overlapping receptive fields is a common

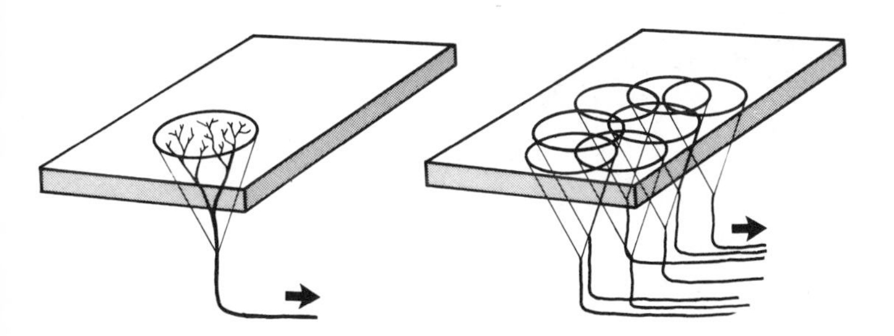

Fig. 4.7 Left, the receptive field of a single idealized cutaneous fibre; right, showing the overlap between neighbouring receptive fields

one in all kinds of sensory systems, and one might well wonder what the point of it is. Surely, one would think, it would be better to avoid this duplication by making the receptive fields smaller, resulting in an improvement of the precision with which a stimulus can be localized. Yet it turns out, on closer analysis, that a system in which the receptive fields overlap is not only just as good as one in which they are discrete, but is in many ways much better.

Consider first the question of accuracy of localization of a stimulus. If the fields were discrete, then all the brain could tell about the position of a stimulus is that it must lie somewhere inside a particular receptive field: there is no way in which it could find out *where* it lies within that field. But consider the case of two overlapping fields: if a stimulus lies within the area of overlap, then it will stimulate the two fibres in different proportions, depending on its exact position. So by analysing the pattern of discharge of the two neurones—by comparing the frequency of firing of one with that of the other—the central nervous system could determine the position of the stimulus much *more* accurately than if the fields were discrete (Fig. 4.8). Overlap has the further advantage that it makes the system much less vulnerable to damage: destruction of any one fibre will still leave each area of skin with the innervation of its neighbours (Fig. 4.8**b**). Now it is perfectly true that if one thinks of the receptors as converting the spatial pattern of the stimulus into a kind of 'neural image'—a corresponding pattern of firing amongst the array

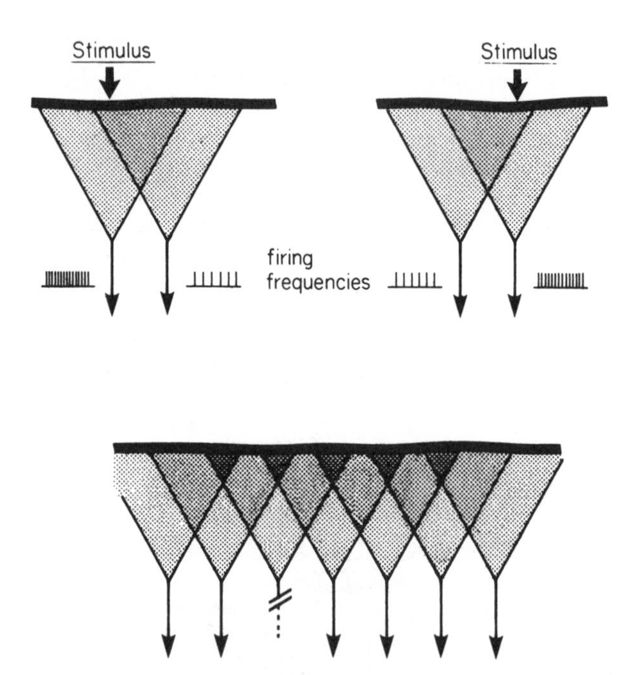

Fig. 4.8 Above, the exact position of a stimulus within an area of overlap can be determined from the relative activities of the corresponding fibres. Below, overlap of receptive fields means that damage to any one fibre does not necessarily produce an area of anaesthesia

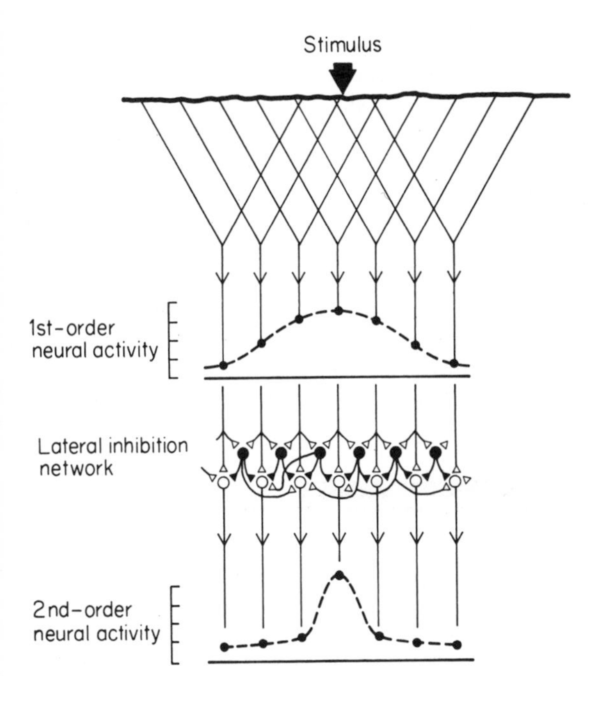

Fig. 4.9 Because of the existence of receptive field overlap, a single localized stimulus to the skin produces a blurred, diffuse, 'neural image' in first-order afferents. But lateral inhibition in a subsequent central relay may then reduce the degree of spread of activity: the most strongly activated fibres will inhibit their neighbours more than they are themselves inhibited. (Black cells and terminals are inhibitory, white terminals excitatory)

of afferent fibres—then it must follow that the effect of having large receptive fields will be to *blur* this neural image. Sharp stimulus boundaries will be converted into a rather fuzzy gradation between those fibres that are firing maximally and those that are not firing at all (Fig. 4.9). However, there is a simple way in which the brain can mitigate the effects of this kind of neural blur, called *lateral inhibition.*

Lateral inhibition

Imagine that at the level at which the incoming fibres first relay on to ascending, second-order, neurones—in this case, in the gracile and cuneate nuclei—they excite local interneurones as well; and suppose that the interneurones in turn send inhibitory connections to neighbouring second-order cells (Fig. 4.9). What will happen now is that each fibre will tend to reduce the activity of the second-order cells driven by neighbouring fibres. But a cell in the middle of the neural image will have a larger effect on its neighbours than they will in turn have on it, because it is more strongly stimulated. The result will be a kind of *image-sharpening* that will tend to compensate for the blurring effect of the original overlap of the receptive fields. This mechanism of *lateral inhibition* is of fundamental importance in understanding the processing of neural information, and occurs in every kind of sensory system. It is also found universally within the central nervous system itself, since blurring can occur not only through having large receptive fields, but also whenever there is convergence and divergence in the projection from one neuronal level to another.

In many ways, lateral inhibition is analogous to adaptation, but in the spatial rather than in the temporal domain: it makes neurones sensitive to a change in activity across a *pattern* of neural activity, as opposed to a change in any one neurone as a function of *time*. Consequently a uniform stimulus occupying an extended area produces more excitation at its edges than it does in the middle, just as adaptation causes a burst of activity at the onset of a steady stimulus (Fig. 4.10). One can sense this quite easily in one's own skin:

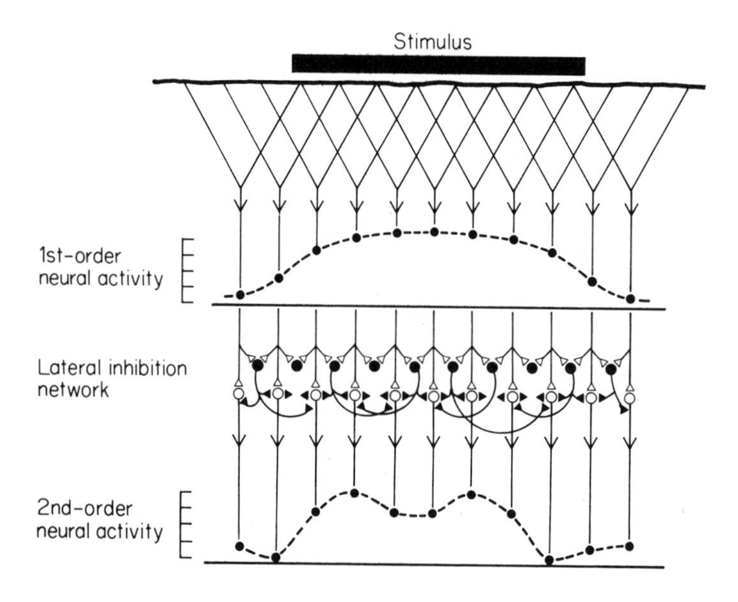

Fig. 4.10 Effect of lateral inhibition on the neural image of an extended stimulus, showing enhancement of response to the edges

stepping into a bath when the water is almost too hot to bear, the maximum discomfort is localized not so much in the foot, but rather at the line formed by the surface of the water around the leg, where the spatial rate of change of temperature is greatest. Or if you put your finger into a beaker of mercury (since mercury is poisonous it is advisable to wear a light rubber glove), what you feel is a tight constriction round the meniscus, even though the pressure is of course greatest at the fingertip. Like adaptation, lateral inhibition helps reduce the *redundancy* of neural signals. An analogy may help to make clear why this is.

Imagine a central weather bureau whose function is to use reports from all over the country to compile up-to-the-minute charts of the changing patterns of weather in the region as a whole. One way of obtaining the necessary information would be to get the local weather stations to ring up every five minutes and describe local conditions. But this would not be a very sensible arrangement; quite apart from the cost of the enormous number of telephone calls that would be required, the central office would have to employ a very large staff simply in order to receive them. Clearly the weather at any one place does not in practice vary much from one 5-minute period to the next, so that most of the phone calls in such a system would be redundant, consisting simply of the message 'same as before'! The first rationalizing step would be to instruct the local stations to ring only when a *change* in the weather has occurred; to act, in fact, like completely adapting sense organs. The next improvement would be to recognize the existence of *spatial* redundancy in their reports; in general, the weather experienced by any one station is likely to be much the same as that experienced by its neighbours. So by telling the local stations to call only when they are aware that they are on the *edge* of a particular condition (such as a cold front), the number of calls, and the staff required to process them, could be reduced still further. We shall see later that this process is particularly prominent in the visual system, and helps to explain why outline drawings are as effective as they are in evoking the appearance of what they represent, even though topologically they are so different: since the visual system in effect converts everything it sees into an outline drawing anyway, it doesn't mind if what it normally throws away isn't there!

However, the one thing that lateral inhibition cannot do—contrary to a popular misconception—is to improve *acuity*. Acuity is a measure of how well a sensory system can transmit in its neural images the spatial detail that is presented to it. A common test of cutaneous acuity is the *two-point discrimination test*: the skin is stimulated at two points simultaneously—a pair of dividers does this very well—and the distance between the points is gradually increased until the subject has the sensation that there *are* two points and not one. Cutaneous acuity varies greatly from one part of the body to another, almost in proportion to the size of its representation in the somatosensory cortex, from a few millimetres on the fingers to nearly 50 on the calves. This variation essentially reflects the size of the cutaneous receptive fields, for if two points of stimulation are separated by less than the receptive field size, no amount of subsequent neural processing in the form of lateral inhibition or anything else will enable one to distinguish the neural image from that produced by a single point. But as we have seen, localization is in general more accurate than acuity, because of the extra information provided by the overlap of receptive fields; consequently one finds the apparently paradoxical situation that with the dividers set to a distance less than the local acuity, one can nevertheless still tell, when only *one* of the points is applied, which one it was.

Finally, it is worth mentioning that the neural circuit for lateral inhibition presented in Fig. 4.9 is only one of several possible arrangements that have broadly similar effects. The one that is shown there is a *feedforward* system—the source of the inhibition is the incoming fibres, and the inhibition itself is of the

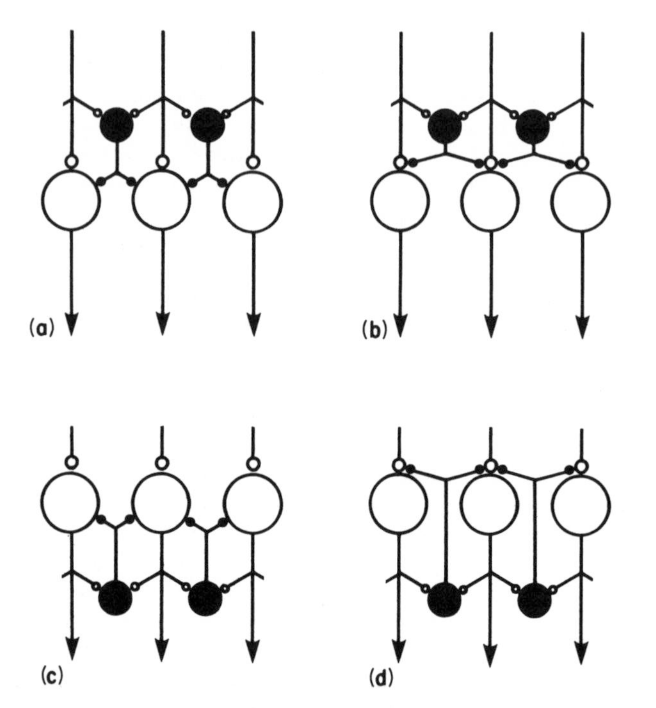

Fig. 4.11 Four possible types of lateral inhibition. (**a**) Feedforward, post-synaptic; (**b**) feedforward, presynaptic; (**c**) feedback, post-synaptic; (**d**) feedback, presynaptic

post-synaptic type. In fact, it appears that the lateral inhibition actuall observed in the dorsal column nuclei is presynaptic in nature (Fig. 4.11**b**) and results in depolarization of the afferent terminals. At higher levels of th somatosensory system, in the thalamus and cortex, lateral inhibition appear to be mainly post-synaptic. A further possibility is that it may be not th incoming fibres but collaterals of the outgoing ones that excite the inhibitor interneurones: this is called *feedback* lateral inhibition, and is found for instanc at the level of the thalamic relay of ascending somatosensory pathways; it functional properties are slightly different. Finally, feedback inhibition ma originate not from the outgoing fibres themselves but from the higher levels t which they project; the cortex can be shown to inhibit both thalamic anc dorsal column relays in this way, as well as the spinothalamic pathways at th level of the cord itself.

Responses from smaller afferents

The cutaneous fibres of groups Aδ and C that are associated with light touch pain and temperature, show response patterns that are a little more comple than those of the larger fibres. *Warm* and *cold* fibres, of Aδ size, fire tonically a a rate that is a function of temperature, with a peak for warm fibres aroun

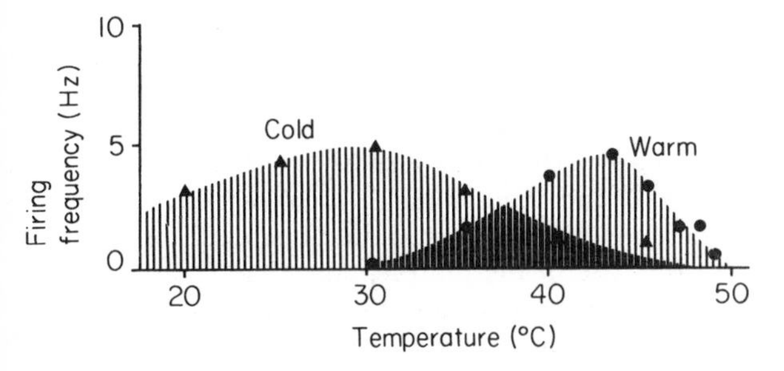

Fig. 4.12 Tonic firing frequencies of cold and warm fibres from monkey skin in response to different steady temperatures. (Data from Kenshalo, 1976)

45°C and for cold receptors around 30°C (Fig. 4.12). Both also show incomplete adaptation: sudden warming of the skin results in a transient increased discharge of warm fibres, whose activity then settles down to a new level, while sudden cooling has the same effect on cold fibres. However, it appears that the cold receptors also respond transiently to *warming* above some 45°C, giving rise to the familiar sensation of *paradoxical cold*: a hot object, when briefly touched, may often give the immediate impression of being intensely cold. These adapting properties of the thermoreceptors dominate one's sense of skin temperature. Thus a swimming pool that seems appallingly cold when one first dives in soon seems quite comfortable, and a bowl of warm water may feel simultaneously cold to one hand and hot to the other, if previously the two hands had been held respectively in hot and cold water. The receptive fields of thermoreceptors are very small and, far from overlapping, are actually separated from one another by large areas of skin that do not respond at all. Thus one may find warm and cold *spots* on the skin; on the hand the cold spots are about 5 – 10 mm apart and the warm spots some 15 mm: temperature is not of course a cutaneous sense for which accurate localization is particularly important. Thermoreceptive properties similar to what is described above are also found amongst C fibres, more of them responding to cooling than to warming; other C fibres respond to light touch in the same general way as A fibres. The remaining Aδ and C fibres serve the sense of pain, produced by noxious stimuli.

It is a common experience that there are two qualities of pain sensation, often called *pricking* pain or first pain, and *burning* pain or second pain. If one stubs one's toe against something, the feeling is a sort of immediate 'Ow!' followed by a more drawn-out 'Ooooh!', and these two kinds of pain are thought to be the result of stimulating the A and C fibres respectively. The main evidence for this comes from experiments in Man in which the conduction of peripheral nerves is partially blocked either by anoxia or by local anaesthetics. Anoxia, which can most easily be produced by inflating a cuff round the arm, affects the largest fibres first and the C fibres only after a considerable delay. The subject loses pressure and position sense first; then, as the Aδ fibres begin to be affected, temperature sense and pricking pain; and lastly burning pain and itch. The sequence of block for local anaesthetics is different: the C fibres are the first to suffer, and the largest A fibres the last; as a result it is then burning pain and itch that are the first to go, followed by temperature and pricking pain, and lastly pressure. Recordings from single afferents have shown that the Aδ pain fibres are specifically sensitive to mechanical deformation of the skin (but at a much higher threshold level than 'ordinary' mechanoreceptors), and that although their receptive fields are quite large, *within* each field the sensitivity is limited to specific 'pain spots' similar to those found in the case of thermoreception. The C fibres on the other hand are of various types: some show a similar high-threshold response to mechanical stimulation, while others respond specifically to extreme cold or heat. Both these and the Aδ fibres responding to noxious stimuli are thought to originate in the free endings of the skin. Pain may also be experienced by certain kinds of stimulation of the viscera, particularly severe distension or constriction; yet the digestive tract is said to be quite insensitive to some stimuli—notably cutting and burning, and some chemical stimuli—that are painful when applied to the skin. Visceral pain is often felt not in its 'true' position, but *referred* to the region of body surface that shares the same dorsal root: thus pain is felt in the groin in response to a stone in the ureter, and in the left arm in angina pectoris. *Itch* is not well understood: the blocking experiments described above indicate that the information generating the sense of itching is carried in C fibres; but specific itch fibres have never been found. It may, like tickle, simply represent the sensation produced by a particular pattern of stimulation of the C fibres, perhaps as the result of the release of histamine from damaged tissue, an extremely powerful stimulus for itch when injected locally.

Central responses

Thalamic responses to lemniscal and anterolateral afferents are not particularly interesting: they show the modality specificity that would be expected from the fibres that project to them, in contralateral receptive fields that may sometimes be larger than those of neurones at lower levels in the somatosensory system. In somatosensory area SI of the cortex, responses are again not qualitatively different from those in the thalamus. One striking feature of the distribution of responses over the cortical surface, apart from its large-scale organization in the form of the sensory homunculus already described, is the fact that the cortical organization appears to be in the form of a mosaic of *columns* a few hundred micrometres in diameter, such that responses from cells at any depth within a particular column are confined both to a particular modality and also to a localized region of the skin. In general, each column is surrounded by neighbours of different modality but similar location, and there are mutually inhibitory connections between columns that presumably accentuate differences in their activity, by a kind of lateral inhibition. In the second sensory area, SII, we begin to find evidence of more complex kinds of analysis of afferent information. Units here are on the whole bilaterally activated, and the receptive field for one side of the body is approximately the mirror image of that for the other. Many of the cells show specific responses to stimuli that *move* across the skin in particular directions.

As one examines more and more outlying regions of the somatosensory cortex, one begins to find cells that may respond to more than one stimulus modality, and also to painful stimuli, a property not usually reported in the main part of SI and SII. Electrical stimulation of the somatosensory cortex in conscious human subjects tends to produce tingling, 'electrical' sensations rather than the illusion of actual tactile stimulation; pain is only rarely reported. Lesions here result in raised tactile thresholds, in reduced two-point discrimination, and a general impairment in the finer somatosensory judgements such as estimating weights. In Man, lesions affecting the further, parietal, regions of the somatosensory cortex are sometimes associated with *astereognosis*, an inability to 'put together' somatosensory information in judging such things as the shape of an object held in the hand, even though primary somatic sensibility—as measured by tests such as the two-point threshold—may be relatively unimpaired.

Pain

The central pathways for pain start with the anterolateral system, section of which causes a complete peripheral analgesia for both pricking and burning pain. At higher levels the two types of pain show slightly differing distributions; ascending fibres concerned with pricking pain go to the somatosensory thalamus and from thence to the cortex, particularly area SII, while the pathways for burning pain appear to be both older and more diffuse, involving the more central thalamic regions, with their rather general projections to the cortex, the ascending reticular formation, periaqueductal grey hypothalamus. In Man, interference with the thalamus generally has more effect on pain than with the cortex. Electrical stimulation of the ventrobasal region may produce sensations of pricking pain, and of the central regions a general sense of intense unpleasantness. Lesions of the thalamus can have widely varying effects ranging from relief from pre-existing chronic pain to the production of unendurable spontaneous pain, while the sensation of pain is usually unaffected, or at most slightly reduced, by cortical damage.

The cental pathways for pain are in fact rather complex and poorly understood, partly because the sensation of pain is itself a complex one, involving a much larger *affective* component than, say, pressure. In fact one might almost go so far as to call pain an emotion that is simply triggered off by certain patterns of cutaneous stimulation in certain behavioural circumstances, in much the same way that, for example, erotic sensations may be produced by other patterns in other circumstances. The relation between the type or intensity of a stimulus and the degree of pain that is felt is a highly variable one, that depends to a large extent on the emotional state of the subject and on any implications or meaning that the pain may have. We have all had the experience of injuring ourselves inadvertently, and of not feeling pain until we actually *see* what we have done. In states of excitement, as frequently reported by soldiers severely wounded in battle, there may be a general insensitivity to injuries that would certainly be painful under normal circumstances. Conversely, some patients, perhaps when particularly apprehensive, may show exaggerated responses to quite mild stimuli: in the

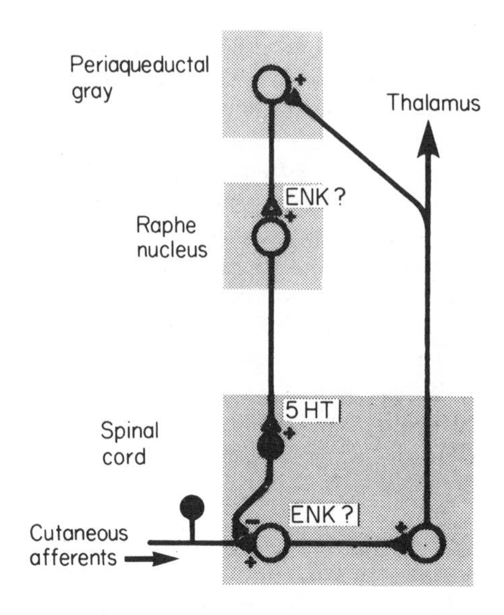

Fig. 4.13 A system associated with the neuropeptide enkephalin (ENK) that may serve to modify the transmission of afferent impulses signalling pain

dentist's chair, we may respond violently to almost any unexpected dental stimulus. Some of these fluctuations in response to pain are almost certainly the result of the changes in transmission in some neural pathways caused by the release of the natural opiates, the *endorphins* and *enkephalins*. The functions of these neuropeptides, that are widely distributed as transmitters throughout the nervous system, and to some extent as hormones as well, are not yet fully understood. Some of them produce marked analgesia when injected either intravenously or into particular regions of the brain. One such area is the periaqueductal gray; it is thought that an excitatory pathway exists from this region to the raphe nucleus of the medullary reticular formation, which in turn sends descending fibres down into the spinal cord which ultimately inhibit the transmission of afferent pain impulses through a spinal interneurone that itself releases enkephalin (Fig. 4.13). It may well be that one of the ways in which this system is activated is by the pain fibres that project to the periaqueductal gray, providing a negative feedback system by which pain might modify its own transmission. A similar gating mechanism may underlie the analgesic effects of stimulation of the larger cutaneous afferents, described below.

One of the clearest pieces of evidence that there is a degree of separation between peripheral neural discharges and the objective sense of the existence of a noxious stimulus on the one hand, and actually *feeling* the pain on the other, comes from patients who have undergone frontal leucotomy to relieve intractable pain. As described later in Chapter 13, on questioning they may indicate to the doctor that they sense the pain, yet from their attitude and mood it is evident that they do not, in any normal sense, 'feel' it. Pain is also influenced to a larger extent than other sensory modalities by other modes of skin stimulation, being reduced for example by warmth and by mechanical stimulation such as rubbing, or for that matter by acupuncture; conversely, in certain circumstances specific damage to the larger cutaneous afferents may result in an increased sensitivity to painful stimuli. All these observations suggest that pain does not have the simple relationship to stimulation of a particular type of Aδ or C fibre that something like the sensation of warmth does, and indicates that it must rather be the result of much more complex central integration. Clearly, a better understanding of the mechanisms involved would be of considerable clinical benefit.

5

Proprioception

This chapter is concerned with mechano-receptors that provide us with information about ourselves: about the positions and movements of our limbs, the forces generated by our muscles, and our attitude and motion relative to the earth. The main use to which this information is put by the brain is of course in the control of movements; consequently a discussion of the *functions* of these proprioceptive modalities is left until Chapters 10 and 11, which deal with the control of muscle length and of posture.

Muscle proprioceptors

Two distinct kinds of proprioceptors are found in voluntary muscles, specialized for providing information about two quite different quantities: *muscle spindles* that respond to muscle *length* and rate of change of length; and *Golgi tendon organs* that signal muscle *tension* or force. Both of these are in essence stretch receptors: their difference in function comes about because of their different situation in the muscle as a whole (Fig. 5.1). Whereas the spindles are in *parallel* with the main contractile elements in the muscle, so that their stretching is simply a measure of the degree of stretch of the muscle itself, the tendon organs are situated in the muscle tendons, in *series* with the contractile elements and the load, so that their stretch is proportional to the tension exerted by the muscle.

Spindles

Muscle spindles are found in practically all the striated muscles of the body, but are greatly outnumbered by the striated muscle fibres themselves: in cat soleus, there is only one spindle for every 500 or so ordinary fibres. Each consists of a fluid-filled capsule some 2–4 mm long and a few hundred micrometres in diameter, whose ends are attached to the exterior sheaths of neighbouring muscle fibres (Fig. 5.2). Inside is a small number of modified muscle fibres called *intrafusal* fibres (the 'fus-' root means 'spindle'), each having contractile ends, and a region in the middle that is not contractile, but contains the nuclei. Two main types of intrafusal fibre are found, differing in the way in which these nuclei are distributed. *Nuclear chain* fibres are thinner, and their nuclei are lined up in a row along the central portion like peas in a

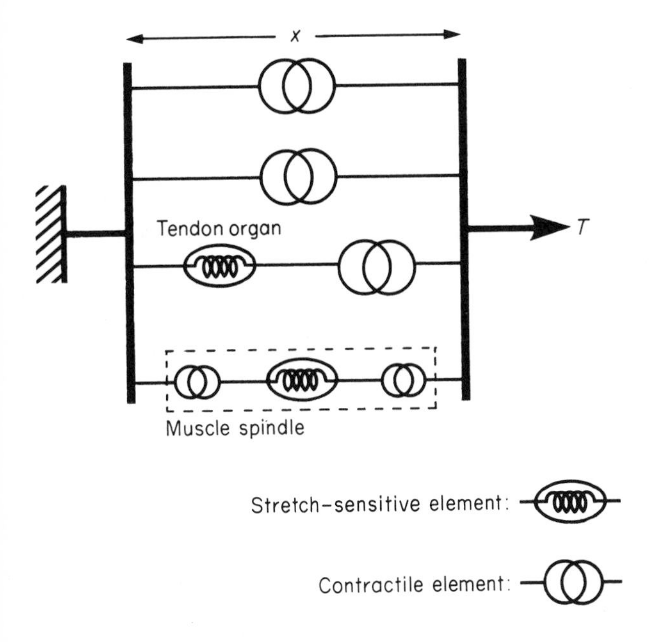

Fig. 5.1 Schematic representation of contractile and stretch-sensitive elements in muscle. The contractile elements within the spindle are innervated separately from the main (extrafusal) contractile elements of the muscle, and make only a negligible contribution to total muscle tension, *T*. Thus tendon organs respond essentially to muscle tension, and spindles to length *x*, but in a manner modified by the activity of their own contractile elements.

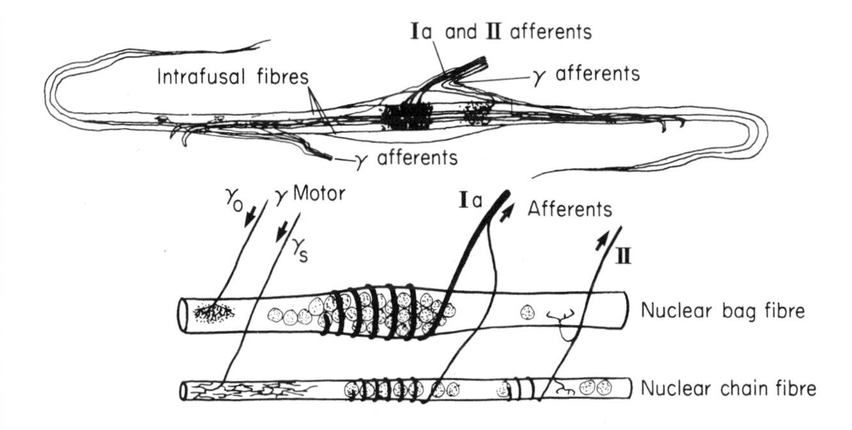

Fig. 5.2 Above, a typical mammalian muscle spindle (simplified from Barker, 1948). Below, schematic representation of central region of a nuclear bag and nuclear chain fibre, showing the afferent Ia and II innervation, and the two kinds of γ fibres. (After Matthews, 1964)

pod; *nuclear bag* fibres have a pronounced bulge in the middle in which the nuclei are bunched together. A typical spindle has about four of the chain fibres and three of the others. Two kinds of afferent or sensory fibres innervate the spindle: the larger, *primary* fibres, belonging to group Ia, send branches to the central portions of both types of fibre and have annulospiral endings; the smaller *secondary* fibres are of group II and terminate partly as annulospiral and partly as flower-spray endings mainly on the nuclear chain fibres, more peripherally than the Ia endings. These two nerve fibres respond very differently to muscle stretch. The secondary fibres are in a sense the simpler: their signals are more or less directly proportional to the degree of stretch of the spindle at any moment, so that their frequency of firing, whether the muscle is suddenly stretched to a new length, stretched more slowly, made to shorten, or alternately stretched and relaxed in a sinusoidal manner, mirrors quite accurately the instantaneous value of the muscle length (Fig. 5.3). These fibres are thus essentially non-adapting or *static*. The Ia fibres, in

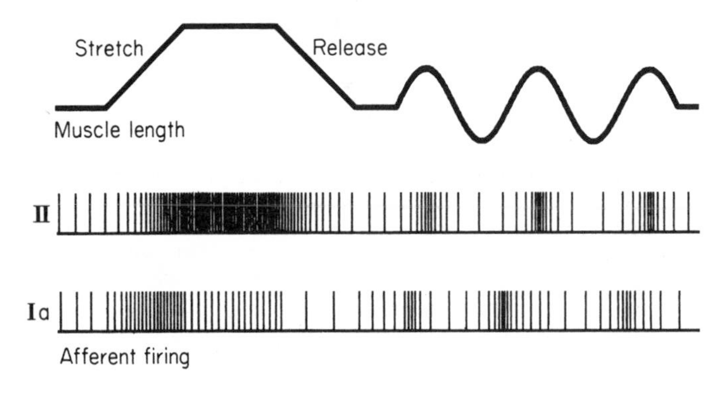

Fig. 5.3 Idealized responses of primary (Ia) and secondary (II) fibres to various patterns of muscle stretch. (After Matthews, 1964)

contrast, are *dynamic* and show very pronounced adaptation. During a sudden stretch they fire maximally during the period of stretching and only at a reduced rate when the muscle is held at its new length; during a slow stretch they again respond most during the movement, at a frequency nearly proportional to the rate of stretch; and during sinusoidal stretching their maximum firing is not at the moment of maximum stretch but near the point of maximum *rate of change* of stretch. In other words, they respond partly in proportion to muscle length and partly in proportion to its rate of change, or velocity.

Now we saw in Chapter 3 that adaptation in sensory receptors can in general be due to two distinct processes: there may be energy filtering, in which static information is wholly or partly thrown away before it even reaches the transducer element itself; or there may be membrane adaptation, in which even a steady conductance at the ending results in a fall-off in firing frequency. Whereas in the Pacinian corpuscle both of these mechanisms contribute almost equally to the adaptation that is observed, in the case of the muscle spindle it appears that nearly all is of the energy-filtering kind, and not very different from what is produced by the concentric lamellae of the Pacinian corpuscle. The contractile portions of the intrafusal fibres behave as if they were very much more viscous than the central portion (this is particularly true of the nuclear-bag fibres), so that the mechanical properties of the fibres as a whole may be represented by a mechanical model like that of Figure 5.4. In a brief stretch, there is no time for the viscous elements to lengthen, and the stretch is entirely taken up by the central region, where the annulospiral endings are; but if the stretch is maintained, the viscous elements gradually yield, releasing the strain on the middle portion, and thus resulting in a lower frequency of firing. The difference between the non-adapting properties of the secondary fibres and the marked adaptation of the primaries seems partly to be due to the fact that the former go only to chain fibres, which are less viscous, and also to the fact that they innervate more peripheral parts of them.

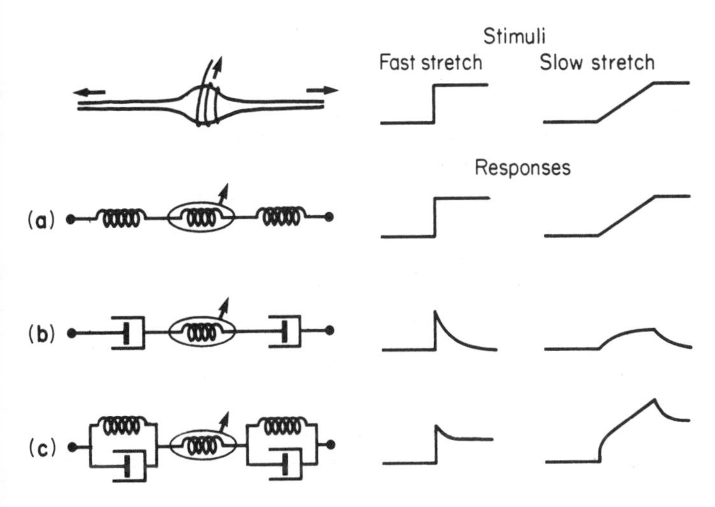

Fig. 5.4 Left, three simple models of the mechanical properties of spindle fibres: (**a**) peripheral region purely elastic; (**b**) purely viscous; (**c**) mixed (viscoelastic). Right, the response of each model, assumed to be proportional to stretch of the central region, to a sudden stretch and to slow stretching. Actual Ia fibres behave most like (**c**), Group II fibres like (**a**)

Besides the two kinds of afferent fibre, spindles also receive a *motor* innervation from the *γ fibres*, (belonging to group Aγ) and around 6μm in diameter. There are two types of fusimotor fibre, called γ_s and γ_d—static and dynamic—and they innervate respectively the nuclear bag and nuclear chain fibres (Fig. 5.2), causing contraction of the peripheral regions. For any given muscle length, such a contraction must of course stretch the sensory elements and thus increase the firing of the afferent fibres, and in fact the effect of γ stimulation is in general much the same as if an extra stretch had

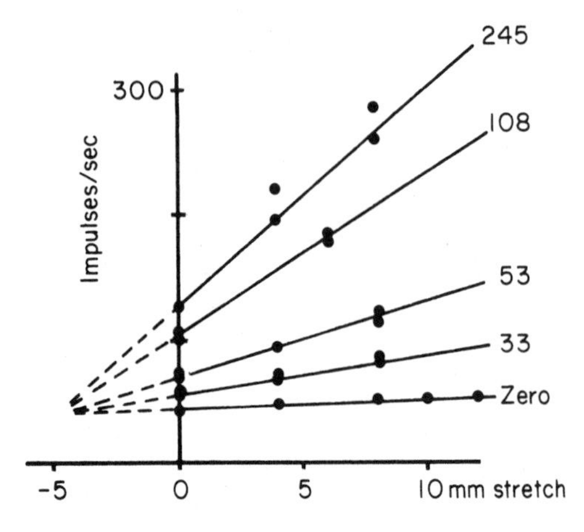

Fig. 5.5 Effect of γ fibre stimulation on afferent response to stretch, in an eye muscle from the goat; the rates of stimulation are shown on the right. (Data from Whitteridge, 1959)

been applied to the muscle as a whole, though they may also increase the sensitivity of the endings to stretch. Figure 5.5 shows some recordings of the firing frequency of stretch receptor afferents in response to different degrees of stretch, when the corresponding γ fibres were also stimulated at different rates: the interaction between external stretch, and the internal stretch produced by the γ activation, can be clearly seen. The static and dynamic γ fibres produce slightly different effects on primary and secondary afferent responses, as would be expected from their differing distribution to the bag and chain fibres. Dynamic γ fibres increase the sensitivity of group Ia fibres but have no effect on group II fibres, whereas the static ones increase the sensitivity of both the secondaries and primaries to static stretch, but actually decrease the primary sensitivity to rate of stretch. Thus the central nervous system can, through the γ efferents, control not just the sensitivity of the spindle afferents, but also in a sense their adaptational properties. The way in which this control is actually used by the motor system is left for consideration in Chapter 10.

Golgi tendon organs

The *Golgi tendon organs* have received much less attention from experimenters. As mentioned above, they are effectively in series with the muscles, lying in the tendons, and there are about ten muscle fibres for each receptor; they are innervated by afferents of group Ib. At one time their importance was underestimated because they seemed to have such high thresholds: large forces had to be applied to the tendon as a whole before they could be induced to fire. But it is now clear that this was because in these circumstances the total tension applied is in effect shared out amongst the tendinous fascicles, so that each tendon organ feels only a small fraction of it: they respond in fact very briskly to the modest tensions generated by the actual muscle fibres to which they are joined. Because they register tension rather than muscle length, during *active* movements their discharge generally has a reciprocal relationship to that of the muscle spindles: extrafusal activity simultaneously increases the tension in the tendons and decreases muscle length; but during *passive* movements, both kinds of response are normally in step with one another.

The central pathways of both these sensory modalities are quite similar. Fibres enter the dorsal roots in the usual way, and the majority synapse in a spinal nucleus called Clarke's column with fibres that ascend in the homolateral *posterior spinocerebellar tract* to an extremely important region in the control of movement, the *cerebellum* (Fig. 5.6). However, Clarke's column fades out above C8 or so, and more rostral afferents turn upwards and ascend to the *accessory cuneate nucleus* of the medulla, from which fibres run in the *cuneocerebellar tract*. Another route by which both cutaneous and proprioceptive information may reach the cerebellum is via the *spino-olivary tract* to the inferior olive, which in turn projects through climbing fibres to the cerebellar cortex (see Chapter 12). Apart from ascending to the cerebellum, branches of fibres from muscle proprioceptors are involved in various reflex mechanisms within the spinal cord, notably the *stretch reflex* (which in its simplest form consists of a

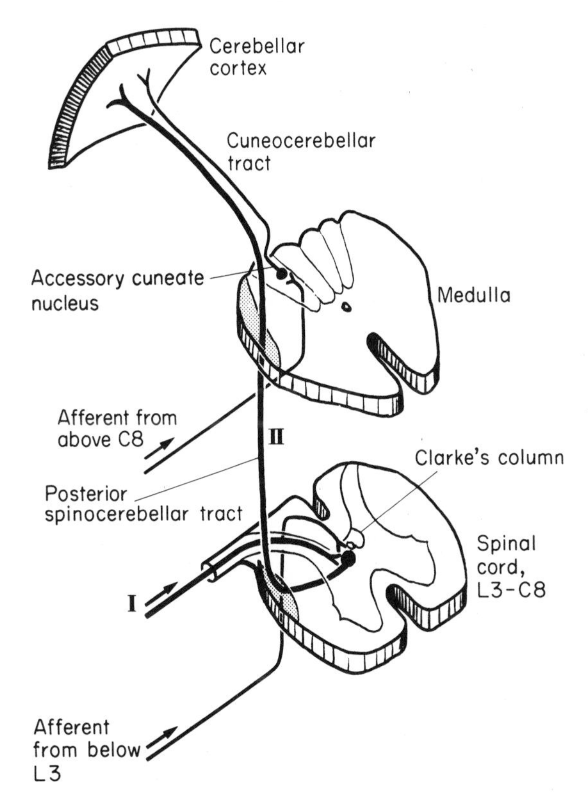

Fig. 5.6 Schematic representation of the principal spinocerebellar pathways. An additional route, not shown, is via the inferior olive and climbing fibres (I, II: first- and second-order fibres)

monosynaptic excitation of a motor neurone by a Ia afferent) and the *clasp-knife reflex*: these are discussed in Chapter 10. There is also some projection of muscle proprioceptors to the cerebral cortex, via the ventral posterior thalamus.

Joint receptors

One might think that muscle spindles, in signalling muscle length, would be ideally suited to telling us about the position of our limbs, yet a number of experiments have established that this function is for the most part not carried out by muscle spindles at all, but rather by specialized endings in and around the *joints*. Thus if we inflate a cuff round the wrist that paralyses the afferent fibres from the fingers, but leaves the muscles that move them unaffected—they are of course situated in the arm—we find a subject with his eyes shut finds it very difficult indeed to sense the positions of his fingers. Some sense of *change* of position may however remain, which is probably mediated by the Ia fibres: selective stimulation of these fibres by vibration applied to the muscle gives rise to illusory sensations of movement. The receptors predominantly responsible for this conscious proprioception are found in the ligaments and capsules of joints, and are of a variety of morphological types: some are similar to the encapsulated endings found in the skin. Of their afferent fibres, which are of group II, some show complete

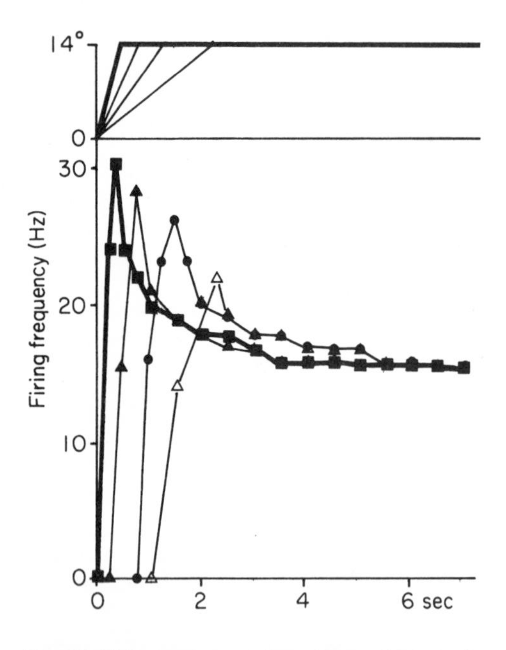

Fig. 5.7 Firing frequency of an afferent from the knee joint of a cat in response to flexion through a fixed angle at the different rates shown, showing incomplete adaptation. (Data from Boyd and Roberts, 1953)

adaptation and are thus more sensitive to rate of change, but most show incomplete adaptation and thus signal limb position as well (Fig. 5.7). The patterns of response are complicated to a certain extent by the fact that few if any of the receptors are able to respond over the whole range of movement that a joint is capable of: their 'excitatory angle' is typically less than half the entire possible range, thus increasing sensitivity to changes in position within that range. This means that information about the position of a limb is partly coded by frequency of firing, but also by *which* neurones are firing. Afferent information from joints follows the same route as that from Pacinian and other corpuscles in the skin (see Fig. 4.5): fibres ascend in the ipsilateral posterior columns, relay in the cuneate and gracile nuclei, cross and proceed via the medial lemniscus to the ventral posterolateral thalamus and thence to the somatosensory cortex; some also contribute to the spinocerebellar pathways.

The vestibular apparatus

The vestibular apparatus forms part of the *labyrinth* of the inner ear. As its name suggests, it is a complex structure, that has partly evolved from the lateral line organ in fishes. This is a system of tubes lined with ciliated sensory cells and in communication with the surrounding water; the cells are stimulated by the flow of water through the tubes, as a result either of something moving in the outside world and setting up fluid currents, or of the fish's own motion through the water. In the course of time, this system sealed itself off from the outside world, and its two functions came to be carried out by two separate organs: the *cochlea*, signalling movement of the surrounding air in the form of sound waves (discussed in Chapter 6), and the *vestibular apparatus*, signalling movement of the head itself. The vestibular part of the labyrinth is divided functionally into two components: the *semicircular canals*, of which there are three on each side of the head, and the *otolith organs*, of which there are two on each side, the *saccule* and *utricle* (Fig. 5.8). The ciliated sensory cells are very similar in all parts of the vestibular apparatus, and will be described first; it is the accessory structures that enclose them that makes them specifically responsive to different types of stimuli.

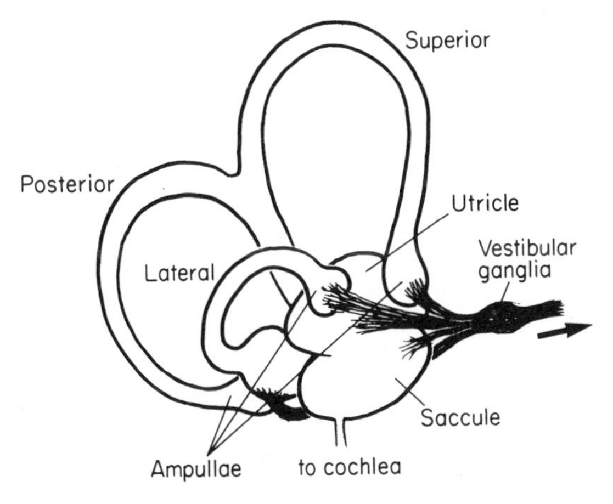

Fig. 5.8 Gross structure of the vestibular part of the labyrinth, viewed laterally (somewhat schematic)

The sensory cells

The sensory epithelia of the vestibular apparatus are made up of a mosaic of sensory cells and supporting cells, the former being divided into two morphological types: a flask-shaped *type I* cell, and a roughly cylindrical *type II* cell. Each sensory cell has a characteristic pattern of cilia projecting from its upper surface, consisting of a single flexible *kinocilium*, whose root is near the edge of the receptor, and between sixty and a hundred *stereocilia*, relatively thin and stiff, and arranged in a regular array in the more central area. The latter are graded in size rather like a set of organ pipes, the longest ones being nearest to the kinocilium (Fig. 5.9). The kinocilium is a much more elaborate

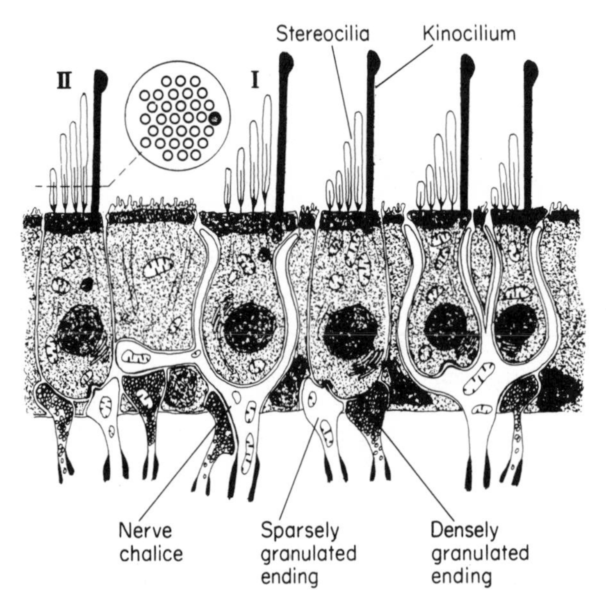

Fig. 5.9 Diagrammatic cross-section of sensory epithelium of vestibular system, showing type I and type II cells with their innervation. Inset, a typical arrangement of the cilia seen in horizontal section

structure than the stereocilia, having the 'nine plus two' arrangement of longitudinal filaments characteristic of motile cilia, and basal bodies. The asymmetrical arrangement of kinocilium and stereocilia defines a direction of polarization for each cell, and it is found that bending in the direction of the kinocilium leads to excitation, while bending in the opposite direction gives inhibition. The exact mechanism is still uncertain, but it is likely that the deformation caused by bending alters the ionic permeability of the receptor cell.

The sensory cells are innervated by branches of the *vestibular nerve*: type I cells are almost completely enclosed in a nerve *calyx* or chalice, and there are often regions of close apposition, which together with the large synaptic area suggest that the transmission process here may be partly electrical as well as chemical in nature. Type II cells generally receive more than one ending; the terminals are smaller in size, and some of them are efferent rather than afferent. There seems to be much more convergence from type II cells on to their afferent fibres than in the case of the type I cells, and more still in the case of the efferent fibres: there are only some 200 fibres in a cat's vestibular nerve going *to* the receptors, in contrast with the 12 000 or so afferents. Sensory and efferent fibres travel together in the vestibular division of the VIIIth nerve to the region of the *vestibular nuclei*, of which the lateral nucleus is the origin of the efferents. Most of the afferent fibres terminate within the nuclei, but some carry on through and project ultimately to the cerebellum. Most afferent fibres fire spontaneously, and one finds that a stimulus that bends the kinocilium in one direction accelerates the rate of firing, and in the other decreases it. Different cells tend to have different spontaneous firing frequencies, and thus function in different parts of the total possible range of stimulation, in a manner reminiscent of joint receptor afferents.

Otolith organs (utricle and saccule)

The utricle and saccule are a pair of hollow sacs containing *endolymph*, a fluid whose ionic composition is of the intracellular type (with a high K^+/Na^+ ratio), and continuous throughout the whole of the labyrinth. Surrounding the endolymphatic sac is a second sac containing *perilymph*, whose composition is low in $[K^+]$ and more like that of a typical extracellular fluid. The receptor cells of the otolith organs are confined to a special region of the sac called the *macula*, and their projecting cilia are embedded in a jelly-like mass, the *otolith*, whose density is increased by the incorporation of large quantities of calcite crystals called otoconia (Fig. 5.10). If the head is *tilted*, this mass moves relative to the macula, bending the cilia and thus resulting in stimulation of the afferent nerve fibres. The cells also respond to *linear accelerations* of the head, since in this case the otoliths tend to get left behind and hence cause the same sort of bending. Thus the afferent signals from the otolith organs are

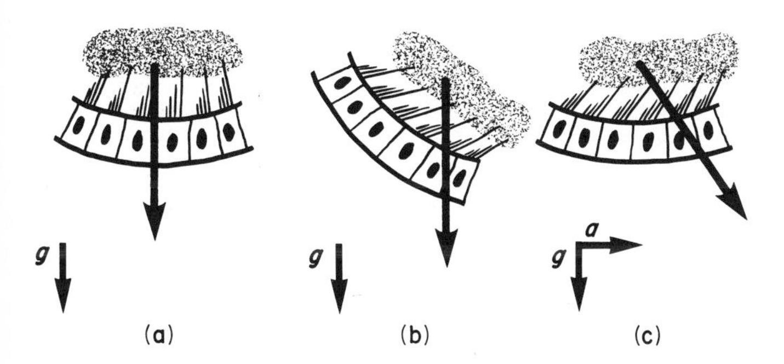

Fig. 5.10 Schematic representation of action of macular receptors (**a**) at rest, (**b**) with head tilted, and (**c**) under horizontal linear acceleration. The last two conditions are indistinguishable in their effects on ciliar bending

dependent simply on the vector sum of the acceleration due to gravity and any linear acceleration that may be occurring at the same time: in other words, on the *effective* direction of gravity (Fig. 5.10). There is necessarily no way in which the brain can distinguish between head tilt and linear acceleration—since 'gravity' is itself of course simply a kind of linear acceleration—nor is it particularly desirable that it should. From the point of view of controlling posture (as for example in trying to stand upright in a bus that suddenly accelerates) it is the *effective* direction of gravity that matters.

Information about the direction of the acceleration vector is available because the maculae of utricle and saccule lie in different planes—in the utricle roughly horizontal, in the saccule roughly vertical—and also because in each macula there is a systematic variation over the macular surface of the direction of polarization of the hair cells (Fig. 5.11); recordings from individual otolith fibres show that each one fires maximally at a particular orientation of

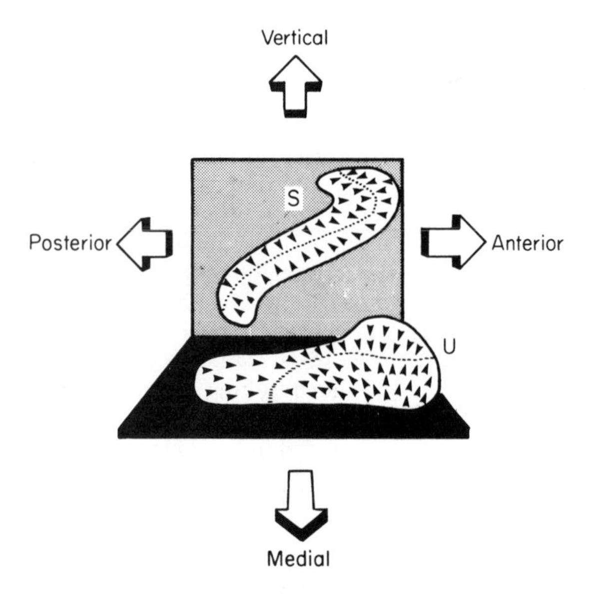

Fig. 5.11 Stylized representation of the orientation of the utricular (U) and saccular (S) maculae in the head. The small arrows indicate the approximate direction of polarization of the receptors at different points on the surface

the head, and that there is a fall-off in frequency as the head is tilted away from that position. Thus the pattern of discharge from the whole population of receptors provides information about the direction of acceleration. As might be expected, most units show little adaptation, and for the most part faithfully signal head position without decrement over indefinitely long periods of time. Some do, however, show adaptation, firing most rapidly during changes of head position, but their adaptation is not complete and their tonic discharge still gives a measure of head position: it is the semicircular canals that primarily signal changes in the attitude of the head.

The semicircular canals

Each canal consists of a looped tube containing endolymph and having a swelling at one point along its length, the *ampulla*, into which projects a crest, the *crista*, which is covered with sensory hair cells (Fig. 5.12). Their cilia are embedded in a jelly-like structure called the *cupula*, forming a kind of flap

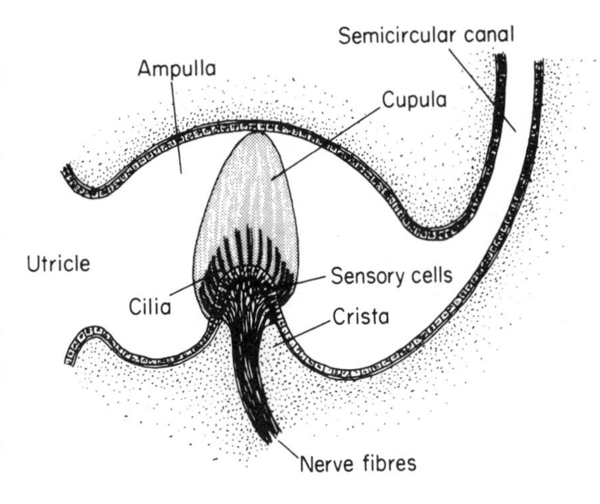

Fig. 5.12 Diagrammatic section through a single canal in the region of the ampulla, showing the cupula, the hair cells of the crista and their sensory innervation

which can swing backwards and forwards in response to movement of fluid along the canal, thus bending the cilia and causing neural excitation. Unlike the otolith, this jelly does not contain calcareous granules, and is in fact of exactly the same density as the endolymph surrounding it: it is very important that this should be so, as otherwise the hair cells would respond to gravity in the same way as the macular receptors. What they do respond to, in fact, is *rotation* of the head; when the head is turned, the fluid in the canals tends to get left behind and pushes on the trap-door-like cupula, thus bending the sensory cilia. Because the three canals on each side of the head are arranged in more or less mutually perpendicular planes (Fig. 5.13) they are able to signal rotations about any axis in space. In most animals, with the head in the normal upright position, the horizontal canals are parallel with the ground, and the superior and posterior canals lie at about 45° to the sagittal plane. All the cells in each crista are orientated in the same direction: thus turning the head to the left stimulates fibres from the left horizontal canal but decreases the frequency of firing of those from the right (Fig. 5.13), and similar mutual antagonisms exist between the superior canal of one side and the posterior canal of the other. Corresponding to this arrangement, one finds cells in the vestibular nuclei that are excited by a particular canal but inhibited by its opposite number on the other side, thus effectively combining the signals from the two sides in a 'push–pull' manner. For this reason, rotation is not a good stimulus to use if

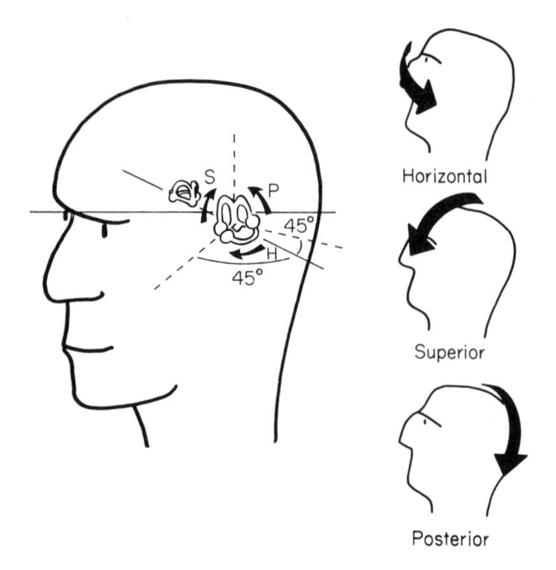

Fig. 5.13 Left, approximate orientation of the semicircular canals in Man (S superior; H horizontal; P posterior): the arrows show the direction of fluid movement that excites each canal. Right, directions of head movement that are stimulatory for each of the canals on the left side of the head

we want to test each of a subject's vestibular organs separately; instead, we need to use a stimulus that only acts on one side at a time. In clinical practice, *caloric* stimulation of the canals is sometimes used; the outer ear is irrigated with warm water, which appears to set up convection currents in the endolymph of the canals and thus produces unilateral vestibular stimulation.

The adaptational properties of the canals are very important, and are almost entirely of the 'energy-filtering' type. If the cupula were completely unrestrained, and moved in unison with the fluid of the canal, then the bending of the cilia of the sensory cells would simply signal rotational position of the head; the endolymph would be acting rather like the gyroscope in an inertial guidance system. A mechanical model of such a system is illustrated in Fig. 5.14**a**. A large railway truck, representing the head, has a smaller truck that is free to move on top of it, representing the cupula and the inertial mass M of the endolymph. If the big truck moves a distance x, the small truck will remain (in absolute terms) exactly where it was before, resulting in a relative displacement between the two—in effect, what is signalled by the hair cells—of x, the new *position* of the big truck.

But consider now what would happen if the little truck, instead of being completely free to move, were coupled to the big one by an elastic element, a spring (**b**). The system is now no longer a position detector; what we have done is to make it into an *acceleration* detector, or accelerometer. For if the truck moves off with acceleration $\ddot{x}$, the spring will experience a force equal to $M.\ddot{x}$, and will therefore stretch by an amount that is simply proportional to the acceleration. Does the cupula in fact behave as a rotational accelerometer? It is

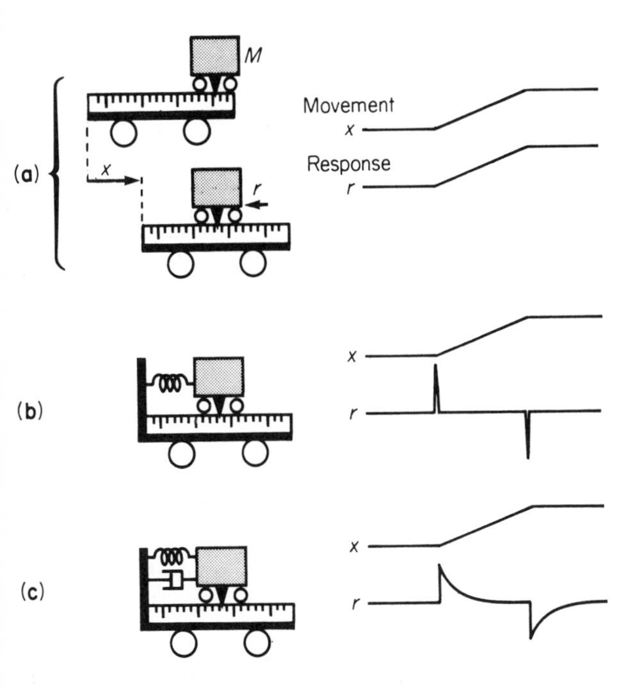

Fig. 5.14 Mechanical model of action of cupula in semicircular canals. (**a**) a small truck of mass *M* representing endolymph inertia rests freely on a larger truck representing the head. The reading *r* of the pointer provides a measure of the position *x* of the big truck. (**b**) If the two trucks are coupled elastically, *r* indicates not position but *acceleration,* $\ddot{x}$. (**c**) Actual cupular movement is as if the coupling were partly elastic and partly viscous

perfectly true that it is indeed elastic, and if pushed to one side exerts a restoring force that tends to bring it back to the middle. But it turns out that the cupula and the endolymph are both very *viscous* (Fig. 5.14**c**) and the effect of this viscosity is to slow down the mechanical response. Consider a subject on a swivel chair that is suddenly set into rotation at constant angular velocity. A true accelerometer would give a brief response only at the instant at which the rotation started, since this is the only time at which acceleration occurs; but recordings of the firing frequency of vestibular units from the canals show that in these circumstances the period of response is very much drawn out by the damping effect of the viscosity, lasting for some 20 seconds or so after the acceleration has stopped. Under *natural* circumstances—continual rotation of the head in one direction is not after all very common in real life!—the frequency of firing mirrors much more closely the instantaneous rotational *velocity* of the head than its acceleration (Fig. 5.15). In fact the semicircular canals are not really rotational acceleration detectors at all, but rather *velocity detectors*. It is only under very peculiar laboratory conditions that their acceleration sensitivity (which can be thought of as an adapting velocity response, just as velocity sensitivity is equivalent to an adapting positional response) manifests itself. A consequence of this adaptation is that a subject

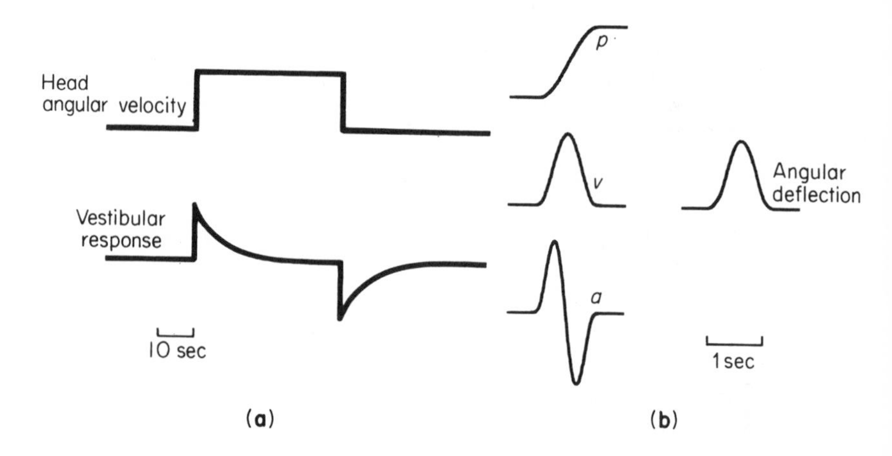

Fig. 5.15 Responses of the semicircular canals. (**a**) During a head rotation at constant angular velocity, canal response declines exponentially over some 20–30 seconds; on stopping the movement, there is an opposite response that again declines in the same way. (**b**) Left, time-course of head angular position (*p*), velocity (*v*) and acceleration (*a*) during a natural voluntary head movement. Cupular displacement (right) is more like the velocity curve than either of the others. (Courtesy Dr. T. D. M. Roberts)

who has been set spinning at constant velocity has a gradually decreasing sense that he is actually rotating; even with his eyes open, after 20 seconds or so he has the strong impression that he is actually sitting still, and that the world is spinning round *him*, as anyone who has had a ride in a fairground 'rotor' will know. Furthermore, if the chair is then suddenly stopped, his cupula, which had previously resumed its resting position, will now be pushed in the opposite direction by the tendency of the endolymph to retain the previous motion of the head; he will then have the very strong impression of rotating in the opposite direction, although he is in fact stationary. Some postural consequences of these adaptational properties are discussed in Chapter 11.

6

Hearing

To appreciate the working of the ear it is necessary first to have at least some acquaintance with the properties of the sound waves to which it responds, a topic that regrettably often seems to get squeezed out of the school physics curriculum.

The nature of sound

Sound is generated in a medium such as air whenever there is a sufficiently rapid movement of part of its boundary—perhaps a moving loudspeaker cone, or the collapsing skin of a pricked balloon. What happens is that the air next to the moving boundary is rapidly compressed or rarefied, resulting in a local movement of molecules that tends to make the regional pressure differences propagate away from the original site of disturbance, at a rate that depends on the density and elastic properties of the medium: this velocity is around 340 m/sec in air, and about four times as great in water. If the original sound source is undergoing regular oscillation—as in the case of the prongs of a tuning fork—the sound is propagated in the form of regular waves of pressure. The *wavelength* of these waves—the distance from one point of maximum compression to the next—is equal to their velocity divided by their *frequency*, the number of vibrations made every second by the source. *Pitch* is the sensory perception that corresponds with frequency, just as colour corresponds with the wavelength of a light, but there is not always an absolutely direct relationship between the two.

Strictly speaking, not all vibrations of this kind are sound; to be audible, the waves must have a frequency lying somewhere between about 20 and 20 000 Hz. The simplest of all vibrations are those in which the variations of pressure along the wave are sinusoidal as a function of time, in other words proportional to sin $(2\pi ft)$, where f is the frequency and t is time. In such a case we may describe the wave completely by means of its frequency and its *amplitude*, the value of the additional pressure at the peak of compression (Fig. 6.1). Sound waves also of course carry energy: the rate at which energy is delivered per unit area, the *intensity* of the sound, is proportional to the square of its amplitude, and also to the density of the medium and the square of the frequency: it obviously takes more energy to vibrate something backwards

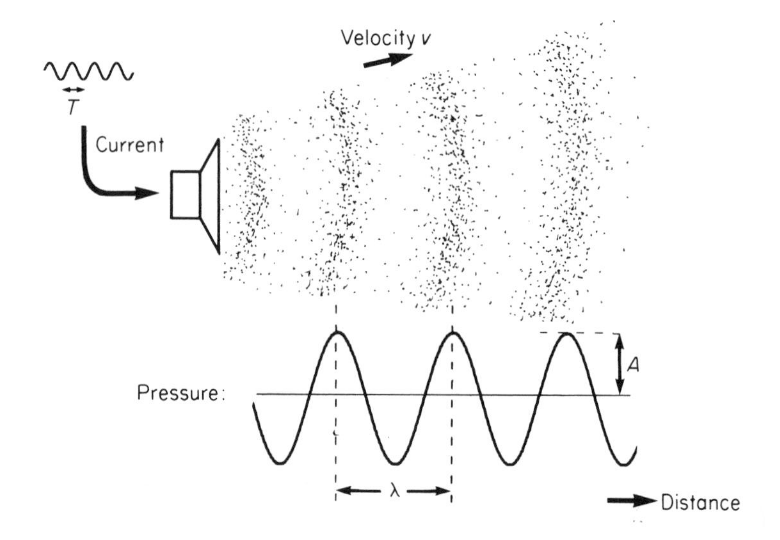

Fig. 6.1 A loudspeaker driven by a sinusoidal current of period T, or frequency $f = 1/T$. The resultant waves of compression and rarefaction of the air travel with velocity v and are of wavelength λ, where $v = \lambda f$. The amplitude A is the difference between the mean pressure and the peak pressure

and forwards very fast than if it is done more slowly. Because the range of intensities to which the ear can respond without damage is a very large one indeed—a factor of some 10^{14}—it is convenient to use a logarithmic scale to describe sound intensities. For this purpose, a standard reference level is used (its value being near the threshold of hearing under ideal conditions, 10^{-16} W/cm^2), and the log of the ratio of the actual intensity to this standard gives the intensity of the sound in Bels (named after Alexander Graham Bell, an inventor of the telephone). In practice, the unfortunate custom has grown up of dealing in tenths of a Bel, or *decibels* (dB), so that the formula becomes:

$$\text{Intensity in decibels} = 10 \log_{10} \frac{\text{Intensity of unknown}}{\text{Intensity of standard}}$$

Figure 6.2 shows the intensity of some common kinds of sound, expressed in decibels. Because intensity is proportional to the *square* of the amplitude, multiplying the latter by a factor of ten results in a *20* dB increase in intensity. Finally, although the measure defined above is an absolute scale of intensity, decibels can also be used to express the ratio of two intensities: thus if one sound has one-tenth the intensity of another, it can be described as having a relative intensity of -10 dB, or as being *attenuated* by 10 dB.

The smallest intensity that can just be perceived, the auditory threshold, depends very markedly on frequency; a graph of measurements of threshold as a function of frequency is called an *audiometric curve* (Fig. 6.3), and often has clinical diagnostic value. The minimum threshold of around 10^{-16} W/cm^2 corresponds to an amplitude of about 0.0002 dyne/cm^2, and a movement of

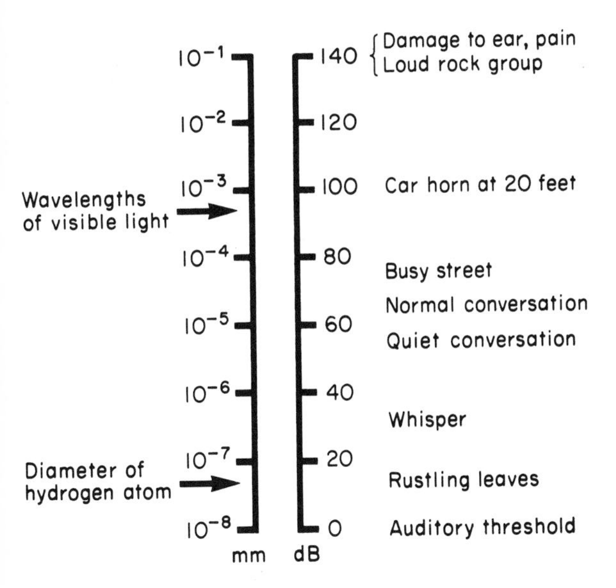

Fig. 6.2 Approximate intensities of various sounds, measured in absolute decibels (dB) and also in terms of the amplitude of the corresponding movements of the molecules in the air (0 dB = 0.0002 dyne/cm^2)

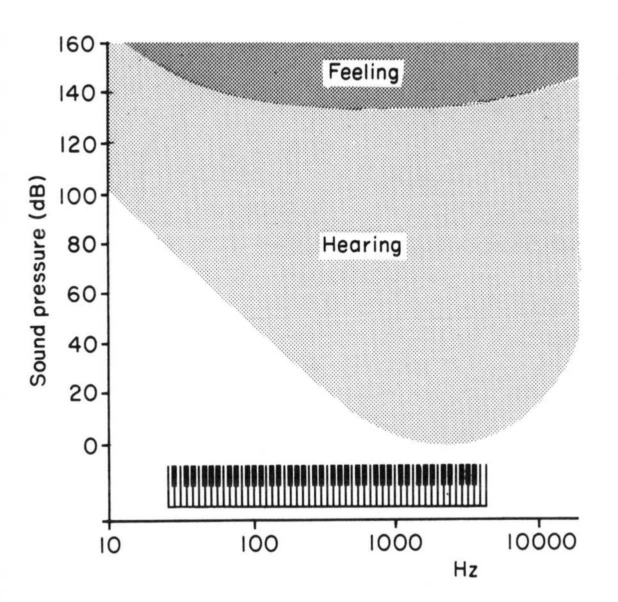

Fig. 6.3 Audiometry curve: the shaded region shows the intensities of sounds that may be heard at different frequencies, averaged over many subjects. (Data from Dadson and King, 1952). The keyboard may help relate the frequency-scale to ordinary musical pitch.

the air molecules considerably less than the diameter of a hydrogen atom! To see what this astonishing sensitivity means in practical terms, in theory a 10 watt loudspeaker of moderate efficiency situated in London and sending out a 1 kHz tone ought to be audible in Cambridge, 50 miles away! (In fact, of course, not only would much of the sound be absorbed by intervening structures, but prevailing background noise will tend to drown the incoming signal: maximum sensitivity can only be obtained when all other sound sources are silenced. Nevertheless, there are well authenticated reports from World War I of heavy shelling at a particular location being heard over wide areas in Europe.)

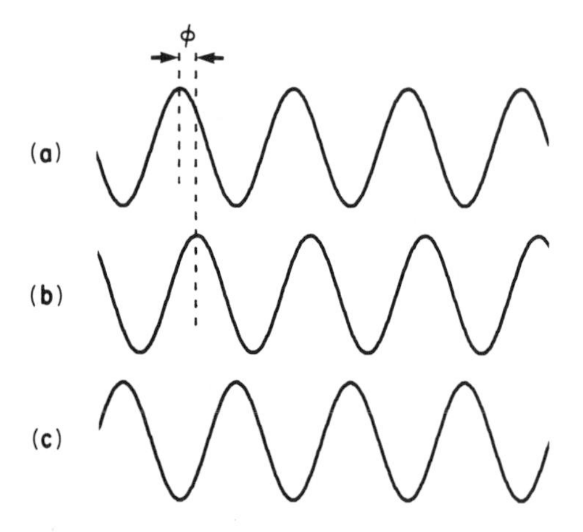

Fig. 6.4 The sine wave (**b**) has the same frequency and amplitude as (**a**), but is shifted in phase by an angle ϕ (in this case a phase advance of 60°). (**c**) A sine wave shifted by 180° relative to (**a**), or in *antiphase* to it

One other parameter that is necessary to describe a sinusoidal vibration in certain circumstances is its *phase*. A sinusoidal wave of constant amplitude and frequency that is sampled simultaneously at two fixed points in space—as for example by the two ears—will by definition have the same amplitude and frequency at each point, but the compressions and rarefactions at the one point may not occur at the same moment as those at the other. In general, one wave will appear to be displaced in time with respect to the other, and the phase difference between the two is a measure of the fraction of a whole cycle by which one appears to be shifted (Fig. 6.4). It is conveniently expressed as an angle, so that a phase shift of 180° brings the waves into antiphase, the peaks of one then corresponding to the troughs of the other, and a further 180° shift, making 360° (or 0°) in all, brings them back into coincidence.

Sound spectra

Now pure sinusoidal waves are actually rather uncommon in real life: musical instruments, for example, produce waves whose profile, though repetitive, is

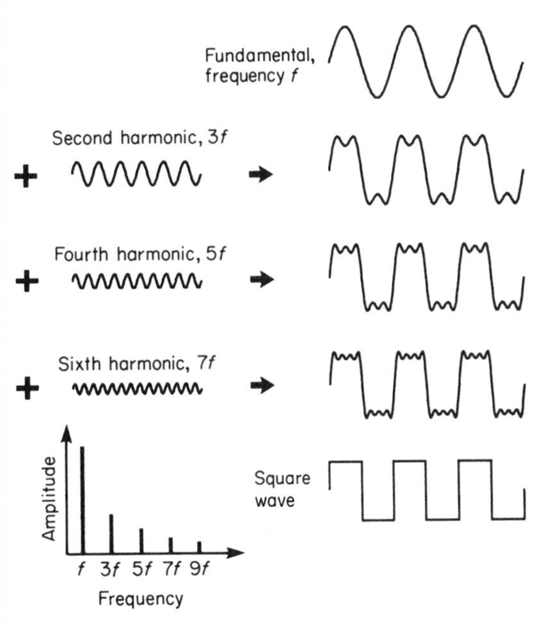

Fig. 6.5 Fourier synthesis: in this case, the gradual approximation to a square wave achieved by adding together successive sine waves of frequency f, $3f$, $5f$ etc; their amplitudes are proportional to 1,1/3, 1/5, etc., as shown in the amplitude spectrum at bottom left

not sinusoidal (Fig. 6.6). However, there is a theorem due to the French mathematician Fourier, that any repetitive waveform can be thought of as being made up of the sum of a series of simple sinusoidal components, whose frequencies are integral multiples of the frequency of the original wave, called the *fundamental* frequency. Thus a square wave of frequency f (Fig. 6.5) can be synthesized by adding together sine waves of frequency f, $3f$, $5f$, $7f$ and so on, with amplitudes in proportion to 1, 1/3, 1/5, 1/7, etc. We can represent a 'recipe' for a Fourier synthesis of this kind in the form of a Fourier *spectrum* that shows in graphical form, as a function of frequency, the amplitude and phase of each of the components that make it up; in practice, the phase information is often omitted, for reasons that will become apparent later. Figure 6.6 shows the sound spectra of a number of different kinds of musical instrument: in each case, the line of lowest frequency shows the amplitude of the fundamental, and those of higher frequency show the amplitudes of the *harmonics*. The Fourier spectrum and the shape of the waveform itself are thus in a sense interchangeable: each contains the same information as the other, for if we know the spectrum, we can add all the components together and recreate the original waveform, and conversely it is possible by a process of Fourier analysis to translate any given waveform into its equivalent spectrum. Furthermore, it turns out that Fourier analysis can even be applied to waveforms that are *not* repetitive: unpitched sounds like that of a boiling kettle, or transients of the kind produced by dropping the teapot. To see why this is so, consider what happens as we continuously lower the frequency of a repetitive waveform of given shape. Since the individual components of the spectrum are spaced out on the frequency axis by f, the fundamental frequency, it follows that the smaller f is, the closer and closer become the lines of the spectrum. In the limit, when the frequency of the wave becomes zero—in other words when its wavelength is infinite, and it never repeats itself at all—the spectral components are infinitely close together: so the spectrum, instead of being a sequence of discontinuous spikes, is now a smooth curve. Thus repetitive waveforms give rise to spectra with discrete harmonics, whereas unrepetitive waveforms produce continuous spectra. In either case, it turns out that the quality of a sound—for example the *timbre* of a musical instrument—is much more closely related to the overall shape of its spectrum than to the shape of its waveform. Musical instruments with lots of high harmonics sound bright and strident (trumpets, clarinets, Fig. 6.6); those with most of their energy in the fundamental sound smoother and more rounded.

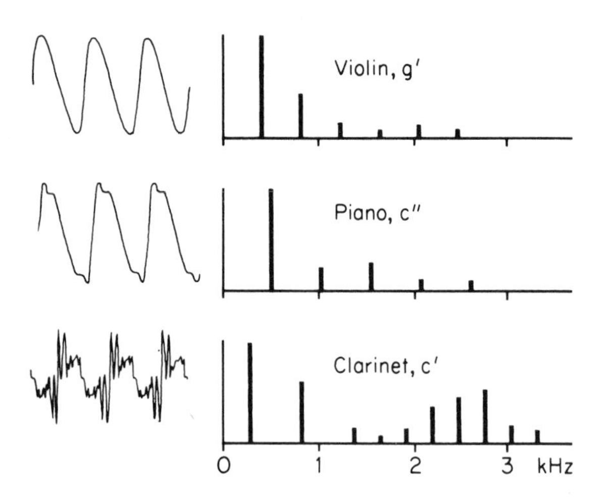

Fig. 6.6 Waveforms (left) and amplitude spectra (right) for different notes played by different musical instruments. (After Wood, 1930)

Sharp sounds, hisses, clicks, all have continuous spectra with prominent high frequencies (the only difference between a hiss and a click is in fact in the phase relationships of its components), whereas thumps, roars and rumbles have their energy concentrated at the low-frequency end.

One of the most striking examples of the use of sound spectra comes from studying the human voice. The larynx, isolated from the rest of the voice-producing apparatus of head and throat, is essentially not very different from the double reed of an instrument like the oboe. As the air passes between the vocal cords, they are alternately forced apart and spring back, producing a repetitive series of compressions and rarefactions whose frequency can be modified by altering the shape and tension of the cords themselves; the

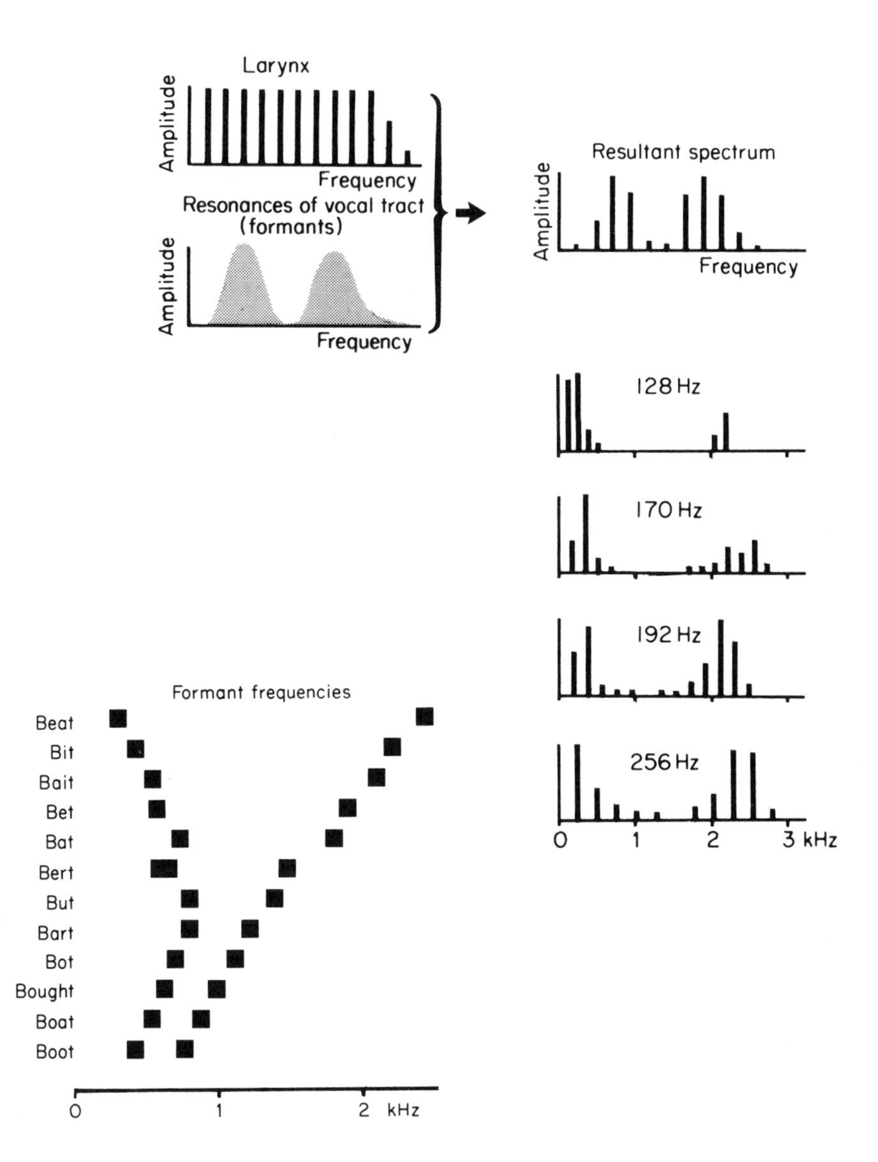

Fig. 6.7 The human voice. Above, the spectrum of vibrations of the larynx is modified by the resonances of the vocal tract (shaded) to produce the spectrum of the sound finally emitted (right): the two peaks are the first and second formants of the sound. Below left: the approximate positions of the formants for a number of different English vowel sounds. Right, the vowel sound in 'EAT' sung at the various frequencies indicated: although the spacing between the spectral lines increases as the frequency rises, the overall shape, defining the vowel quality, remains largely unchanged. (After Wood, 1930)

resultant spectrum is roughly of the form shown in Fig. 6.7. But in real life, the sounds it produces have to pass through a number of hollow cavities—the throat, nose and mouth—before they reach the outside world; these cavities tend to resonate, absorbing certain frequencies and reinforcing others, so that the original sound spectrum becomes distorted (Fig. 6.7). The two or three main resonance peaks in this spectrum simply reflect the fact that the tongue effectively divides the mouth cavity into separate compartments, and each compartment acts as an independent resonator, at a frequency that depends mainly on the shape of the tongue and the degree of jaw opening. What is striking is that different vowel sounds are associated in a closely reproducible way with particular positions of these resonant peaks (called *formants*), and these characteristics are largely independent of the speaker, the pitch of his or her voice, and whether the vowel is spoken or sung: yet the *waveforms* produced by different speakers pronouncing the same vowel are generally completely different from each other. In other words, it seems that vowels are recognized, independently of the quality of the voice or its pitch, by the overall shape of the spectrum and not by the shape of the wave itself. It is the frequency of the fundamental that produces the sense of the pitch of the voice, and very often conveys information in its own right (as in the different emphasis in '*I* love you', 'I *love* you' and 'I love *you*'), particularly of an emotional nature. Finally, it seems to be the fine structure of the spectrum, the little bumps and hollows on it that are the results of idiosyncrasies of the way our own particular mouths and throats are constructed, that enable us to differentiate one speaker from another. It turns out, as we shall see, that almost the first thing the ear does to the sound waves it receives is to Fourier-analyse them, and the pattern of activity of the fibres of the auditory nerve is—at least for medium and high frequencies—in effect a representation of the spectrum of the sound that is heard: only at low frequencies is information about the actual shape of the waveform available to the brain.

The structure of the ear

The visible external ear, or *pinna*, has little effect on incoming sound except that of 'colouring' it by superimposing little idiosyncratic resonances on it in the high-frequency region, that are dependent on the direction from which the sound is coming and can, as will be described later, provide quite accurate information about the position of a sound source. The auditory canal, the *external meatus*, similarly does little in Man except to add its own rather broad resonance peak around 3000 Hz. In other animals with less vestigial pinnae, they may have a more significant part to play in gathering sound, and if movable can provide a great deal more information about where a sound is coming from.

At the end of the meatus, the sound impinges on the *tympanic membrane* ('eardrum' in English) that separates the outer ear from the middle ear. On the inner side, the tympanic membrane is attached to the *malleus* ('hammer'), the first of a set of three tiny bones, the *ossicles*, whose function is to transform vibrations of the eardrum into vibrations of the fluids that fill the inner ear (Fig. 6.8). The malleus is joined to the *incus* ('anvil') which in turn bears on

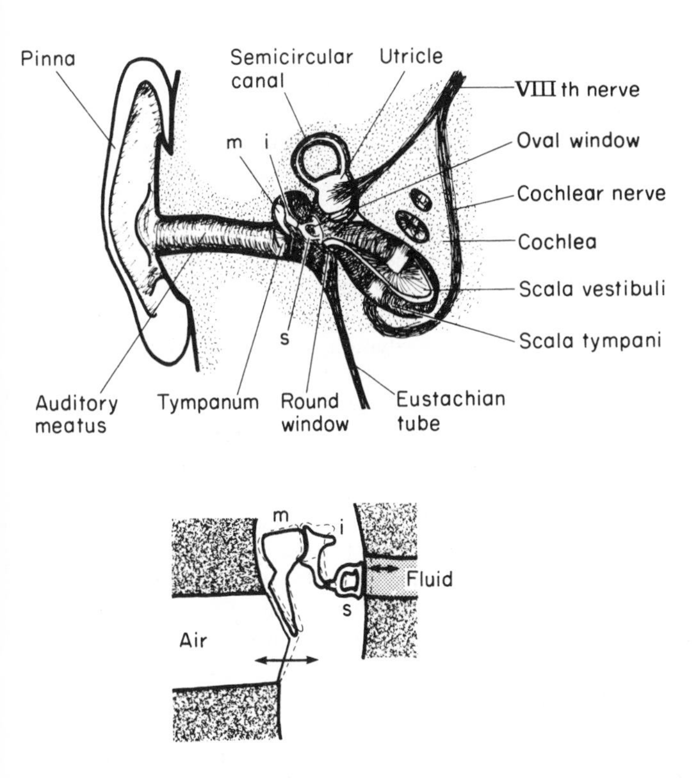

Fig. 6.8 Above, diagrammatic section through the ear. Below, showing the conversion by the chain of ossicles of low-pressure waves in the air into high-pressure, small-displacement waves in the cochlear perilymph: m, malleus; i, incus; s, stapes

the *stapes* ('stirrup'), whose footplate rests on the *oval window*, a membrane separating the middle and inner ears (Fig. 6.8). This chain of ossicles acts as a kind of lever system, converting the movements of the eardrum, which are of comparatively large amplitude but small force, into the smaller but more powerful movements of the oval window; this increase in the pressure of the vibrations is further enhanced by the fact that the tympanic membrane is much much larger in area than the oval window. The reason why this transformation is necessary—it is not an amplification, for like all passive systems it does not increase the *energy* of the waves that are transmitted—is because the fluid of the inner ear is very much denser than air. If a sound wave in air strikes a dense medium like water, the pressure changes of the air are too small to make more than a slight impression on the fluid, and most of the sound is reflected back. To ensure the most efficient transfer of energy from air to fluid, we need some way of increasing the pressure changes in the sound wave to match the characteristics of the new medium, and this *impedance matching* appears to be the primary function of the middle ear; without it, only some 0.1 per cent of the sound energy reaching the eardrum would reach the inner ear. A second function of the middle ear is that it is capable of acting as a kind of gate: it contains muscles—the tensor tympani and stapedius—that effectively disable the transmission system when they contract, thus providing a mechanism that might protect the inner ear from sounds powerful enough to cause damage. However the reaction time of such a response to intense sounds is such that it cannot in fact provide much protection against things like loud bangs, since by the time the muscles contract the damage has been done. But in discos they may help by acting as automatic ear-plugs.

The inner ear is simply the labyrinth, described in the previous chapter; the part of it that is concerned with sensing sounds is the *cochlea*, in effect an elongated sac of endolymph some 35 mm long, shaped as if it had been pushed sideways into a corresponding tube of perilymph, rather like the sausage in a hot dog (Fig. 6.9): the upper half of the perilymph is called the *scala vestibuli*, the lower is the *scala tympani*, and the endolymphatic sausage is the *scala media*. At the far end of this structure, the two perilymphatic regions join up through an opening called the *helicotrema*; and finally the whole thing is rolled up into a conical spiral, giving it the shape of a snail shell. In cross-section (Fig. 6.9) one may see that the scala vestibuli and scala media are separated by only a very thin membrane, *Reissner's membrane*, whilst the boundary between scala media and scala tympani is much more complicated and contains several layers of cells, including the receptors themselves, resting on the *basilar membrane*. The oval window faces onto the scala vestibuli, while a similar structure, the *round window*, separates the scala tympani from the air of the middle ear. (This air, incidentally, is in communication with the outside atmosphere through the *eustachian tube* which joins it to the pharynx; this tube is normally closed, but opens briefly during swallowing and yawning, causing a characteristic modification of one's hearing: one may get relief in this way from the ear-drum pain sometimes experienced in air travel as a result of pressure differences between the atmosphere and the middle ear.)

Now endolymph and perilymph are virtually incompressible fluids; so any movements of the oval window must result in corresponding movements of the round window. In other words, sound energy has no option but to pass from the scala vestibuli to the scala tympani, either by crossing through the endolymph, or by travelling right to the end of the cochlea and traversing the helicotrema. In the former case, it will cause the basilar membrane, with all its elaborate superstructure, to vibrate as well. Now the cochlear receptors—the *hair cells*—are peculiarly well adapted to respond to exceedingly small up-and-down movements of the basilar membrane. Their cell bodies are attached at the bottom to the membrane itself, and at the top to a rigid plate called the *reticular lamina*, which is in turn attached to a strong and inflexible support called the *arch of Corti*, that also rests on the basilar membrane. Consequently any movement of the basilar membrane results in an exactly similar displacement of the hair cells. These hair cells are very like the vestibular hair cells described in the previous chapter; they possess a number of stereocilia, arranged this time in a characteristic W-formation and graded in size, but the auditory hair cells of adults do not appear to have kinocilia. The cilia project through holes in the reticular lamina, and in the case of the outer hair cells the tips of the cilia are lightly embedded in the lower surface of the *tectorial membrane*, a flap of extracellular material analogous to the cupula and otolith,

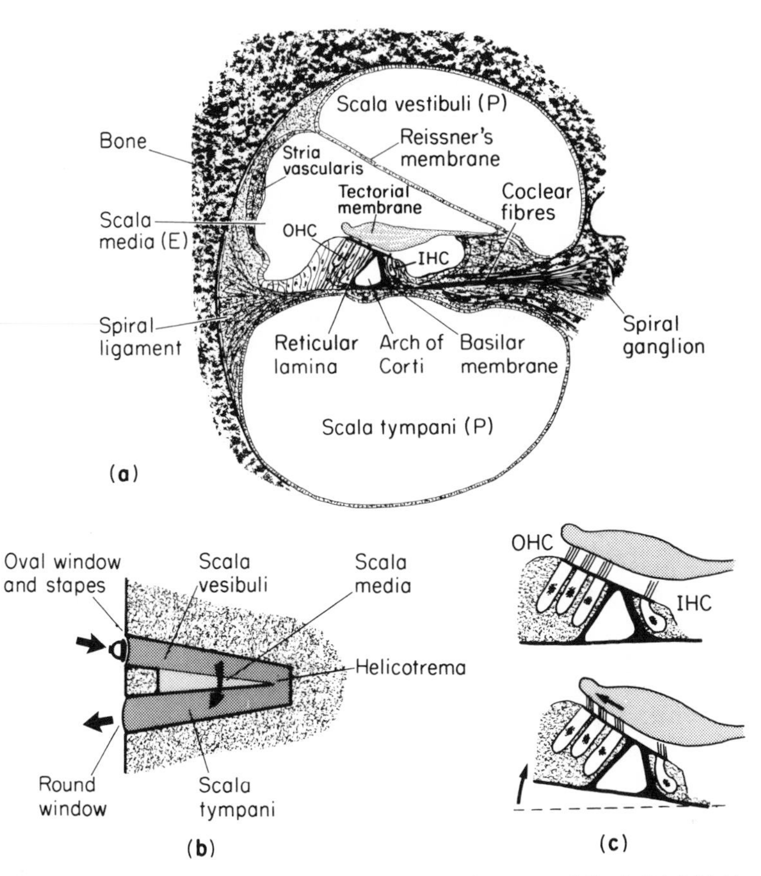

Fig. 6.9 (**a**) Section through the cochlea, showing the organ of Corti. (**b**) A highly stylized representation of the relationship between scala vestibuli, scala media and scala tympani, and the path of the sound through them. (**c**) Showing how vertical deflection of the basilar membrane causes shear between the reticular lamina and the tectorial membrane, thus bending the cilia of the hair cells. (For clarity, the movement of the basilar membrane is of course vastly exaggerated) (OHC and IHC are outer and inner hair cells, respectively; E is endolymph, P is perilymph)

that extends from the inner boundary of the scala media and lies along the top of the inner and outer hair cells. The cilia of the inner hair cells do not appear to make contact with the tectorial membrane, but lie just short of it. A consequence of the geometry of this arrangement is that any up-and-down movement of the basilar membrane will result in horizontal sliding, or shear, between the reticular membrane and the tectorial membrane. This in turn will bend the cilia of the hair cells (thus presumably stimulating them), through an angle which will be enormously greater than the original deflection of the basilar membrane (Fig. 6.9). As in the case of vestibular hair cells, this bending alters the ionic permeability of the receptors, giving rise to generator currents that eventually lead—probably with an intervening stage of chemical transmission—to stimulation of the auditory fibres. An electrode in the vicinity of the cochlea can be used to pick up the summed receptor potentials of the hair cells, giving a recording that is closely related to the form of the original sound-waves called the *cochlear microphonic* potential. If amplified and used to drive a loudspeaker, the result is a quite faithful reproduction of the sound entering the animal's ear.

But although it is fairly clear how the cochlea may act as a *transducer* of sounds into nervous energy, nothing has yet been said about whether it also carries out any *analysis* of the sound at the same time, in the way that for instance the cones of the retina begin the analysis of colour by responding preferentially to light of different wavelengths. In fact it turns out that different regions along the length of the cochlea are especially responsive to different sound frequencies, providing a rough kind of Fourier analysis of the type described earlier in this chapter. Before going on to discuss the electrophysiological responses of auditory nerve fibres, it is useful first to consider the mechanical properties of the cochlea that enable it to do this.

Fourier analysis by the cochlea

The physicist and physiologist Hermann Helmholtz (1821–94) was the first to appreciate the way in which sound quality was related to its frequency spectrum. Observing that the basilar membrane gets wider as one approaches the helicotrema, he suggested that it might be the cochlea itself that carried out this Fourier analysis. His notion was that individual transverse fibres of the basilar membrane might act rather like the strings of a harp, tuned to different frequencies and resonating in sympathy whenever their own particular frequency was present in a sound. (If you open the lid of a piano and sing loudly into it with the sustaining pedal depressed so that the strings are undamped, you will hear it sing back to you as particular strings are set into sympathetic vibration: and if there were some device that signalled which strings were vibrating and which weren't, you would have a kind of Fourier analyser). However, direct measurements of the mechanical properties of the basilar membrane show that it cannot in fact behave in this way. There is too much longitudinal coupling of the membrane: it is as if neighbouring strings in the piano were tied together; and in any case it is easy to show, simply by cutting it and observing that it does *not* twang back, that there is hardly any tension in the basilar membrane at all, and certainly not enough to make it resonate in the way Helmholtz suggested.

Nevertheless, it is clear from a number of pieces of evidence—for instance, that lesions of the basal end of the cochlea are associated with the specific loss of high-frequency auditory sensitivity—that different regions of the cochlea really are sensitive to different frequencies, in a systematic way. The true mechanism by which this comes about was not established until Georg von Bekésy, some 40 years ago, applied his superlative experimental skill to a demonstration of the way the basilar membrane *actually* responded during stimulation by sounds of different frequencies. What he found was that at any particular frequency, as one explored from the base towards the apex, the amplitude of the vibration of the membrane increased up to a maximum and then fell off again more sharply. The *position* of this maximum along the cochlea was dependent on the frequency, and the lower the frequency the nearer it was to the helicotrema: by 50 Hz or so the maximum had more or less reached the end of the cochlea. There were also phase differences at different points along the membrane: the greater the distance from the basal end, the greater the phase lag between the movement of the membrane and that of the oval window. Consequently when one looked at the basilar membrane as a whole, the appearance was of a *travelling wave* progressing towards the apex, growing larger as it went until the point of maximum response was reached, and then abruptly decaying to zero (Fig. 6.10).

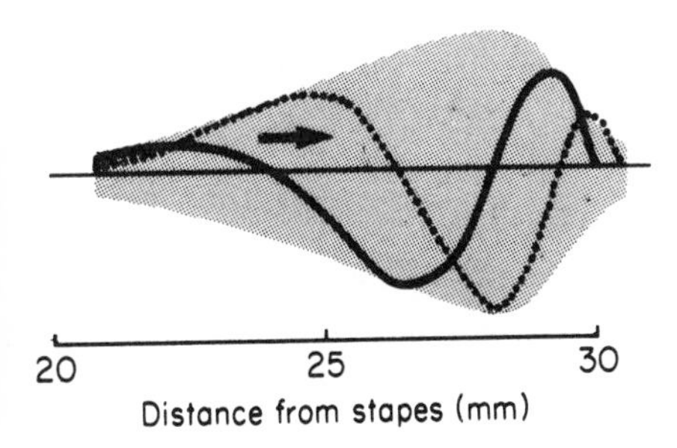

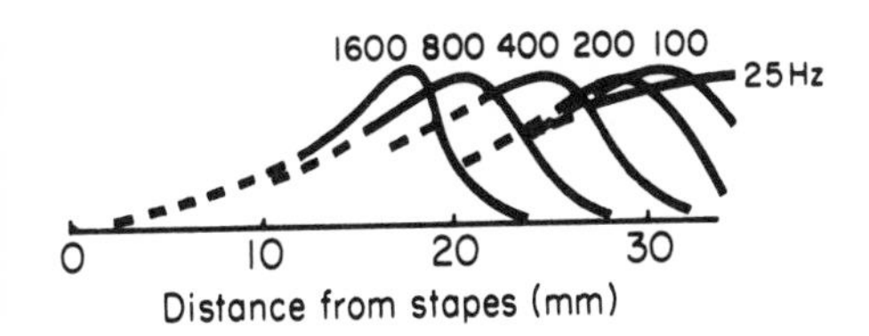

Fig. 6.10 Above, 'snapshot' of displacement of basilar membrane in response to a sine wave, at different distances from the stapes (solid line); the dashed line shows the displacement that would be seen a few instants later. The whole wave moves from stapes to helicotrema, with an envelope (shaded) that is a function of the frequency. Below, peaks of the envelopes associated with the various frequencies indicated. (After von Békésy, 1960)

The mechanical properties that give rise to this behaviour are rather complex. In essence what happens is something like this: the sound waves entering the scala vestibuli at the oval window can only get out again through the round window, but there is a choice of routes by which they can get there. They might for example cross straight through the basilar membrane at the basal end: the membrane is stiffer here, but on the other hand this would be a short route involving less movement of perilymph. Alternatively, the sound waves might prefer to travel further along the cochlear duct before crossing into the scala tympani, involving a longer mass of fluid to have to move, but an easier crossing because of the decrease in stiffness of the basilar membrane at increasing distances from the base. Now the relative hindrance offered by the stiffness of the membrane and the inertia of the periplymph depends very much on what the frequency of the sound is; the energy required to move a mass backwards and forwards increases enormously with increasing frequency, so that at high frequencies the path of least resistance is for the sound to cross as quickly as possible from scala vestibuli to scala tympani. At very low frequencies, the opposite is true: it is not much trouble to move the entire mass of perilymph backwards and forwards, and by doing so the sound can make the easier crossing at the apical end, where the stiffness is least. In other words, the preferred route will represent a compromise between the relative disadvantages of membrane stiffness and perilymph inertia, and the structure as a whole will sort out vibrations of different frequencies into different positions along the membrane, and thus act as a kind of Fourier analyser. One can find out how well it carries out this task by looking at the responses of the auditory fibres themselves.

Responses from auditory fibres

The synaptic connections between primary auditory fibres and the hair cells are broadly similar to those found in the vestibular system; corresponding to the distinction there between type I and type II cells, there are clear differences in innervation between the inner and outer hair cells of the cochlea. Each inner hair cell has afferent connections from some dozen or so radial fibres (Fig. 6.11), each of which appears to terminate on a single receptor. By contrast, the spiral afferents that innervate the outer cells run along the cochlea for a millimetre or so, and send afferent terminals to large

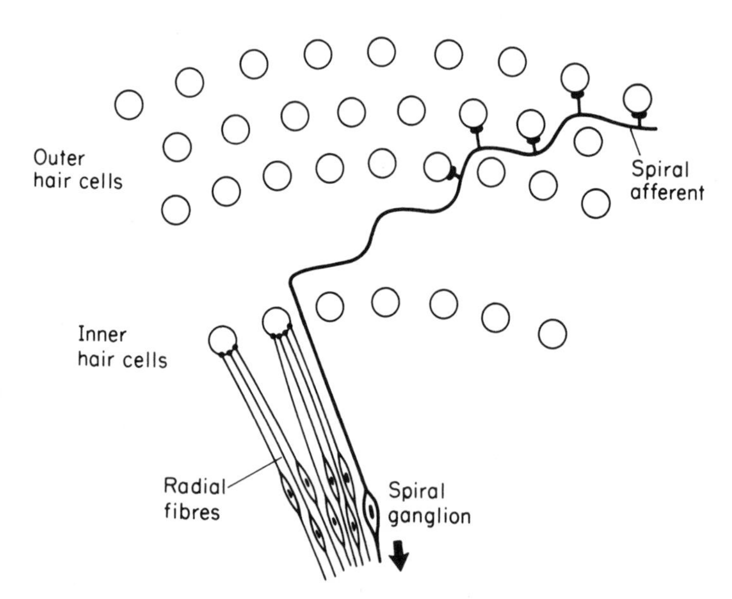

Fig. 6.11 Afferent innervation of outer hair cells (thick line) and inner (thin lines), showing convergence in the former case and divergence in the latter. (After Spoendlin, 1968)

numbers of receptors; about 10 per cent of the auditory nerve fibres come from outer hair cells and 90 per cent from inner. Thus there is a great deal of convergence from outer hair cells onto the afferent fibres, but little or none from the inner hair cells. There is a rough analogy here with the rods and cones of the retina: the outer hair cells have (like rods) a low threshold for stimulation—partly because their cilia actually stick into the tectorial membrane—and are grouped together in large receptive fields; whereas the inner hair cells, like cones, have reduced sensitivity but a more discrete connection with the brain, suggesting that here too the inner hair cells may have better 'acuity' (in this case to small differences of frequency). The hair cells also receive an efferent innervation, originating from a nucleus in the brainstem called the *superior olive*; these fibres terminate on the afferent endings of outer hair cells, and perhaps of inner as well.

There are two different sorts of electrical response that can be measured in the cochlea; with microelectrodes one can record action potentials from the auditory nerve fibres, and with larger electrodes in various areas within and around the cochlea one may in addition record several kinds of slower potential. The hair cells have resting potentials that are some 80–90 mV negative to the scala tympani, while an electrode in the scala media records a standing potential of some 80–90 mV *positive* with respect to perilymph: this is simply the result of the different ionic composition of endolymph and perilymph, and has the desirable consequence that the voltage difference across the top end of the hair cells is roughly twice what is usually found across neural membranes, implying that any given conductance change will give twice the usual generator current, thus presumably increasing sensitivity. One may also record *cochlear microphonic* potentials with large electrodes almost anywhere in the vicinity of the cochlea; these are thought simply to be the summed generator potentials of large numbers of hair cells, and appear to be more-or-less proportional to the local displacement of the basilar membrane; they thus follow the shape of the sound wave itself quite accurately, though with different degrees of frequency-filtering at different distances from the oval window. The auditory nerve fibres show similar properties to primary vestibular fibres, having a spontaneous resting discharge whose frequency is increased when the basilar membrane moves towards the scala vestibuli and decreased when it moves in the opposite direction, with an S-shaped relationship between displacement and firing frequency. Different fibres have response curves lying on different positions along the displacement axis, and the effect of stimulation of the efferent fibres seems to be to reduce their sensitivity by shifting the curves to the right. This effect is rather small under experimental conditions, being some 10 dB, and it is difficult to believe that this is all they are for.

Because of the frequency analysis performed by the mechanical properties of the basilar membrane, individual auditory fibres show a marked frequency selectivity (Fig. 6.12). A mysterious feature of this selectivity is that the cells are much more sharply tuned—the range of frequencies is much narrower—than one would expect from the response of the membrane itself to pure tones. It seems as though there must be some extra mechanism—a *second filter*—that makes the responses more selective than they would otherwise be.

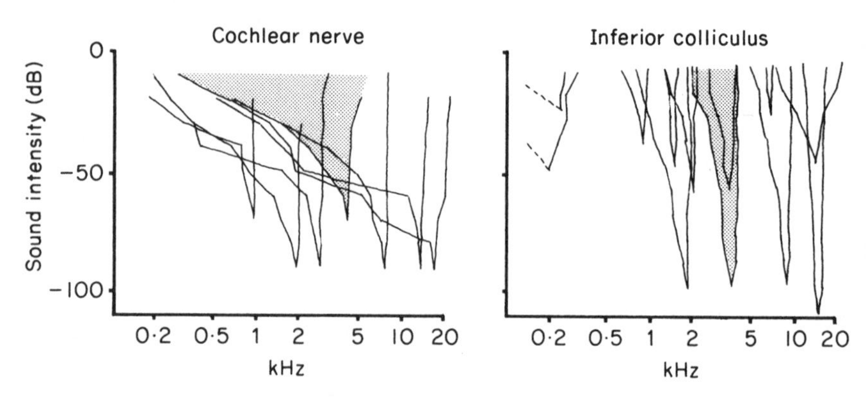

Fig. 6.12 Threshold response curves for individual units in cochlear nerve (left) and inferior colliculus, as a function of frequency, in the cat. (After Katsuk, 1961)

Recently it has been shown that individual hair cells respond in a frequency-selective manner even to *electrical* stimulation at auditory frequencies (which of course bypasses the mechanical filter provided by the basilar membrane), suggesting that the second filter is an intrinsic property of the receptors themselves. A further contributing mechanism is the existence, as in all sensory systems, of *lateral inhibition*. If one determines the tuning curve for a single auditory unit using a single tone, and then adds to this a second tone of different frequency, one finds that in the regions immediately neighbouring on the original tuning curve the response of the cell is actually *reduced* by the extra sound. In other words, each fibre has—in terms of *frequency*—a central excitatory area and an inhibitory fringe, which serves further to sharpen its selectivity.

The mechanisms of frequency analysis described so far operate essentially at medium and high frequencies; indeed at frequencies above a kilohertz or so there is no other way that information about frequency *could* be transmitted to the brain except by peripheral analysis and recoding, since individual nerve fibres are incapable of firing more frequently than about a thousand times a second at the very best, and hence cannot reproduce the pattern of the sound waves reaching the ear. But this is not the case at low frequencies, and in any case we saw in Fig. 6.10 that the frequency analysis produced by the basilar membrane begins to become ineffective at low frequencies because the maximum of activity has nearly reached the helicotrema. Recordings from single auditory units show that as the frequency of a stimulating tone is decreased, there is an increasing tendency for firing to be *phase locked* to the stimulating frequency; even if the frequency is too high for any one fibre to be able to fire once in every cycle, it may do so every two cycles, or every three or more (Fig. 6.13). So although no single fibre will be firing at the frequency of the stimulus, the average activity over the whole set may nevertheless be modulated at this frequency. In practice, phase locking is not quite as rigid as this: even at low frequencies where the fibres would be perfectly capable of following the imposed frequency, unless the stimulus intensity is very great

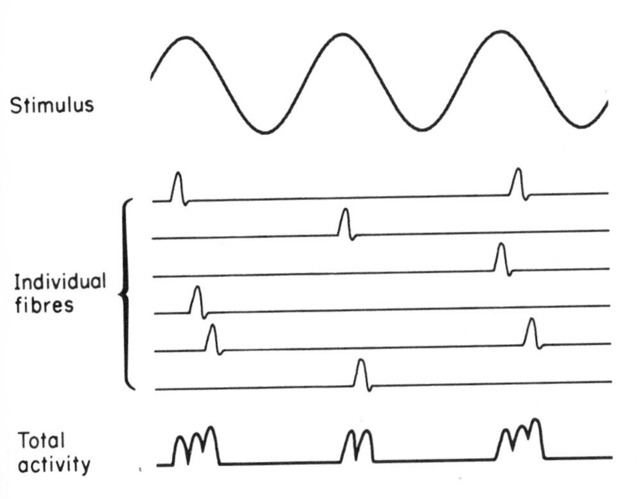

Fig. 6.13 The principle of phase locking: although no one fibre out of the whole ensemble fires in every cycle of the sound wave (top), nevertheless the modulation of the *total* activity reflects the frequency of the original stimulus

what one observes is simply that there is an increased *probability* of firing during one part of the cycle rather than another. At all events, phase locking provides a method of conveying auditory information to the brain without peripheral frequency analysis, and has the advantage that it retains information about the phase of incoming sound, information which is thrown away at high frequencies, above some 5 kHz. It seems that the perceived pitch of a sound depends essentially on the periodicity of the afferent neural activity—i.e. on its *temporal* pattern—whereas the sensation of timbre or quality, which requires the detailed perception of the high-frequency power spectrum of the sound, is almost certainly coded by the relative activity of fibres from different parts of the cochlea, i.e. by their *spatial* pattern of activity. Loudness is presumably simply a matter of the total amount of auditory activity; as auditory intensity is increased, there is both an increase in the firing of any one fibre, and also an increase in the total number of fibres that are firing at all, through recruitment.

Central pathways and responses

The central auditory pathways are complex (Fig. 6.14) and poorly understood. Primary auditory fibres synapse first in the ipsilateral *cochlear nuclear complex*, a group of three nuclei, each having a systematic *tonotopic* representation of the basilar membrane, so that neighbouring areas correspond to neighbouring frequencies. There is a certain amount of overlap of the projection of individual auditory units to neurones of the cochlear nucleus, but frequency selectivity is not reduced as a result because of the existence of further sharpening through lateral inhibition at this level, rather as in the dorsal column nuclei. The cochlear nuclei project to neurones of both

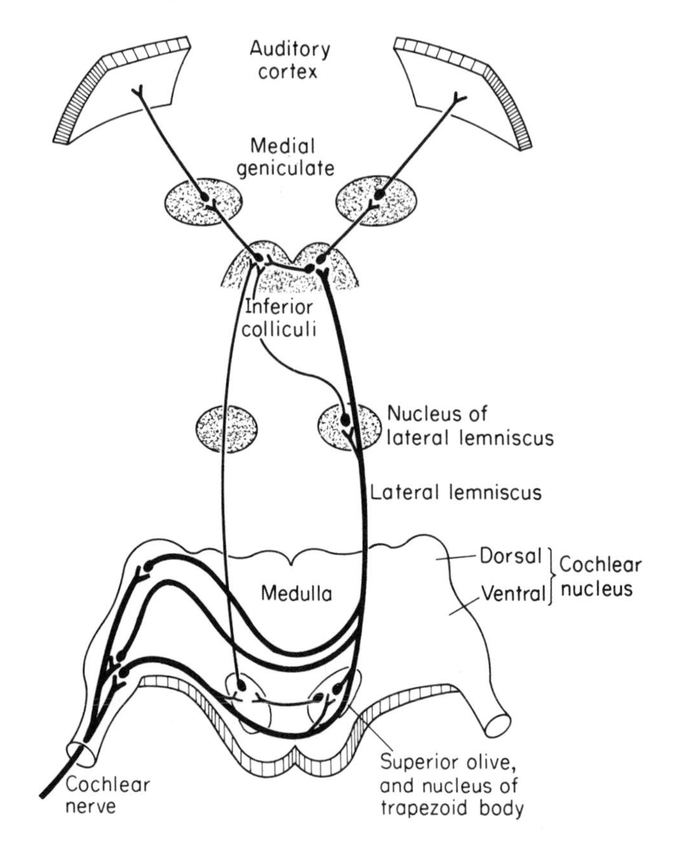

Fig. 6.14 Schematic diagram of ascending auditory pathways: first- and second-order fibres are shown thicker than the others. (Simplified, after Carpenter, 1976, with additional data from Brodal, 1981)

ipsilateral and contralateral *superior olives*, which ultimately project upwards to the *medial geniculate nucleus* of the thalamus—whence there is a direct relay up to the *primary auditory cortex*—via the *inferior colliculus*. The latter is an important integrating area for orienting responses to sound, such as turning the head or ears, and in lower animals is also the highest auditory level.

Nothing very spectacular seems to happen to the auditory information as it traverses these various neuronal levels: the orderly arrangement of cells in terms of their characteristic frequencies is preserved in most of these regions, and on the whole their tuning curves get sharper as one ascends to the medial geniculate (Fig. 6.12). But at the cortex this trend is reversed: although primary auditory cortex is still tonotopic, certainly for high frequencies, neighbouring auditory regions are less obviously so, and many of the units found there have novel properties that are very much more complex than simple frequency selectivity: they may respond to *changing* frequencies, or react preferentially to such specialized, 'real', stimuli as clicks, hisses and voices. Many units are binaurally driven, and respond only when the sound is coming from a particular direction, usually contralateral. What is mysterious at present is how these seemingly sophisticated responses can emerge so suddenly at the cortical level, with little evidence lower down of any kind of preliminary analysis of the kind characteristic, as we shall see later, of the visual system.

Spatial localization of sound

There are two components to localization—distance and direction—and these are carried out in very different ways by the auditory system. Since the energy of a sound wave decreases with the square of the distances it has travelled, one could in principle judge *distance* if one knew in advance the power of the sound source and the degree of absorption of the intervening structures; but in practice intensity can give only very approximate information. Very much more useful is the fact that not all frequencies suffer equal attenuation with distance: in an ordinary sort of environment, shorter wavelengths tend to be reflected or absorbed by physical objects, whereas long wavelengths simply ignore them. For this reason, the further one is from a sound source, the more of its high frequencies are lost: as a marching band approaches, it is the bass drum and then the tubas and euphoniums one hears first. Or again, when listening to a radio play one has a clear sense of how far the actors are from the microphone from the ratio of high to low frequencies in their speech sounds: close-to, the consonants—particularly sibilants like 'S'—are predominant, whereas with increasing distance it is the lower-frequency components—mostly vowels—that are heard most prominently.

Locating the *direction* of a sound source is a much more precise business, and is carried out by at least three separate mechanisms. Contrary to popular belief, one can localize sounds quite accurately using one ear alone: as was mentioned earlier, the peculiar pattern of bumps and whorls that decorate our pinnae add a coloration to all the sounds one hears, a pattern of small peaks and troughs in one's frequency sensitivity curve that is dependent on the angle at which the sound waves impinge on the ear. In the course of growing up one presumably learns that particular kinds of coloration are associated with particular directions, and in the adult this mechanism has been shown to provide localization of sound accurate to a few degrees. Using both ears produces only a slight improvement, to perhaps 1–2°. The extra information provided by *binaural* listening is of two distinct kinds: differences in interaural *intensity* and in interaural *phase*. Intensity differences come about because the head casts a 'sound shadow' that screens the ear to a certain extent from sounds coming from the opposite side (Fig. 6.15**a**). For the head to cast a shadow of this kind, it needs to be at least of the order of magnitude of the wavelength of the sound itself; thus screening of this type can only cause significant effects at frequencies *higher* than some 2–3 kHz. You can explore this effect for yourself by using a transistor radio as a source of sounds of different frequencies, covering one ear and listening to the changes in the intensities of low- and high-frequency components as you move the radio around your head. Intensity differences alone, even at high frequencies, do not permit sounds to be localized very accurately unless one is also allowed to

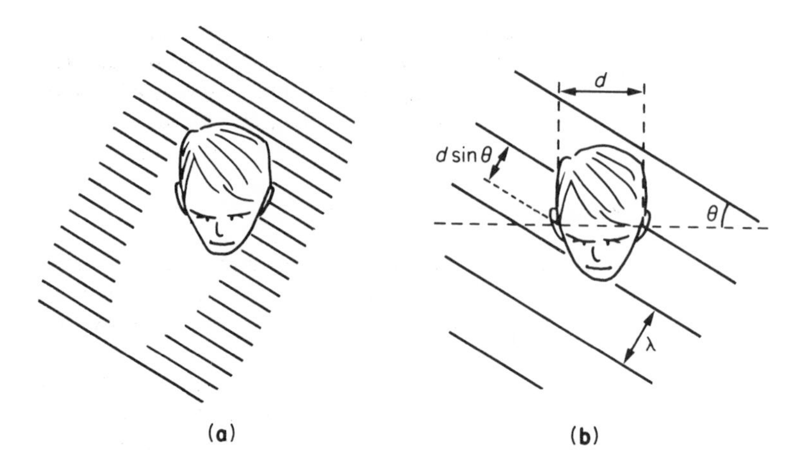

Fig. 6.15 Two binaural methods of localizing sounds. (**a**) Sounds of sufficiently high frequency cast a sound shadow on the far side of the head: the wavelength λ must be less than the order of magnitude of the head diameter, *d*. (**b**) A sound coming from a direction at an angle θ to straight ahead is associated with a phase difference between the ears of $\frac{d}{\lambda}\sin\theta \times 360°$, equivalent to a time difference of $\frac{d}{v}\sin\theta$ milliseconds, where v is the velocity of sound in km/sec and d is expressed in metres

move one's head to find the direction for which the intensity is most nearly equal in the two ears.

Phase differences arise because a sound coming from one side takes slightly longer to reach one ear than the other (Fig. 6.15(**b**)). Since by using phase information alone a subject may detect movement of a sound source of only 1 – 2°, one can calculate that the brain must be sensitive to interaural time differences of the order of 10 μsec, or about one hundredth part of the duration of an action potential! However, although phase differences can give accurate information about the direction of sounds, pure tones cannot be localized in this way if their frequencies are higher than some 1 – 2 kHz. The reason for this is that once the wavelength of the sound is less than the distance between the ears, ambiguities can occur, in the sense that a given phase relationship could be the result of more than one possible source direction (Fig. 6.16). In any case, we have already seen that information about phase is simply not transmitted by auditory nerve fibres at high frequencies. Thus the two fundamental binaural mechanisms of location are—rather conveniently—exactly complementary to one another: at low frequencies, only phase can be used, and at high frequencies, only intensity. Neither of these binaural mechanisms can do more than tell you the angle between the direction of a sound source and an imaginary line joining the two ears (Fig. 6.17); in particular, they cannot distinguish between a sound lying immediately behind the head, immediately in front, or somewhere overhead in the sagittal plane. For a complex sound of known frequency composition, this extra information can be provided by the monaural mechanism of directionally selective coloration described earlier. When this is impossible,

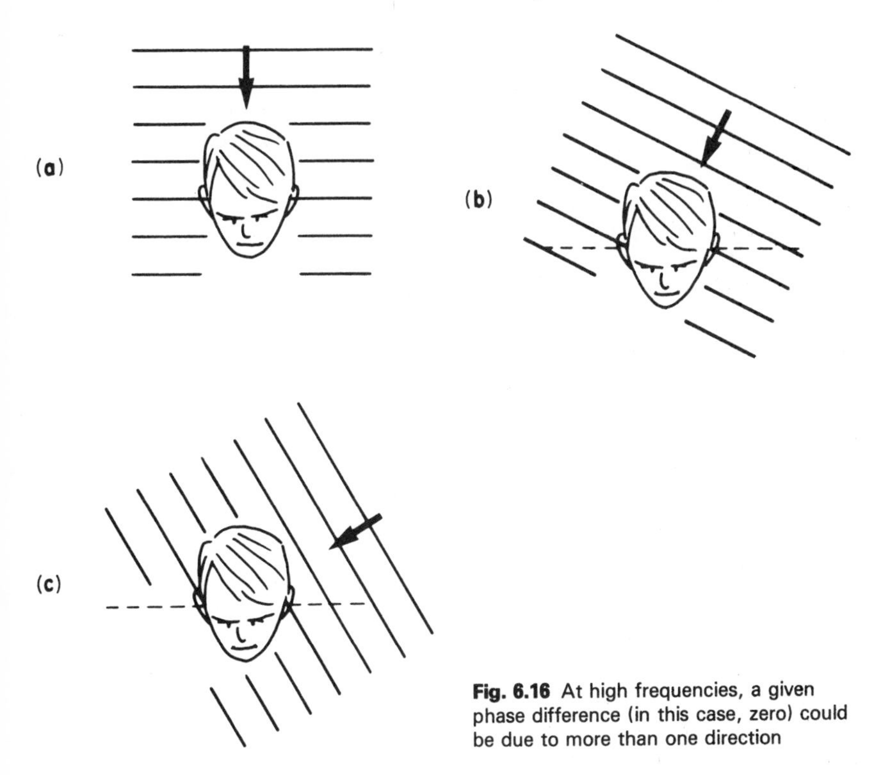

Fig. 6.16 At high frequencies, a given phase difference (in this case, zero) could be due to more than one direction

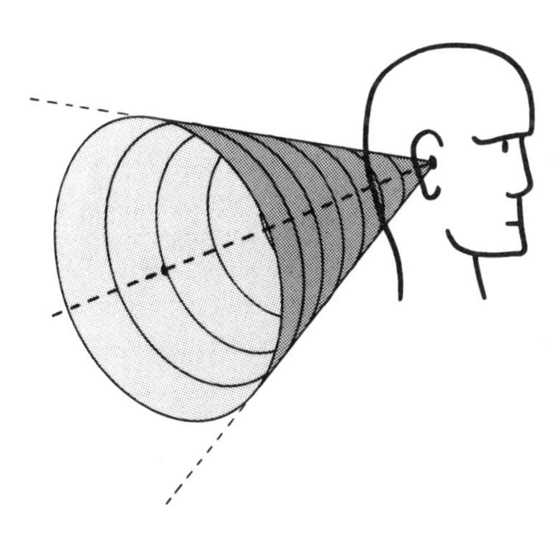

Fig. 6.17 A sound source producing a given phase delay, even at low frequencies, may lie anywhere on the cone which is the locus of all points lying at a given angle from the axis formed by the two ears. (This is not strictly the case if the source is close to the head)

then moving the head can provide an extra 'fix' on the sound that will enable its exact three-dimensional direction to be established, as when a dog cocks his head on one side when trying to locate the source of a sound.

These mechanisms are obviously of very great importance in the design of stereo audio systems. Ordinary stereo heard through a pair of loudspeakers is not very realistic for a number of reasons. First of all, it can provide an impression only of right–left localization and not of vertical localization; but more important, it messes up the normal time delays between the ears that are vital in low-frequency localization: each ear hears the sound from *both* speakers, so that a single recorded click reaches the brain as four separate clicks, two to each ear. This problem can obviously be got round by listening through headphones; and although this can certainly give a greatly improved sense of localization, one is then up against a different problem. In order to achieve good balance between different instruments in an orchestra or band, sound engineers like to use a vast array of microphones scattered about in different locations, and then mix them all together to form the two stereo channels. As a result, the phase relations between the same sound on the two channels are more confused than ever, and a single click is likely to end up as many dozens of clicks by the time it reaches the listener. Since most people listen through loudspeakers which muddle up the phase relationships anyway, this is not thought to matter much; but it means that recordings of this type do not work very well even through headphones. A very great improvement is the use of *dummy head* microphones. Here each channel is recorded through its own single microphone, which is placed in a dummy head, carefully designed with detailed modelling of the external ears in such a way that it produces the same kind of directional coloration that a real head would. If one now listens to the recording through headphones, the effect is extraordinarily realistic, partly because both amplitude and phase information is preserved, and also

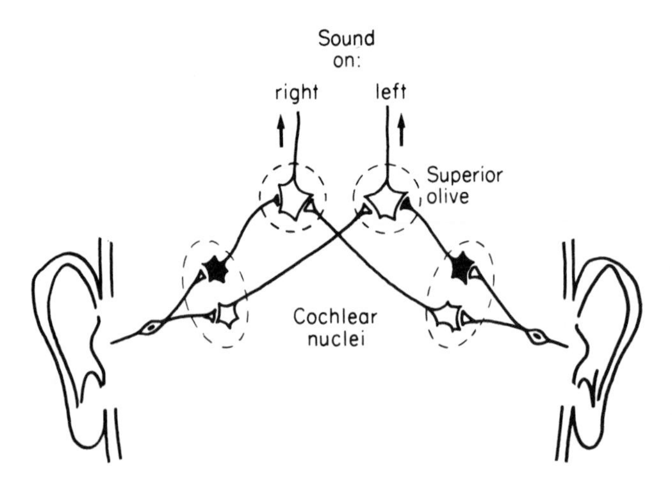

Fig. 6.18 A possible neural mechanism for binaural sound localization. A sound that is either louder on the right, or arrives first on the right, will excite the left superior olive neurone more than the right. (After van Bergeijk, 1962)

because for the first time it is possible to perceive the vertical localization of a sound. The one snag, which is true of all headphone systems, is that moving the head moves the sound image with it, reducing the illusion.

The neural mechanisms of sound localization are poorly understood. We have already seen that some auditory cortical units are specifically responsive to sound from particular directions, but how this is achieved is uncertain. One possible model is shown in Fig. 6.18. Here, neurones of the superior olive (the lowest point in the auditory system at which there is an integration of information from both ears) are excited by afferents from the contralateral ear and inhibited by those from the other. Consequently, if the sound at one ear either arrives earlier than that at the other, or has greater intensity, the olivary neurones on the opposite side will be excited more than the ipsilateral side, resulting in a sense of right–left localization. With such a system one might expect to be able to cancel out a time delay on one side by increasing the corresponding intensity; this kind of *time–intensity trade* can in fact be demonstrated quite easily in the laboratory by arranging for a subject to hear a tone of the same frequency in each ear, but of different phase, and asking him to adjust the relative loudness of the two stimuli until they sound as if coming from straight ahead.

7

Vision

Light is a form of energy propagated by electromagnetic waves travelling at an immense velocity—some 300 metres per microsecond—and carried in discrete packets called quanta or *photons*. Only a very small range of all the wavelengths of electromagnetic radiation known to physicists are visible (Fig. 7.1). The longest waves that we can just see, forming the red end of the spectrum, are some 0.7 μm in length, and only rather less than twice as long as the shortest waves at the blue end. In nature, most electromagnetic radiation is generated by hot objects: the hotter they are, the more of this energy is radiated at shorter wavelengths. The peak of the spectrum of light from the sun—an exceedingly hot object—corresponds quite closely with the range of wavelengths seen by the eye. Of man-made sources of light, many, like the

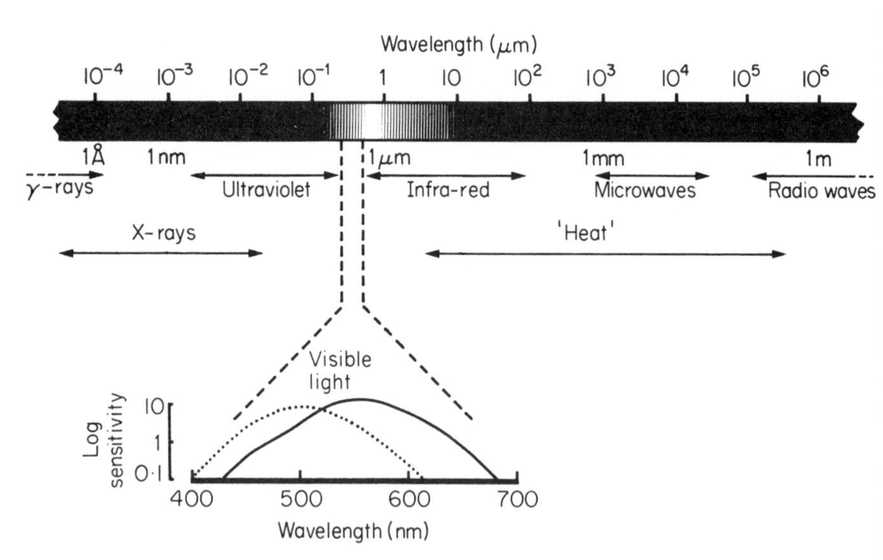

Fig. 7.1 Above, the electromagnetic spectrum: vertical shading indicates the approximate energy distribution of radiation from the sun. Below, the visible portion of the spectrum expanded, showing the relative sensitivity of the human eye to different wavelengths in the dark-adapted (dotted curve) and light-adapted (continuous) state

ordinary incandescent electric lamp, radiate as hot bodies and have a smooth and broad emission spectrum: others are quite different, and emit light only at a few discrete wavelengths. The sodium lights used for street lighting, for example, are effectively monochromatic, their energy being concentrated in a very narrow band in the yellow region. Domestic fluorescent lamps have a spectrum consisting of a number of emission lines superimposed on a continuous background. *Colour* is a function of the relative energy in different parts of the spectrum.

Photometry

The spectrum of a light is in a sense a description of its *quality*: describing *quantities* of light (photometry) is unfortunately a much more complicated business. There are two related kinds of photometric measurements: first, how much light is *emitted* by a source of radiation, and secondly, how much light is *received* by an illuminated object—for example, a piece of paper lying under the open sky. In the first case, we can talk about the *luminance* of a source, measured in candelas per square metre (cd/m^2). The candela is a measure of the rate of emission of light by an object: an ordinary 60 watt bulb is equivalent to about 100 candela. The amount of light received by an object per unit area is its *illuminance*, and is measured in lux. This unit is defined as the degree of illumination of a surface one metre from a source of one candela radiating in all directions. Full sunlight may provide about 100 000 lux.

Now objects in the real world tend to scatter back some of the light that falls on them, so that in general an illuminated surface is also a luminous one, and the ratio of luminance to illuminance in these conditions is a measure of the surface's whiteness or *albedo*. If we shine one lux on a perfectly white object that is also a perfect diffuser, it will have a luminance of about 0.32 cd/m^2, and such a surface is said to have an albedo of unity. Ordinary white paper has an albedo of about 0.95; paper printed with black ink, about 0.05. The photometry of coloured objects, which scatter back light of a different spectral composition from that which illuminates them (so that their albedo is a function of wavelength) is much more complex and requires special definitions and methods of measurement.

Figure 7.2 gives some idea of the range of luminances found in nature. It can be seen that the brightest lights tolerated by the eye without damage are some 10^{15} times more intense than the dimmest that can just be perceived. In practice, however, *at any one moment* the actual range of luminances to which the eye is exposed is very much smaller than this. Because the albedos of natural objects vary only from about 0.05 to 0.95, the range of luminances that can be seen as one looks for example round a uniformly illuminated room is only about 20:1, and this ratio is of course unaffected by changes in the overall level of illumination. Black objects seen in daylight look black because they lie at the bottom end of the range of luminances in the environment, even though *absolutely* they may radiate more light than white objects seen under dimmer artificial light at night. Although not very bright in absolute terms, the latter look white because they are at the top end of the range of luminances in the vicinity (Fig. 7.3).

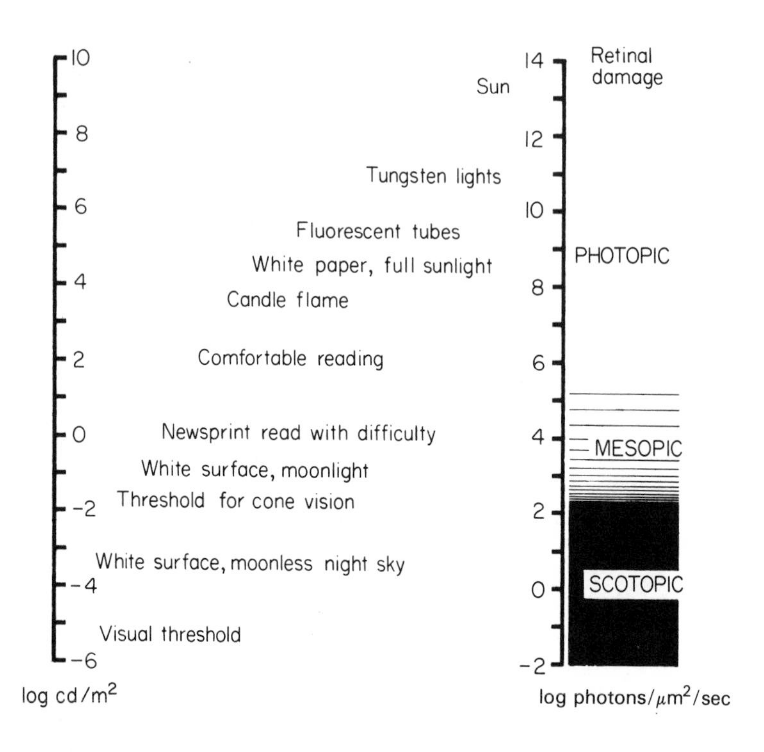

Fig. 7.2 The range of luminances (left) and retinal illumination (right, approximate) found in the natural world

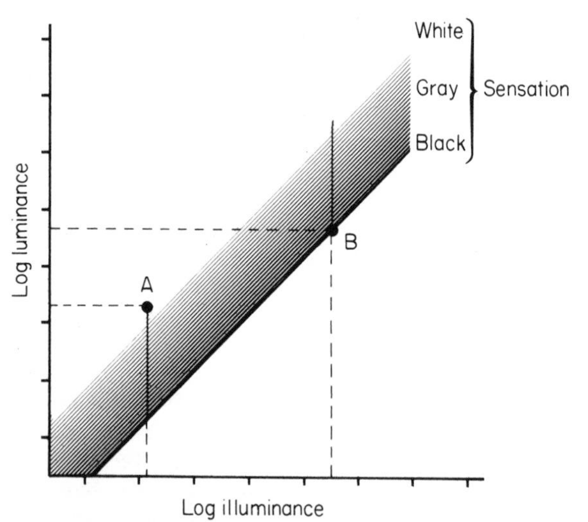

Fig. 7.3 The eye's sliding scale of brightness. A piece of white paper that is dimly lit (A) looks white because its luminance lies at the top of its local scale, even though this luminance may be *less* than that of a piece of black paper that is brightly lit (B). The latter looks black because it is at the bottom of its local scale

Light and dark adaptation

The eye operates in other words on a sliding scale of brightness that can be moved up and down the whole 15 log unit range in such a way as to match the prevailing level of luminance; this property is the result of various mechanisms of *adaptation*. It follows that the eye responds not so much to the luminance of natural objects as to their *albedo*: a much more useful sensory quality, since albedo is an intrinsic property of objects, whereas their luminance depends on how much they happen to be illuminated.

But it turns out that we cannot adjust our sliding scale instantaneously, and if we go from daylight to a dark room we find that it takes nearly 40 minutes for the eye to adjust its sensitivity fully to the reduced level of illumination. The simplest way to demonstrate this process of dark adaptation is to measure a subject's *absolute threshold*—the luminance of the dimmest light he can just perceive—at regular intervals during this adapting period. Such curves normally show two distinct components (Fig. 7.4): an initial one that levels off

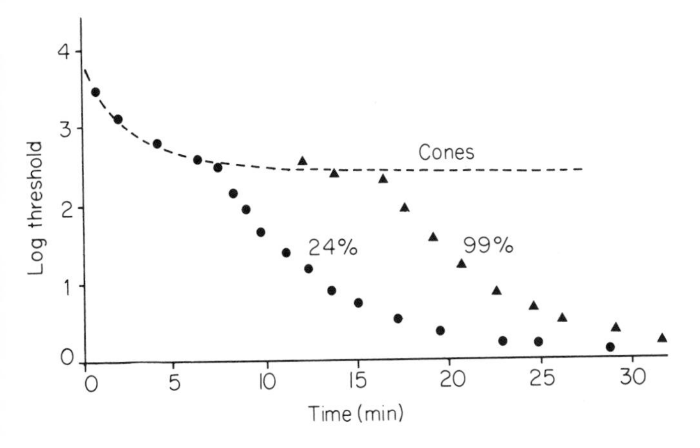

Fig. 7.4 Dark adaptation curves. The points show measurements of absolute threshold at different times after a strong light that bleached 99 per cent (triangles) or 24 per cent (circles) of the pigment in the rods. The dashed line indicates the approximate time-course of recovery of sensitivity of the cones alone. It is clear that recovery occurs in two stages: the first is due to cones, and the second, delayed, stage to rods. (Data from Rushton and Powell, 1972)

after some 8 minutes, and a further, slower, increase in sensitivity that takes another 30 minutes or so to reach completion. This dual response is due to the presence in the retina of two different types of receptor: *cones*, that function at high light levels (the *photopic* region of Fig. 7.2) but cannot respond to luminances lower than some 0.1 cd/m^2, and *rods*, which are much more sensitive and respond throughout the scotopic and mesopic range, but are overloaded by bright lights and cannot contribute much in the photopic region. The cones adapt relatively quickly and are responsible for the first branch of the adaptation curve; the rods more slowly, and provide the final and more gradual component of dark adaptation. Animals that are specialized for night vision have only rods in their retina, which means that they are unable to respond effectively to the enormous range of luminances that can be appreciated by an eye of mixed type such as our own. We shall see later that cones have further properties that are useful in vision, in particular that by responding preferentially to different narrow bands of the spectrum, they provide a mechanism by which the visual system can respond to the *quality* (i.e. the colour) of a source of light, as well as to quantity.

Image-forming by the eye

When parallel rays of light pass into a denser medium with a convex surface, or a less dense medium with a concave surface, they are brought to a focus at a distance that is a function of the radius of curvature and of the ratios of the refractive indices of the two media. In the eye, there are three surfaces of this sort that act together to bring the images of distant objects to a focus on the retina: they are the *cornea*, and the front and back surfaces of the *lens* (Fig. 7.5).

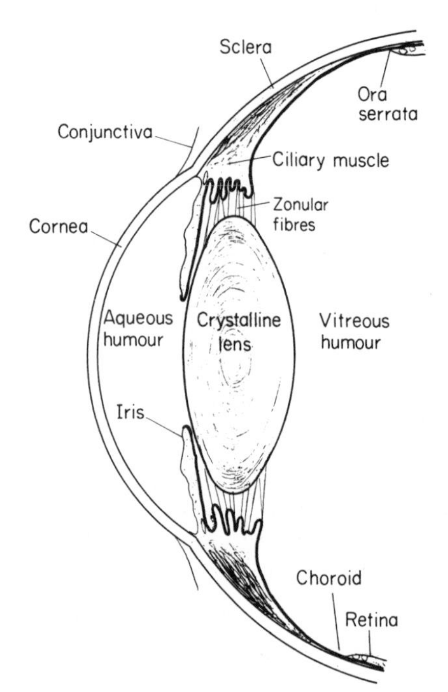

Fig. 7.5 The anterior part of the human eye, shown in parasagittal section

The refractive index of the aqueous humour that separates the cornea and lens is much the same as that of the vitreous humour that fills the rest of the eye, and is about 1.34; that of the crystalline lens is only slightly greater than this, being about 1.42, so that most of the refractive power of the eye is due to the cornea rather than the lens. Ophthalmologists describe the power of refractive surfaces by the reciprocal of their focal length in metres, and these units are called *dioptres* (D). Since the distance from the cornea to the retina in Man is about 24 mm, the total refractive power of the eye when focused on a distant object is some 42 D; of this, about 36 D are due to the cornea alone. But although the lens does not contribute much to the total refractive power, its main use lies in the fact that it can alter its shape and hence change the eye's effective focal length: this function is called *accommodation*. The lens is encircled by a ring of muscle called the ciliary muscle, to which it is joined by the zonular fibres (Fig. 7.5). When the ciliary muscle is relaxed, the zonular fibres exert a radial pull on the edge of the lens that tends to flatten it; in accommodation, the ciliary muscle contracts and thus allows the lens to revert to its natural, more biconvex, shape. In consequence, changes in the degree of innervation of the ciliary muscle can alter the power of the lens by as much as some12 D. As one gets older, however, the elasticity of the lens declines, and by the age of 60 the possible amplitude of accommodation may have fallen to 1 D or so, a condition known as *presbyopia* (Fig. 7.6). The range of accommodation can easily be measured by finding the positions of the *near and far points* of the eye, the nearest and furthest distances at which objects can just be brought into focus. For a normal or *emmetropic* eye with accommodation fully relaxed the far point will be at infinity, and the range of accommodation will be given by the reciprocal of the distance of the near point in metres: for a young subject, the near point will typically lie at around 80 mm. In old people, this distance will be typically of the order of a metre, so that convex spectacles have to be worn if reading matter is to be brought within comfortable holding distance. But few people are exactly emmetropic, and what is normally found is that when the accommodation is fully relaxed the total refractive power is either too great or too small in relation to the distance from the cornea to the retina. If it is too great, images of distant objects are brought to a focus in front

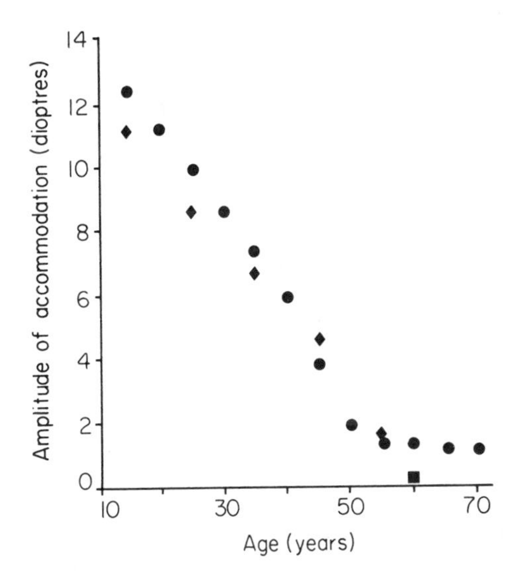

Fig. 7.6 Presbyopia: decline in the amplitude of accommodation as a function of age in three large groups of subjects (Data from Fisher, 1973)

of the retina, and the eye is said to be *myopic*, or short-sighted: under these circumstances the far point will not be at infinity, but nearer to the eye. Such a condition may be corrected by the use of concave or negative spectacle lenses. *Hypermetropic*, or far-sighted, subjects are precisely the opposite, and require convex or positive lenses in order to focus at infinity with relaxed accommodation (Fig. 7.7). In either case, the degree of disability may be indicated by the power and sign of the lens needed to bring the eye back to emmetropia. This is called the spherical correction, and in general is not the same in both eyes. Another common defect is *astigmatism*: here the refractive

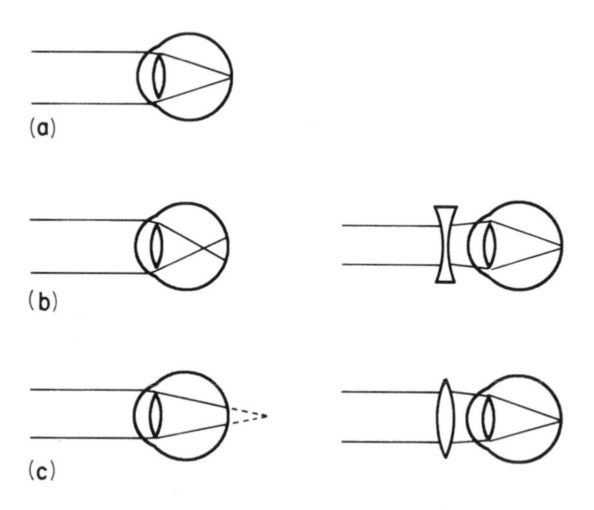

Fig. 7.7 Errors of focusing. (**a**) shows an *emmetropic* eye with relaxed accommodation that focuses parallel rays exactly on the retina. A *myopic* eye (**b**) brings parallel rays to a focus that is too close to the lens: the defect may be corrected with a negative (concave) lens (right). A *hypermetropic* eye (**c**) cannot bring parallel rays to a focus at all: a positive (convex) lens is needed for correction

power of the lens is found to be different in different meridians, generally because of non-uniformities of the radius of curvature of the cornea. If for example it has a smaller radius of curvature in the horizontal plane than in the vertical, the far point when measured with a vertical line as test object will be closer than when a horizontal line is used. Opticians test for astigmatism by means of a target like that of Fig. 7.8, called an astigmatic fan; an astigmatic subject will see some of the lines more sharply than others, and this will tell the optician the angle at which a cylindrical lens should be placed in front of the eye to make the refractive power as nearly as possible equal in all meridians. The power of the cylindrical lens that is needed to do this, together with its meridional angle, make up the cylindrical correction that is the second part of a prescription for spectacles.

Astigmatism and incorrect refractive power are not the only faults that may be found in the eye's optics, and as in many man-made optical systems, the cornea and lens together produce a number of different types of optical

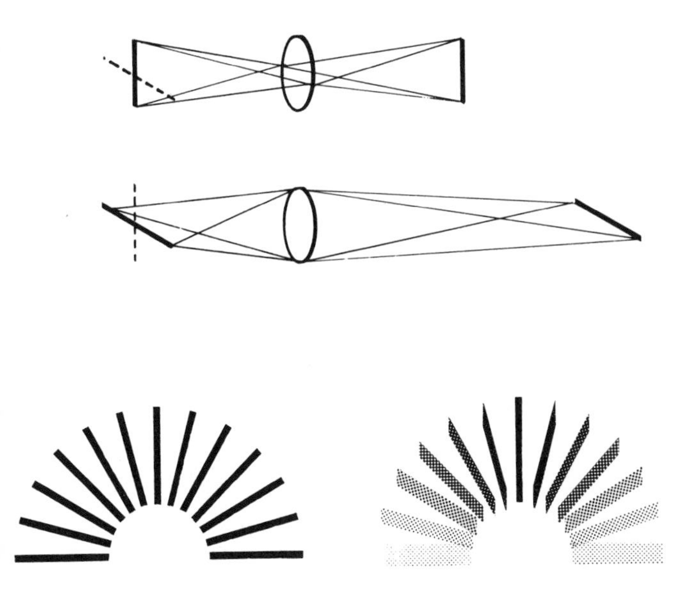

Fig. 7.8 Astigmatism. Above, an astigmatic lens has different focal lengths for line targets in perpendicular directions. Below: left, an astigmatic fan, a target used for testing for astigmatism, and its appearance (right) to a subject with marked astigmatism in the horizontal/vertical directions

aberration. The first of these is due to the fact that the refractive indices of the various optical media of the eye depend on the wavelength of the incident light. In general, the refractive index increases with decreasing wavelength, so that blue light is refracted more than red. This phenomenon is called dispersion and gives rise to defects in the resultant image, in the form of coloured fringes, called *chromatic aberration*. This means that if one looks at a blue object and a red object lying side by side at the same distance from the eye, they cannot both be in focus simultaneously, and a subject who is emmetropic when his far point is measured in red light will be short-sighted if it is measured in blue: his far point will then be only a metre or so away. This forms the basis of a simple clinical test for errors of refraction, consisting of an illuminated screen divided into three portions that are red, green and white: identical test figures are superimposed on each field, and the subject is simply asked which figure he sees most clearly. The emmetrope will see the one on the white background best, while the hypermetrope and myope will see most clearly those lying respectively on the green and red backgrounds.

The second major type of aberration is one that is common to all systems formed of spherical refracting surfaces, and is called *spherical aberration*. This arises because the shape of refracting surface needed to bring parallel rays to a point is not strictly speaking a spherical one at all, but an ellipsoid. For surfaces that are small in comparison with their radii of curvature the difference is slight, and spherical aberrations are often negligible. But in the case of the eye, the aperture is of the same order of magnitude as the radius of curvature of the cornea, and the result is that rays entering near the periphery of the cornea are bent too much, and form a closer focus than those entering near the centre. To some extent Nature has compensated for spherical aberration, first of all by making a cornea that is not exactly spherical but tends towards the desired ellipsoid, and secondly in that the refractive index of the lens is not constant throughout, but graded from a maximum of some 1.42 at its centre to about 1.39 at the edge, thus cancelling out, to some extent, the extra bending of peripheral light rays. The degrading effect of both spherical and chromatic aberration, and of other defects due to irregularities of the refracting surfaces, depends on the aperture of the eye, and this in turn is under the control of the *pupil* or iris.

The control of the pupil

Unlike the lens, the pupil is under the control of two different muscles: one, the sphincter pupillae, lies circumferentially round the pupil, and the other, the dilator, lies radially. The two muscles thus have opposed effects, the first causing contraction of the pupil and the second dilatation, and they are respectively under the control of the parasympathetic and sympathetic systems. It is not entirely clear which of the two branches of the autonomic nervous system is responsible for normal tonic control of pupil size, and one may cause *mydriasis* (enlargement of the pupil) by drugs that either block the action of acetylcholine on the sphincter (e.g. atropine) or simulate the effect of noradrenaline on the dilator (e.g. phenylephrine). The advantages of a large pupil size are first that the eye receives more light (over the normal range of pupil diameters, about 2 – 8 mm, the amount of light caught by the eye varies by a factor of 16), and secondly that the diffraction effects that always occur when light passes through a small aperture are minimized (see p.171). The advantages of a *small* pupil, on the other hand, are an increased depth of field (a greater tolerance of errors of focus) and a reduction in the magnitude of the optical aberrations: the extent of this effect can be seen for oneself if the pupil is dilated with a mydriatic. Thus the ideal size for the pupil is something of a compromise, and depends on the ambient light level. Under bright photopic conditions the eye can take advantage of the excess light by reducing the pupil and improving the quality of the retinal image. In scotopic conditions, however, the eye needs all the light it can get and the quality of the retinal image is of secondary importance: in any case, we shall see later that the rods are not capable of passing on accurate information about the detailed structure of the retinal image. It is important to emphasize that pupil dilatation contributes very little to the enormous changes in sensitivity that accompany dark adaptation, since it can only vary the incoming light by a factor of 16 at most, or 1.2 log units. The control of the pupil is also closely linked to accommodation: when the ciliary muscle contracts in order to focus on a near object, there is normally an associated constriction of the pupil (the near reflex). As these responses are usually also combined with binocular convergence movements of the two eyes, the whole pattern of response (constriction, accommodation, convergence) is also known as the *triple response*.

Under certain clinical conditions, notably in neurosyphilis, one may find that the pupillary response to near objects remains despite loss of the response to bright lights: this condition is known as the Argyll Robertson pupil, and is an important diagnostic neurological sign.

The retina

One might hope to be able to see another person's retina directly by eyeball-to-eyeball confrontation: if both eyes are emmetropic and relaxed, each retina should be clearly in focus on the other (Fig. 7.9). The reason why this doesn't in fact work is that the presence of the observer's eye also prevents light falling on the other's retina, so that nothing can be seen: under normal conditions the pupil of the eye is always dark. The *ophthalmoscope* is a device that gets round this problem by projecting a small beam of light into the subject's pupil at the same time (Fig. 7.9). It also has an arrangement whereby one of a set of negative and positive lenses can be introduced into the optical pathway: the

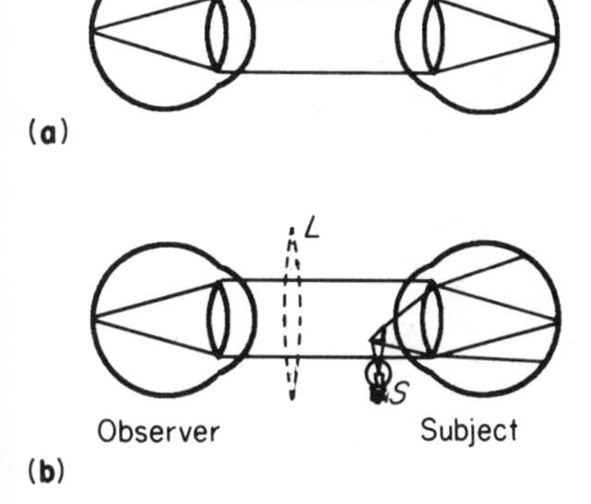

Fig. 7.9 The principle of the ophthalmoscope. (**a**) If an observer looks into a subject's eye and both are emmetropic, the retina of one will be focused on that of the other. But the observer's eye prevents light from reaching the subject's retina, so that nothing can be seen. (**b**) The ophthalmoscope introduces an extra source of light, *S*, to illuminate the subject's retina, and is fitted with a set of lenses (*L*) that correct for errors of refraction

power of the lens that exactly brings the subject's retina into sharp focus is equal to the combined refractive errors of observer and subject. Thus so long as an oculist knows his own correction, the ophthalmoscope provides an objective method for determining what spectacles the subject requires, as well as permitting the examination of the retina for signs of disease.

Two features of the retina are immediately obvious when seen through the ophthalmoscope (Fig. 7.10). About 15° to the nasal side of the optical axis, the blood vessels of the retina are seen to converge to join the central retinal artery and vein. This is also the point at which the nerve fibres of the retina come together to form the optic nerve, and since both blood vessels and nerve fibres lie on the interior side of the retina, in front of the receptors themselves, this area—the *blind spot*—is incapable of responding to light. Although it is some 5° across, one is usually unaware of its existence because the brain tends to fill it in with whatever background colour or pattern immediately surrounds it (Fig. 7.11). The other gross feature of the retina that is visible with the ophthalmoscope is that very close to the centre of the retina is an area about 5°

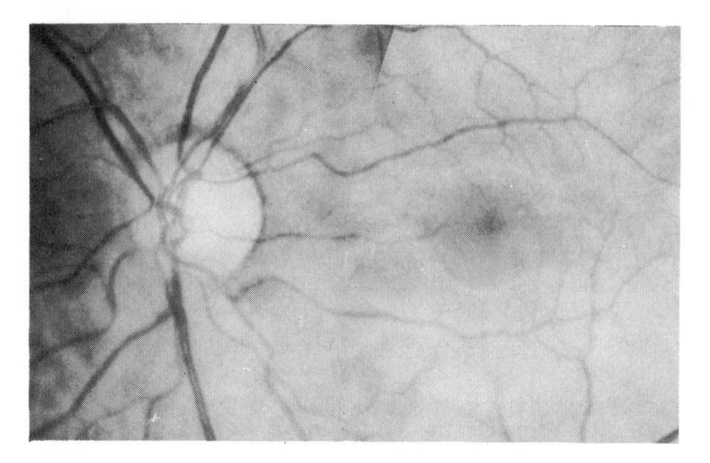

Fig. 7.10 Photograph of a living human retina, showing blood vessels, optic disc and blind spot (left), and macula lutea, with fovea at its centre (right). (Courtesy of J. Keast-Butler, FRCS)

Fig. 7.11 Demonstration of the blind spot. Close the left eye, and fixate the cross with the page at about 25 cm from the eye. The face will disappear, and no discontinuity on the background will be apparent

across that is free of large blood vessels—they arch around on each side to supply it from the edge—and is also distinctly yellower than the rest of the field. This is the *macula lutea* (yellow spot), and at its centre is a very small dot—actually a depression or pit—called the *fovea centralis*. When we look at a small object in the outside world, it is the fovea that is directed to the corresponding part of the retinal image: its angular size is about that of one's finger nail with the hand fully extended. The fovea is specialized for high-quality, photopic vision: it is quite without rods, and the cones themselves are tightly packed to give the maximum information about image detail (Fig. 7.12). Cones in this region are about 1.5 μm across, corresponding to a visual angle of some 20 sec arc. The depression arises because the retinal structures that elsewhere in the retina lie between the receptors and the lens—remembering again that the retina is inside-out in its layered structure—are here displaced to one side so as to cause the minimum scattering of incoming light. The supply of oxygen and nutrients for this

Fig. 7.12 Section through the human fovea. The direction of incident light is from above downwards, passing through the layers of neural elements (which in the central fovea are pushed to one side) before reaching the receptors

region must derive almost entirely from the blood vessels that richly supply the *choroid*, the layer immediately superficial to the receptors and separated from them by the thin pigment epithelium.

The retina is quite different from any of the sense organs we have met so far, in that a good deal of the neural processing of the afferent information has already occurred before it reaches the fibres of the optic nerve. No doubt the reason for this is that the eye is a highly mobile organ, and if each of the 130 million or so receptors sent its own individual fibre into the optic nerve, the latter would have to be some 11 times thicker than at present, and would be a considerable hindrance to rapid movement of the eye; and of course the blind spot would be correspondingly larger as well. The fibres of the optic nerve are in fact at two synapses' remove from the retinal receptors, and particularly as far as the rods in the periphery are concerned, there is considerable convergence of information from large groups of receptors. What happens is that receptors synapse with *bipolar cells,* and these in turn synapse with the million or so *ganglion cells* whose axons form the optic nerve. These two types of neurone form consecutive layers outside the receptor layer—except in the fovea, where we have seen that they are pushed to one side—and are mingled with two other types of interneurone that make predominantly sideways connections. These are the *horizontal cells* at the bipolar/receptor level, and the *amacrine cells* at the ganglion cell/bipolar level. The arrangement of their connections is shown schematically in Fig. 7.13. We shall see that there are marked differences in the electrical behaviour of these different neurones: although ganglion cells and amacrines show spike discharges in response to retinal stimulation, the bipolars, horizontal cells and the receptors themselves do not: they are small enough to be able to interact electrotonically without the need for active propagation.

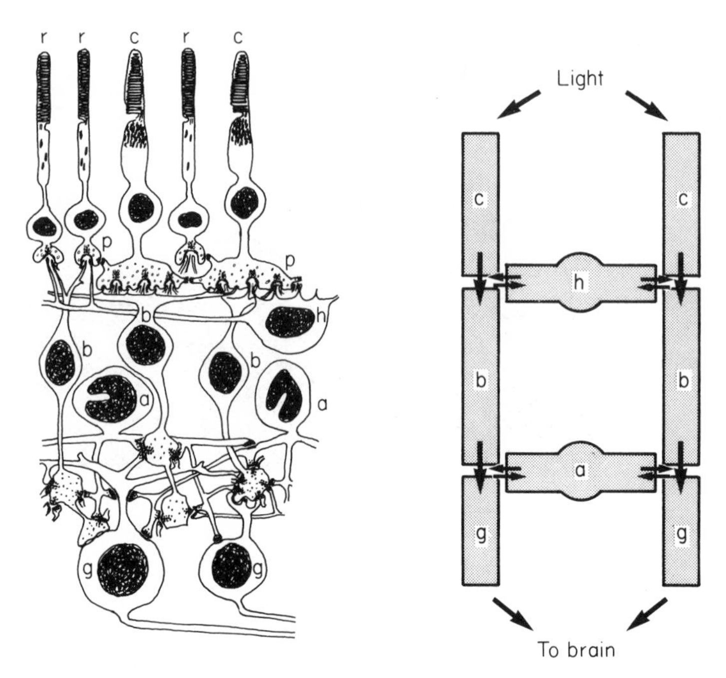

Fig. 7.13 Retinal neurones and their connections. Left, simplified representation of section of primate retina, showing rods and cones (r,c), with their pedicles (p) forming synapses with horizontal cells (h) and bipolars (b). The latter connect with amacrine cells (a) and ganglion cells (g) whose axons form the optic nerve. (After Dowling and Boycott, 1966). Right, highly schematized representation of the principal connections shown on the left

The receptors

Rods and cones both consist of two distinct parts: an *outer segment*, apparently a modified cilium, and an *inner segment* containing the nucleus. The outer segment possesses a high concentration of *photopigment*, associated with a richly folded set of invaginations of the outer surface, which in the case of rods at least have separated off to form a stack of flattened saccules (Fig. 7.13). At the base of the outer segment the remains of the ciliary filaments and centrioles can be seen. The inner segment has mitochondria as well as the nucleus, and its inner end forms the synaptic junction with bipolar and horizontal cells. There is no doubt that the photopigment bound to the membranes of the outer segment discs play a key role in transforming incident light into electrical changes, for if the pigment is isolated from the receptor it is found that its absorption of light of different wavelengths corresponds closely with the spectral sensitivity of the receptors themselves. Retinal photopigment consists of two portions: a *chromophore* called retinal or retinene (a derivative of all-*trans* retinal, better known as Vitamin A), in association with a lipoprotein complex which may be called *opsin*. It is thought to be slight differences in the composition of the latter that give rise to the different spectral sensitivities of rods and cones. The first effect of light on the visual pigment found in rods

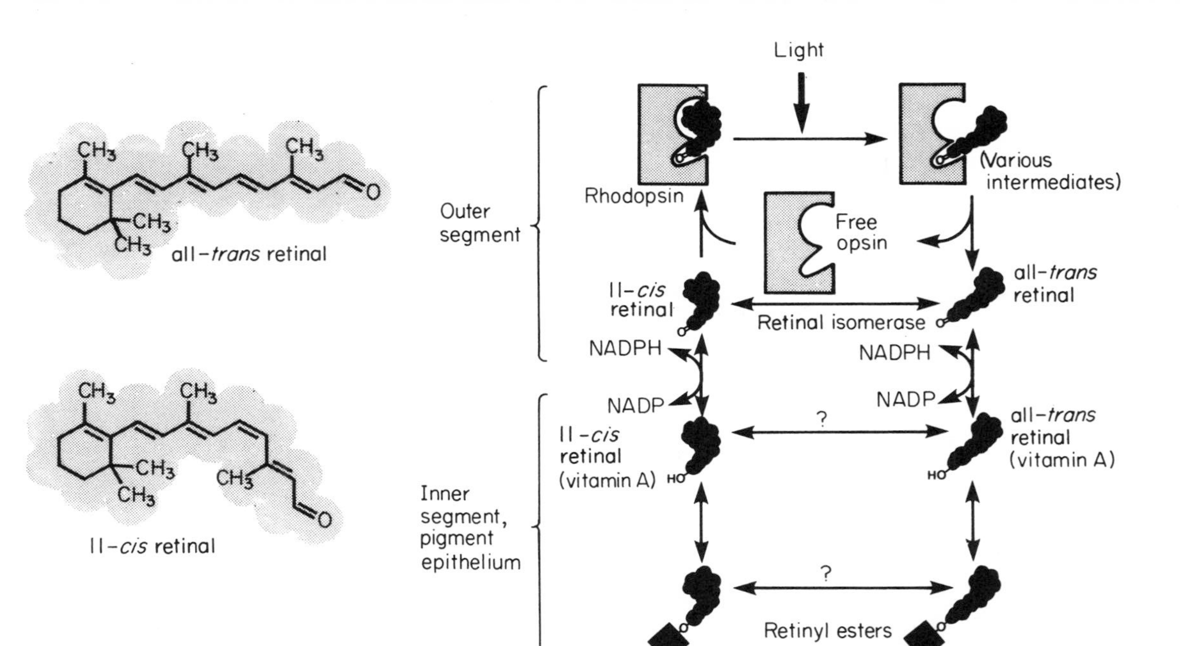

Fig. 7.14 The photochemistry of the rods. Left, the structure of all-*trans* retinal and 11-*cis* retinal. Right, the cyclical sequence of events by which light leads to isomerization of retinal and its dissociation from opsin, followed by the relatively slow processes that lead to the final regeneration of rhodopsin

(rhodopsin) is to cause an isomerism of the retinal from the normal 11-*cis* form to the all-*trans* configuration. This in turn leads to a series of changes in the configuration of the rhodopsin, producing a number of more-or-less short-lived intermediates, that culminates in the complete dissociation of the opsin from the retinal. In vitro, this is the end of the matter, and the pigment is said to be *bleached*. But in the rods, the bleached pigment can be regenerated by enzymes present in the receptors and in the pigment epithelium that lies behind them. The first stage of this process consists of the reconversion of the free all-*trans* retinal back to the 11-*cis* form, a relatively slow process (Fig. 7.14). We shall see later that it is this slow regeneration of pigment that determines the long time-course of recovery of rod sensitivity during dark adaptation that has already been mentioned (Fig. 7.4). Since the retinal is fundamentally derived from vitamin A, a deficiency of this vitamin may result in there not being enough retinal present to convert all the available free opsin back to active pigment, producing a degree of night-blindness because of the resultant insensitivity of the rods. In bright light, most of the rhodopsin is in the bleached form: an equilibrium is reached in which the rate of bleaching equals the rate of regeneration. Estimates of the amount of pigment in the receptors of a living eye during particular stimulus conditions may be made by the technique of retinal *reflection densitometry*, in which one measures the amount and spectral composition of the light scattered back from the retina when a light is shone into the eye. In this way it is possible to track continuously the amount of bleaching that occurs in both rods and cones under relatively natural visual conditions. Another technique that has been used to characterize the cone pigments is that of microdensitometry, in which the spectral absorptions of individual receptors may be measured in a preparation on a microscope slide. As far as we know, the reactions that occur in rods and cones are fundamentally similar, though the regeneration of cone pigment is substantially quicker than in rods, and under photopic conditions a smaller fraction of the cone pigment is in the bleached state than is the case for rods: this is one of the reasons why the cones are able to function at much higher light levels.

Electrical events in retinal receptors

What is not at all clear is exactly how the bleaching of pigment by light is linked to the resulting neural stimulation. One mechanism that has been suggested is that the series of changes in the opsin configuration that results from the initial isomerization of the retinal might in effect open or close ionic channels in the membrane of the outer segment and hence cause electrical currents to flow between the outer and inner segments, and ultimately at the distal synapse. Measurements of the absolute threshold for seeing dim flashes of light when the eye is fully dark-adapted show that a single rod is capable of responding to a single absorbed photon. This astonishing sensitivity is usually taken to imply the existence of some mechanism of amplification within the rods, for it is hard to believe that a single ionic channel in a receptor (out of a total population of some 10^8 or 10^9, as estimated from the amount of pigment in a single rod) would be capable of initiating a significant electrical response.

Cones seem to be intrinsically less sensitive, though even they will respond to perhaps some 8–10 absorbed quanta.

Electrical recording from single rods and cones has now become a commonplace technique of visual research, but as yet has not done much to elucidate the fundamental mechanisms concerned. If the retina is subjected to a flash of bright light, a rapid electrical response (the early receptor potential, or *ERP*) may be recorded from the receptors. Because the ERP occurs with virtually zero latency from the moment of the flash, and is unaffected by changes in the ionic composition of the surrounding medium, it seems likely that it represents the direct effect of light on the distribution of electron density in the pigment molecules, resulting from the initial isomerization of the retinal from 11-*cis* to all-*trans*. In the same way, it is found that the amplitude of the ERP depends not only on the energy of the flash, but also directly on the fraction of pigment that is in the unbleached state; thus its size increases during dark adaptation. Larger and slower potentials generated by the receptor in response to light may also be recorded: these are big enough to be measurable in the electroretinogram or *ERG* that can be recorded either by means of an electrode applied to the surface of the cornea, or with a microelectrode in the retina itself. Unlike the ERP, these potentials are strongly influenced by the ionic composition of the bathing fluid, and are almost certainly due to changes in ionic permeability. By examining the way in which these potentials alter as the electrode is inserted to different depths in the receptor layer, one can show that they are generated in the outer segment, and that when light falls on the eye there is a flow of current from the outer segments to the inner, resulting (in mammals) in a hyperpolarization rather than a depolarization of the receptor terminals. If one plots the size of the receptor potential as a function of the intensity of the flash (Fig. 7.15) one finds a characteristic S-shaped, or saturating, relationship; and the effect of

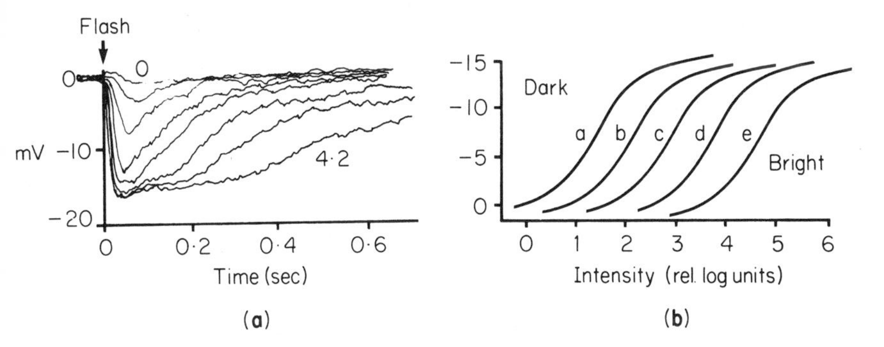

Fig. 7.15 Electrical responses from cones. (**a**) Hyperpolarizations generated by a very brief flash of various intensities ranging from 0 to 4.2 relative log units in steps of 0.6 log unit. (**b**) Showing the effect of pre-adaptation to different backgrounds in such an experiment. Each curve is the result of an experiment like the one on the left, where peak electrical response is plotted as a function of the flash intensity. Curve *a* was measured in the dark, *b–e* under different increasing background adaptation levels over a range of some 3.5 log units. (After Baylor and Fuortes, 1970, and Normann and Perelman, 1979)

different levels of light adaptation is to shift this curve along the intensity axis, providing one of the mechanisms by which the sensitivity of the retina is adjusted to suit prevailing luminance.

Retinal interneurones

The synaptic connections of the receptors with the bipolars and horizontal cells are of an unusual type (Fig. 7.13), in which invaginations of the foot of the receptor receive processes from both of the other types of neurone in a kind of three-way junction. The receptor both influences and is influenced by the horizontal cells, and both affect the bipolar. In the mammalian retina, horizontal cell responses are of the same hyperpolarizing form as those of the receptors, but the bipolar cell is hyperpolarized by the receptors and depolarized by the horizontal cells. The latter therefore provide lateral inhibition, as well as carrying information from receptor to receptor, both rods and cones: this function may also be carried out by the direct synaptic contacts that have also been observed between the receptors themselves. This pooling of receptor signals is also mediated to some extent by the bipolars, which in the peripheral retina may synapse with large numbers of rods and cones and hence provide the first stage of convergence which enables the number of optic nerve fibres to be so much smaller than the number of receptors. In the fovea this convergence is much less evident, and most of the bipolars contact only a single cone. The lateral inhibition provided by the horizontal cells can be demonstrated by electrical recording from the bipolars. Their electrical responses are generally similar in size and time scale to the slow potentials that can be obtained from the receptors themselves, but show very different receptive field properties. If one plots the response of bipolar as a function of the position of a small stimulating spot of light, one finds that there is a central region in which the light causes a hyperpolarization of the bipolar, surrounded by a region in which the light produces depolarization; one may also find cells for which the centre response is depolarization and the surround is hyperpolarizing. The effect of this antagonism between centre and surround is to make the bipolar respond more vigorously to small stimuli in the centre of its receptive field than to uniform illumination that stimulates both centre and surround (Fig. 7.16).

As noted earlier, ganglion cells differ from all the other types of cell except the amacrines in that they respond to light with repetitive spike discharges. Like bipolars, most ganglion cells have receptive fields consisting of a centre region with an antagonistic surround, many of them having in addition the property that they respond only *transiently* to sudden changes of retinal illumination. Sometimes a transient burst of firing is found in response to an increase of illumination (an 'on-response'), and sometimes to a decrease ('off-response'): and sometimes one may find a burst response both at the beginning and end of a period of steady illumination (an 'on-off response'). A cell with an on-response at its centre will normally show an off-response in its surround, and vice versa (Fig. 7.17) and they show on-off responses in intermediate regions. Thus as far as their field properties are concerned, ganglion cells are similar to bipolars: the new feature of transient sensitivity is probably the result of

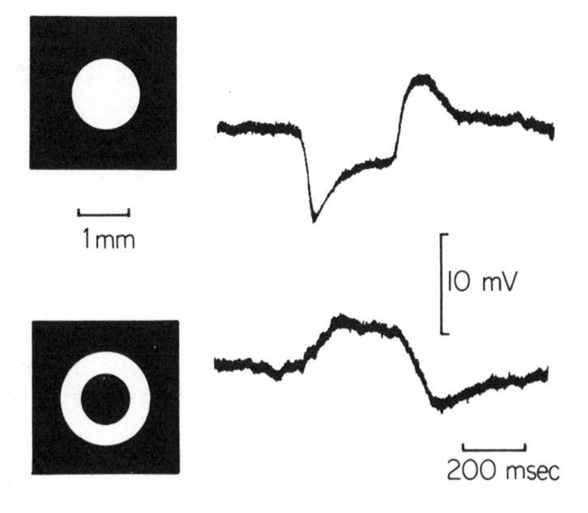

Fig. 7.16 Responses of bipolar cell to disc (above) and annulus of light, showing antagonism between centre and surround (Kaneko, 1970)

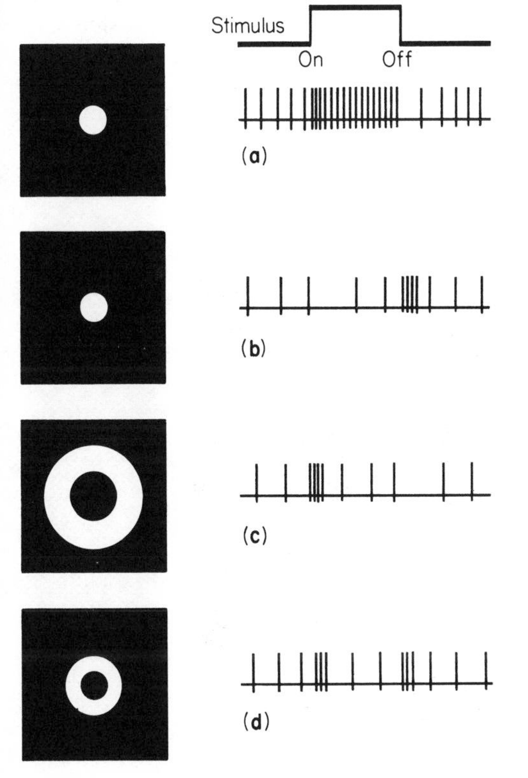

Fig. 7.17 Schematic representation of types of ganglion-cell response: (**a**) sustained; (**b**) off-centre; (**c**) on-surround; (**d**) intermediate on–off

feedback inhibition from the amacrines, whose responses are very similar to those of ganglion cells and have the right sort of connections for mediating lateral and self-inhibition of the type that would explain the time-course of the ganglion cell responses: but little is known as yet about their precise function. A small number of ganglion cells show sustained responses, and can convey information to the brain about steady light levels: clearly some such source of information must project into the optic nerve to explain such tonic responses to steady light levels as the tonic pupil light reflex. Several other types of ganglion cell have been described, differing in such aspects as the size and homogeneity of their receptive fields, and the extent to which they show linear summation.

Central visual pathways

Although both our eyes point forwards and have much the same view of the world, the brain is lateralized in the sense that the left brain is primarily concerned with things on the right side of the body and vice versa, so it is not surprising that the first thing that happens to the optic nerve fibres after leaving the two eyes is that they are sorted out according to which side of the retina they come from, and brought into association with the corresponding fibres from the other eye. This occurs in the optic *chiasm* (Fig. 7.18). A consequence of their rearrangement is that whereas lesions of the optic nerve

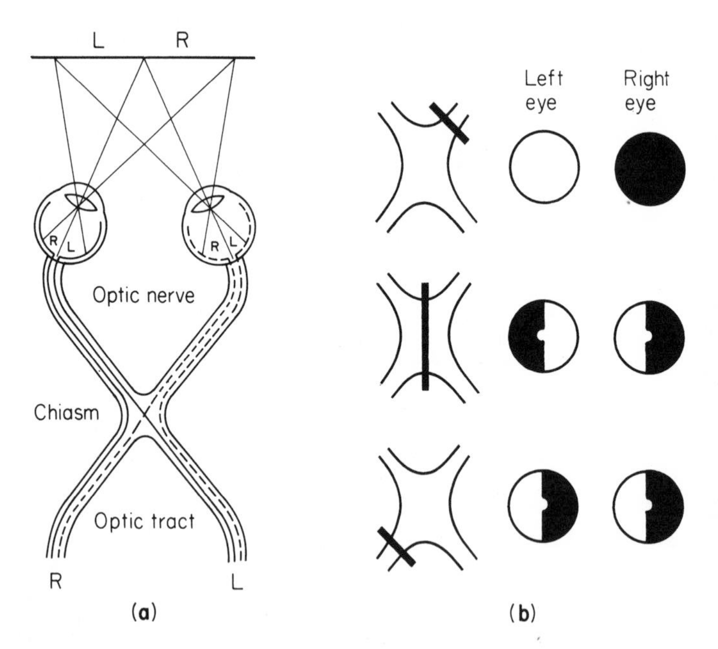

Fig. 7.18 **(a)** Schematic representation of the partial decussation of the optic nerve fibres in the chiasm. **(b)** the effect on the visual fields of each eye of lesions at three different points on the peripheral visual pathway

naturally enough cause blindness in one eye (unilateral anopia or anopsia), a unilateral lesion of the optic *tract* (the continuation of the fibres after the chiasm) results in blindness of the same half field of each eye (homonymous hemianopia); damage to the chiasm itself may give a bitemporal heteronymous hemianopia. Not all the fibres decussate completely, however, and the fovea of each eye is to some extent represented in both cerebral hemispheres. This gives rise to the phenomenon of *macular sparing:* lesions that would be expected to produce an exact homonymous hemianopia often show no loss of vision on the affected side near the point of fixation.

Most of the fibres of the optic tract project to the *lateral geniculate nucleus* (LGN) of the thalamus, the rest passing to the *superior colliculus* and to pretectal regions of the brainstem. The geniculate neurones in turn project through the *optic radiation* to the primary *visual cortex* or striate cortex, in the occipital lobe of the cerebral hemisphere (area 17). Throughout the whole of this system the neurones preserve their relative topographical arrangement, so that fibres that are near one another correspond to neighbouring parts of the visual field. There is, however, a certain amount of distortion, in the sense that central regions have a relatively greater representation than more peripheral ones: this is partly a consequence of the greater degree of retinal convergence found in the periphery. In primates, the LGN has six distinct layers of cells, three of which are entirely associated with the ipsilateral eye and the other three with the contralateral one: the functional significance of this arrangement is not understood. The receptive fields and responses of geniculate neurones are not markedly different from those of retinal ganglion cells, and show the same concentric organization of on- and off-responses. Many LGN cells also show centre–surround antagonisms that are wavelength-dependent. A cell of this type might for example show an on-response in the centre to red light, and an off-response in the surround with green light; or for that matter, vice versa. Yellow-versus-blue cells may be found as well as red-versus-green ones, and some cells show spectral antagonism of this type without having a centre–surround organisation. The significance of these spectrally opponent cells in the perception of colour is discussed later, on p.180. Another general feature of geniculate neurones is that to a far greater extent than ganglion cells their response to light may be modified by nervous activity in other parts of the brain: in sleep, for example, their responses to flashes of light are considerably reduced. This 'gating' of the geniculate cells, allowing some control of what information reaches the cerebral cortex, seems to be a general feature of thalamic relays, and it is important to remember that the thalamus receives many descending fibres from other parts of the brain in addition to its primary sensory projections.

It is a relatively simple matter to record from cells in the visual cortex, and this has been a happy hunting-ground for neurophysiologists over the last twenty years or so. Although some cells in the visual cortex show receptive field properties that are very similar to the roughly circular and concentric fields seen in retinal ganglion cells and in the LGN, most of them (and all of them outside the primary projection area, area 17) show receptive field properties that are quite novel. *Simple* cells are found predominantly in area 17 and have fields that are not circular, but consist of a central strip with antagonistic strips

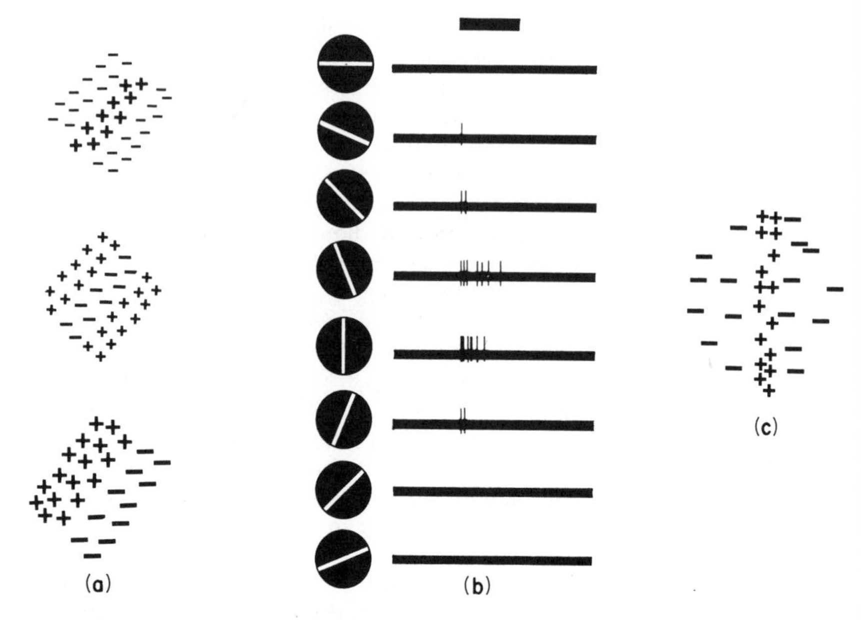

Fig. 7.19 Visual responses of simple cells. (**a**) Receptive fields of three typical cells, showing regions responding to a localized spot of light being turned on (+) or off (−). (**b**) Responses of a single unit to a bar of light presented at the various orientations shown: it is preferentially stimulated by a vertical bar. (**c**) The receptive field of the unit whose responses are shown in (**b**). (After Hubel and Wiesel, 1962)

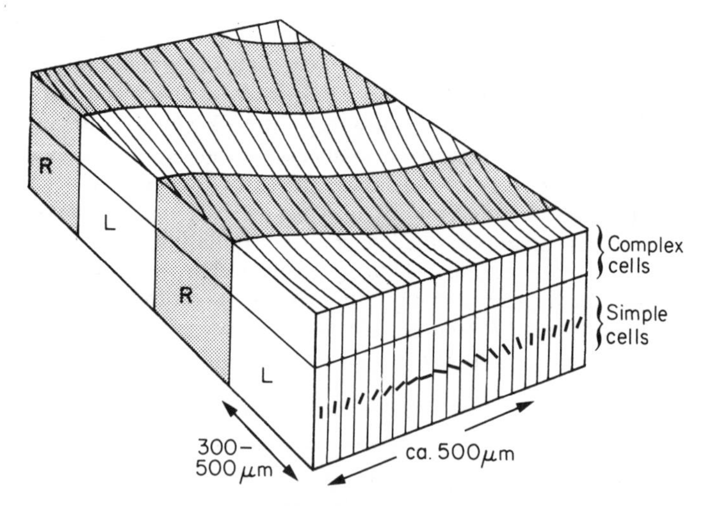

Fig. 7.20 Highly stylized representation of a slab of visual cortex, showing its organization into 'columns', narrow strips in which cells share a common preferred orientation, and roughly perpendicular dominance bands, preferentially driven by one or other eye. Preferred orientation usually changes systematically (as shown) in passing along a set of columns. An analogous 'columnar' organization is found in other neocortical regions

flanking it (Fig. 7.19): the centre may be excitatory or inhibitory, and the orientation of the entire field is different for different cells. Moving or flashed stimuli are in general more effective as stimuli than steady ones. The normal arrangement is that the cells appear to be functionally grouped in *columns* perpendicular to the cortical surface, the cells in a particular column sharing the same preferred orientation and the preferred orientations of the columns (more properly, strips) change in a systematic way as one moves across the cortical surface (Fig. 7.20). *Complex* cells are more common; like simple cells, they respond best to a bar or edge of a specific orientation, but in this case they will respond to such a target placed anywhere within their field of view. It is not possible to map out excitatory and inhibitory areas of their fields as it is with simple cells. Moving stimuli are again more effective, and the neurones often show responses of opposite sign to movement in the opposite direction. *Hypercomplex* cells are even more fussy: not only has the orientation to be correct in order to obtain a response, but the length of the stimulating bar or line must also lie within certain limits (Fig. 7.21). It is easy to imagine how a simple cell might derive its receptive field from the summation of the outputs of two or more geniculate cells arranged in a straight line; and although it is tempting to extrapolate this notion by supposing that the complex cell response might similarly be the result of the summation of the outputs of simple cells, it appears that in fact this is not so, and that both types of response are actually the result of appropriate connections from geniculate afferents. Responses to colour are less evident than in the geniculate, and of the cells that show colour-opponent responses, most are of the simple or concentric type. Finally, it is at the cortical level that we first find convergence

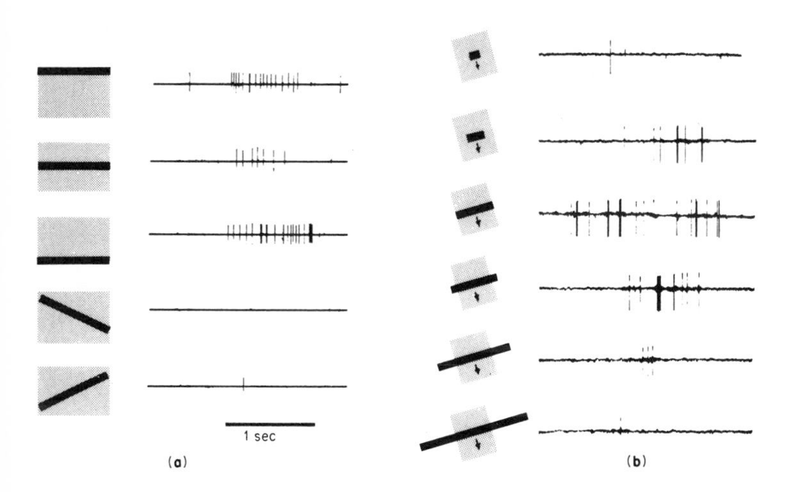

Fig. 7.21 Complex and hypercomplex cells. (**a**) Responses of a complex cell to a dark bar within its receptive field (shaded): it responds if the bar is horizontal, wherever it is within the field, but not if the bar is tilted. (**b**) Responses of a hypercomplex cell to a moving bar of optimum orientation but different lengths: there is clearly an optimum length for evoking maximum activity. (After Hubel and Wiesel, 1965)

of information from the two eyes, and most cortical cells, certainly in area 18 in primates, can be stimulated by either eye (though not necessarily with equal ease). Although some of these binocular neurones have receptive fields that are identically situated with respect to the fovea of each eye, others have pairs of fields that do not exactly correspond in the two eyes: this *retinal disparity* is undoubtedly an important source of information about the distances of visual objects from the plane of fixation, and is later discussed in the context of depth perception, on p.186. The cortical cells so far described lie in area 17 (the striate cortex) and areas 18 and 19 (the prestriate cortex); but from the prestriate cortex a further projection extends to the inferotemporal lobe (areas 20 and 21), lesions of which in the monkey can give rise to visual defects that are more subtle than those associated with lesions in the occipital lobe, such as difficulty in recognizing or appreciating the significance of visual objects. It is only recently that recordings have been made from neurones in these areas, and preliminary results seem to indicate that their receptive properties are even more specific than those of hypercomplex cells, and that some at least may respond to objects as closely defined as the appearance of the monkey's own hand.

It is important to appreciate that these various classifications of specificities amongst cortical cells are to a large extent arbitrary and artificial; as was emphasised in Chapter 4, if an experimenter finds any response at all from a cell, it must necessarily be to a stimulus that he has chosen beforehand. Recent examinations of cortical visual cells suggest a richness of variety that is not well conveyed by conventional descriptions in terms of 'simple cell', 'complex cell' and so forth.

The functions of the fibres that avoid the geniculate and pass to the superior colliculus and other areas of the midbrain are not well established. Neurones in the *superior colliculus* do not show orientation specificity and have roughly circular fields, but are responsive to targets moving in particular directions, especially moving away from the fovea. The cells are arranged in an orderly way on the surface of the colliculus, forming a map of visual space, and one interesting feature of this map is that electrical stimulation of a region corresponding with a particular part of the visual field often initiates an eye movement or head movement which is of the right size and direction to bring that part of the visual world onto the fovea. This suggests that one of the functions of the colliculus might be to move the eyes to objects of interest in the visual field: but since such movements are affected surprisingly little by lesions of the colliculus in primates, it seems that this function has perhaps in the course of evolution been taken over by the visual cortex, a region from which eye movements may also be evoked by electrical stimulation. The projection of optic tract fibres to other parts of the brainstem may also have some function in the control of eye movements: cells can be found in the pontine region that are also particularly sensitive to visual movement, and this region is also one that is closely associated with the generation of saccadic eye movements in response to visual objects. The pathways that must presumably exist for the visual control of the pupil and of accommodation are not absolutely established: the former is probably mediated by pretectal projections from the optic tract, and the latter by descending pathways from the visual cortex.

Mechanisms of adaptation

The fundamental importance of adaptation in the visual system, for enabling the eye first to respond over a large range of light levels, and secondly to register the albedo of objects rather than their luminances, has already been emphasized. It turns out that there are several different mechanisms in the eye that contribute to adaptation; it is convenient to divide them into those that are more or less immediate in their effect, and respond to the actual value of the retinal illumination at any moment, and those that depend on the state of bleach of the retinal pigment and thus reflect not the immediate degree of illumination but rather its past history over a period of the order of the time it takes for the pigment to regenerate: in the case of the rods, this may be a matter of half an hour or more. The first kind of adaptation may be called *field adaptation*, and the second *bleaching adaptation*.

Field adaptation

The simplest way to demonstrate the changes in sensitivity that accompany field adaptation is by means of an *increment threshold* experiment. Here the subject is presented with a background field of steady luminance I, and a test flash of luminance ΔI is suddenly superimposed on it: what is measured is the smallest value of ΔI that can just be perceived against the background. It turns out that over a moderate range of background intensities the ratio $\Delta I/I$ is constant: in other words, the sensitivity to incremental flashes is proportional to $1/I$, the inverse of the light level to which one is adapted. This is known as the *Weber–Fechner* relationship, and the quantity $\Delta I/I$ as the Weber fraction. This proportionality breaks down both at very high and very low luminances (Fig. 7.22). At the high end, the size of flash needed increases out of proportion to the background: this can be explained very well in terms of the kind of saturation of receptor response shown in Fig. 7.15. At low luminances, the value of ΔI levels off to a fixed quantity, ΔI_0, which is the *absolute* threshold (i.e. the 'incremental' threshold for a flash when there is no background present at all). This levelling off can be explained by supposing that there is a continual spontaneous background activity of the retina even when it is not illuminated—retinal noise, or 'dark light'—providing a constant signal that adds to the signals generated by any real backgrounds that may be present. If we call the luminance that is equivalent to this virtual background I_0, then the shape of the low-intensity part of the increment threshold curve can be described as $\Delta I/(I + I_0) = k$, where k is a constant and equal to the absolute threshold divided by the 'dark light'. The reason why there is an absolute threshold at all is simply that even when no real background is present, the test flash still has to be detected against the virtual background produced by the retinal noise.

A useful aid to thinking about mechanisms of visual adaptation is to treat the visual system as a 'black box' for which the input is retinal illumination, I (with the added noise signal, I_0), and whose output is the magnitude of the sensations perceived (Fig. 7.23). The latter is rather a nebulous concept, but the only assumption we make about it is that in the increment threshold

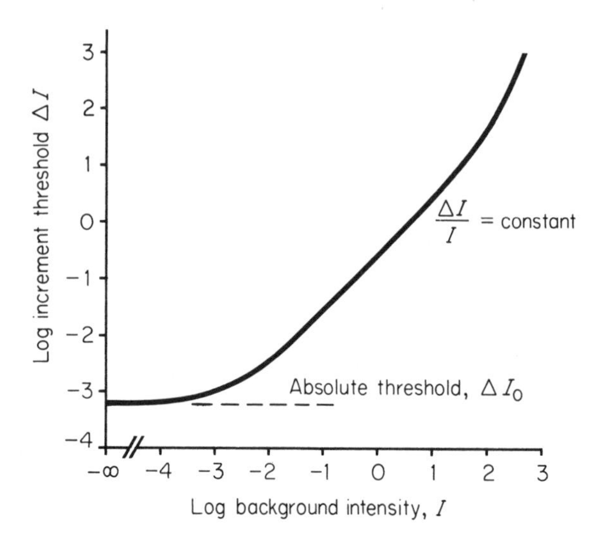

Fig. 7.22 The increment threshold curve. The smallest increment ΔI that can just be seen against a background I is plotted; for very small values of I the curve levels out at the absolute threshold, ΔI_0, while in the mid-range ΔI is proportional to I. At very large values of I, saturation occurs, and ΔI begins to rise more steeply. (Data from Aguilar and Stiles, 1954)

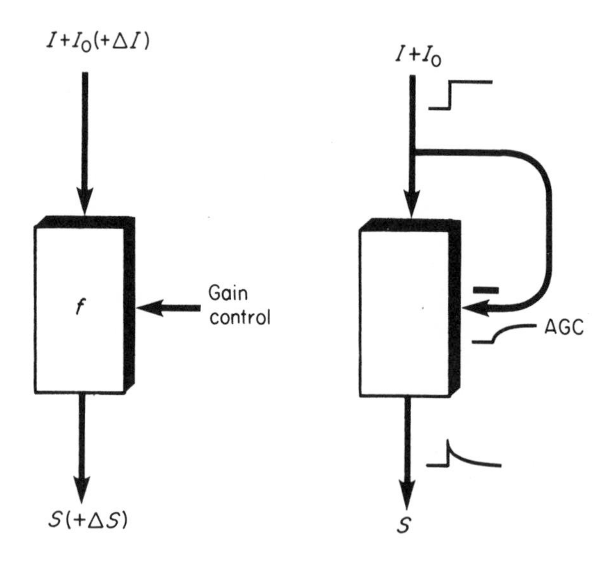

Fig. 7.23 The concept of gain control. Left, a representation of the process f that converts illumination of the retina ($I + \Delta I$, together with the 'dark light' I_0) into a sensory signal $S + \Delta S$. The gain control determines the sensitivity of the system: that is, how large ΔS is for a given ΔI. Right, if the input $I + I_0$ is also used to turn the gain down, the system becomes an *automatic* gain control. If the gain only changes relatively slowly, the output S will show a kind of adaptation

experiment, the change in S, ΔS, is constant when the flash is only just perceived: in other words, that there is a fixed threshold for perceiving changes in S. We can then say that $S = f(I + I_0)$, where f is the function that describes the way in which the box processes its input signals. If f were linear, so that the output was always proportional to the input, then of course the system would show no adaptational properties at all: its sensitivity or *gain*, $\Delta S/\Delta I$, would be constant. What properties must it have if it is to behave like the real visual system, in particular if it is to have the property that $\Delta I(I + I_0)$ is constant? One possibility is for the output of the box to turn down its own gain. Such a device is called an *automatic gain control* or AGC, and is used for example in radio receivers to maintain a roughly constant output to the loudspeaker despite fluctuations in the strength of the received signal: the analogy with visual adaptation is obvious. By making the gain of the box a suitable function of the input, the system can be made to reduce its gain automatically as I increases in such a way that $\Delta I/(I + I_0)$ is constant. If the gain control is sluggish in operation, taking some time to respond to changes in the input, a consequence will be that the system will show transient responses to sudden step-changes of illumination (Fig. 7.23). If the luminance is suddenly increased from a steady low level to a high, the initial response will be large but will fall off as the AGC comes into operation and turns the gain down to a level appropriate for the new luminance. This will generated an output of the classical 'incomplete adaptation' type, reminiscent of the responses of many ganglion cells to step-changes of illumination, except that they show complete rather than incomplete adaptation. Other evidence for complete adaptation in the visual system comes from the study of *stabilized images*. It is possible—for example, by mounting a tiny projector on a contact lens attached to the cornea—to arrange to project an image on the retina in such a way that it always falls on exactly the same receptors, despite the subject's eye movements. Under these circumstances, if the image is fully stabilized, it fades from sight completely in a matter of seconds. (In ordinary voluntary fixation this fading cannot be seen because the involuntary miniature eye movements of fixation continually move the images of objects over the receptors, producing continual small changes of luminance so that adaptation cannot occur.) To account for complete adaptation of this kind, we need to add to the black box, at the output end, an extra mechanism of complete adaptation that responds only to changes in the output and not to steady levels, that lets through ΔS but filters out S.

Bleaching adaptation

Bleaching adaptation is quite different in its properties. It can be demonstrated by means of the same apparatus as for measuring increment thresholds, exposing the subject to an adapting field I which is then turned off *before* the eye's sensitivity is tested with the test flash ΔI. The results are very different from the previous case. Whereas in field adaptation ΔI depends directly on I, now it is found that ΔI is a function not of I on its own, but rather of *how much pigment was bleached* during the adapting period: for fairly short adapting periods this is proportional to the product of I and the time of exposure. The second difference is that the adaptation lasts a considerable period after the adapting field has been switched off, a time that in fact corresponds closely to the time required for the pigment to regenerate (see Fig. 7.4). From our knowledge of the photochemistry of the pigment we might well have expected some such effect: obviously sensitivity must depend in part on the amount of active pigment present, and if say 20 per cent of it is in the bleached state, then we would expect the eye to be 20 per cent less sensitive. But it turns out that the changes in sensitivity that result from pigment bleaching are vastly *greater* than the simple proportionality that would be expected by this argument. By means of the technique of retinal densitometry (p.152) it is possible to measure ΔI in this experiment at the same time as monitoring the proportion of pigment that is bleached. In the case of rods, it turns out (Fig. 7.24) that it is not ΔI, but $log \Delta I$, that is proportional to B, the fraction of pigment bleached, and that very small bleaches produce very large changes in sensitivity: a 20 per cent bleach produces not a 20 per cent reduction in sensitivity, but a reduction by a factor of 10 000! A clearer idea of how this comes about may be obtained by carrying out increment threshold measurements as a function of background intensity after bleaches of different sizes: what one then finds is that though the values of ΔI are unchanged at high background luminances, as I is reduced the curve flattens off sooner to a higher value of ΔI_0 the more pigment is in the bleached state (Fig. 7.25). It is

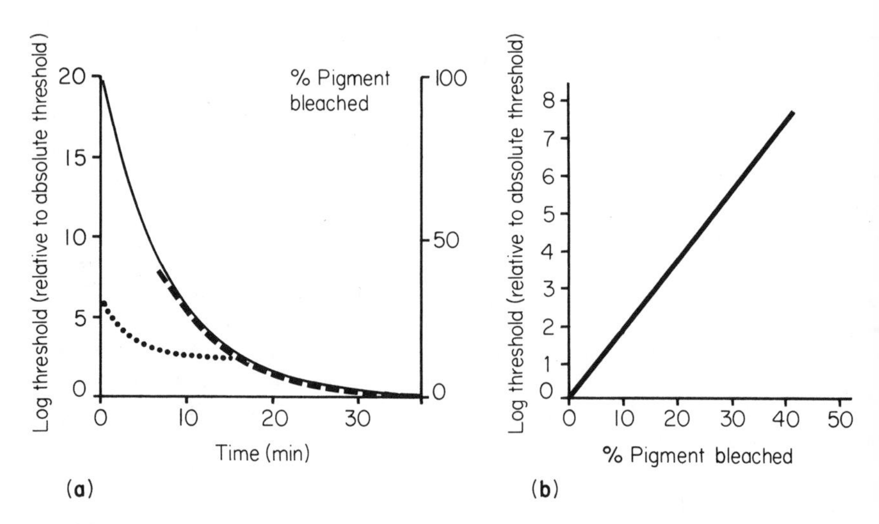

Fig. 7.24 The relation between rhodopsin concentration and sensitivity during dark adaptation. (**a**) Continuous line shows the percentage of rhodopsin in the bleached state at various times during recovery after a full bleach: the curve is closely similar in a normal subject and in a rod monochromat. Also shown are simultaneous measurements of absolute threshold by a rod monochromat (dashed line) and normal subject (dotted), plotted on a logarithmic scale as shown on the left. (**b**) the relation between per cent pigment bleached and log threshold obtained from this experiment in the case of the rod monochromat: it is evident that the relationship is a linear one, and threshold is proportional to 10^{aB}, where B is the percentage of pigment bleached, and a is a constant. (Data from Rushton, 1965)

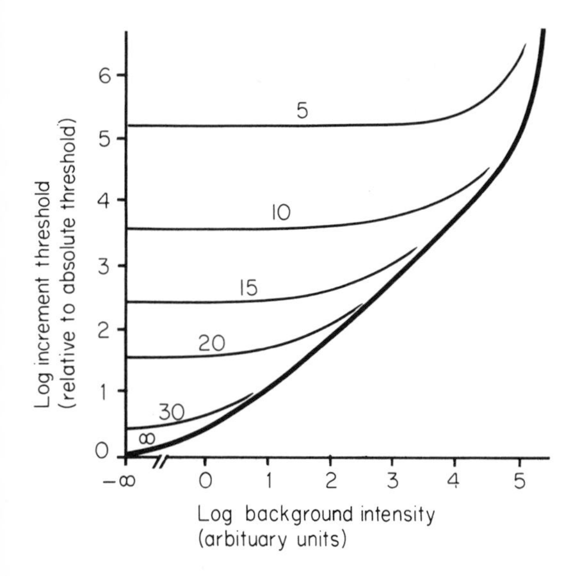

Fig. 7.25 Effect of bleaching on increment thresholds. Increment threshold curves are plotted as in Fig. 7.22, at various times (shown in minutes by the number above each curve) after a very intense bleach of the rods. (Data from Blakemore and Rushton, 1965)

as if the presence of free opsin as a result of bleaching caused an increase in the level of retinal noise or 'dark light', and indeed such curves can be very well explained by supposing that the dark light, I_D, is given by $I_0 10^{aB}$, where a has a value of about 20. In other words, bleaching an area of retina has much the same effect on sensitivity as shining a light of luminance $I_0 10^{aB}$ on it; this imaginary light is called the *equivalent background.* But why do we not see this light? The answer is that sometimes we do: after viewing a light that is bright enough to bleach a significant amount of pigment—an ordinary light bulb does very well—we see initially a bright *after-image* (positive after-image) which *is* in effect I_D; but because it is stabilized on the retina, it is subject to the complete adaptation already described, and after some seconds it fades from view. If we now look at an illuminated surface, the after-image may reappear, but in its negative form: the areas that have been bleached are less sensitive than the rest of the retina, because of the steady dark light that turns down the AGC, and so the bleached areas look darker than their surroundings. When we go from a brightly lit room into a dark one, the reason we cannot see clearly at first is that everything we look at is superimposed on an invisible background consisting of all the after-images that we have accumulated over the past 20 minutes or so. Although short-term adaptation prevents us from seeing the dark light that is generated by bleaching, the pathways that control the size of the pupil do not show this complete adaptation, and during the course of dark adaptation the pupil responds to the dark light in exactly the same way that it would to real lights.

What is not clear in this system is how the function 10^{aB} comes about, and whether the dark light signal is conveyed in the same neural pathways as those from real lights. It is difficult to think of plausible mechanisms by which a receptor's output would be the sum of a signal corresponding to the amount of light falling on it, i.e. proportional to the *rate* of bleaching, and a signal so non-linearly dependent on the *amount* of bleach.

Visual acuity

Visual acuity is a measure of the fidelity with which the visual system can transmit fine details of the visual world: it is the equivalent of the ability of a camera to produce sharp pictures. In a camera there are essentially two stages at which sharpness may be lost: either through optical defects that blur the patterns of light in the image on the film, or by defects in the film itself, such as graininess, that limit the density of detail. These correspond in the eye to the quality of the optics, and to the density of the retinal receptors. But a third factor that must also be considered is the possible degradation of the image that may occur in the course of the neural processing that takes place in the retina and in the brain.

In an ideal system, a point source of light such as a star would produce a point of excitation in the neural output pattern: but the effect of the various types of image degradation mentioned above is to blur this final image so that its excitation is spread out over a finite area. The distribution of excitation in the image of a point source is described by the *pointspread function*. There are two distinct consequences of spreading of this kind. The first is that since the incident energy from the point source is spread out over a larger area, the maximum intensity at the central peak is necessarily reduced, leading to a

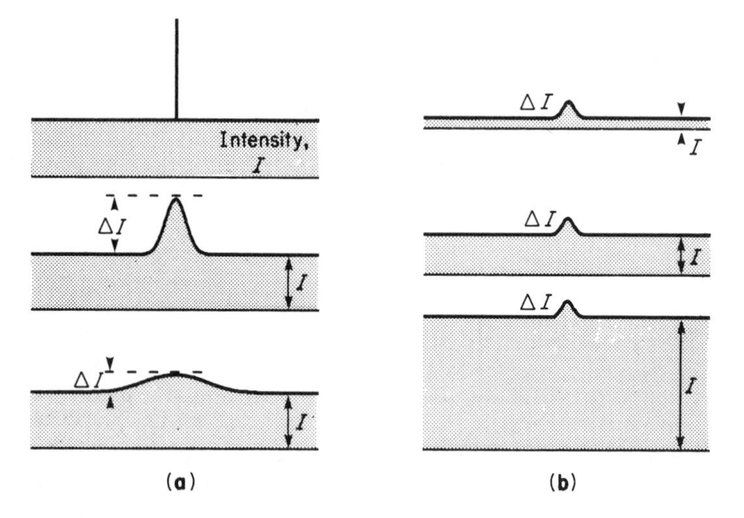

Fig. 7.26 The effect of light spread on contrast. In each case the shaded area represents the variation of local intensity with distance across the retina. (**a**) An ideal point image is shown at top, on a background of intensity I; below it is shown the effect of the lateral spread of light in reducing the contrast, $\Delta I/I$. (**b**) The contrast of the image of a star will fall as the sun rises and increases the background intensity I against which it is seen, even though ΔI is constant

decrease in the *contrast*, $\Delta I/I$, that determines whether it will be seen against its background (Fig. 7.26 **a**). For objects of intrinsically high contrast, such as stars seen against the void of space, this will not matter much, and subjects with poor visual acuity as measured conventionally (see below) are not as bad at seeing stars as one might expect, bearing in mind the fact that such objects subtend an almost infinitely small angle. In the dark, whether one sees a star or not is almost entirely a matter of whether a sufficient number of photons from it fall upon a rod summation pool; under other conditons its visibility depends on whether $\Delta I/I$ exceeds the threshold contrast. With increasing values of I, for example as the sun rises, fewer and fewer stars meet this criterion, and they fade from sight one by one (Fig. 7.26 **b**). When, as in this case, or for example in viewing a thin black line against a white background, the contrast is limited and threshold depends on $\Delta I/I$, visibility will be much dependent on visual acuity (Fig. 7.27). The minimum width of a black line that can be seen on a white background by a normal eye is some ½ sec arc: but white lines of the same width on a black background can be seen very well, because their contrast is higher.

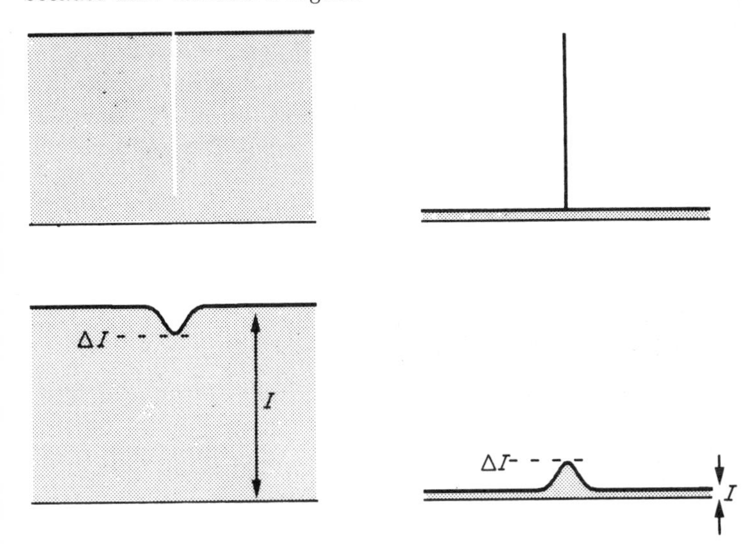

Fig. 7.27 A thin black line on a white background (left) and a thin white line on a black background (right), with their images shown below. Even though the width of the lines and the intensities of the black and white parts of the targets are matched, the visibility of the white line will always be greater than that of the black, because its contrast, $\Delta I/I$, is higher

The second effect of the spreading of the images of points is a spatial one: it means that images of adjacent points on an object will overlap and lead to obliteration of spatial detail. Consider for example a pair of point sources that are close to one another—as for example a double star (Fig. 7.28). As the pointspread function is increased in width, there will come a point at which the distribution of excitation in the image will no longer exhibit a dip in the middle, so that an observer will be unable to see that there are two stars present and not just one. For normal observers the angle of separation for

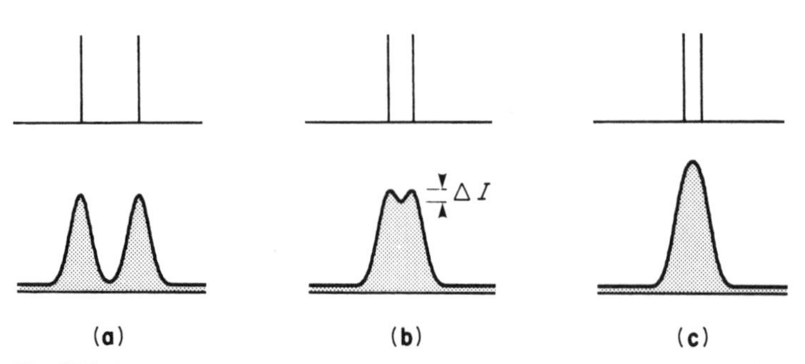

Fig. 7.28 Resolution and contrast. As a pair of point sources are gradually brought together (above), their retinal images (below) begin to overlap. (**a**) is easily resolved, so long as the points can be seen at all; (**b**) will be resolved only if the contrast is sufficiently high; while (**c**) can never be resolved, whatever the contrast

which this kind of resolution can just be performed is an order of magnitude greater than the width of a black line that can just be seen, and is just less than one minute of arc. In a case like that of Fig. 7.28**c**, it is clear that we cannot improve resolution simply by increasing the contrast, and such a stimulus may be described as absolutely unresolvable. But in an intermediate case like Fig. 7.28**b**, whether resolution is possible or not will depend on the contrast of the original object. This interaction between resolution and contrast can best be investigated by using *grating patterns* as test targets. A grating is simply a regular pattern of stripes, such as the square-wave grating shown in Fig. 7.29**a**: it is called square-wave because if one were to plot intensity as a

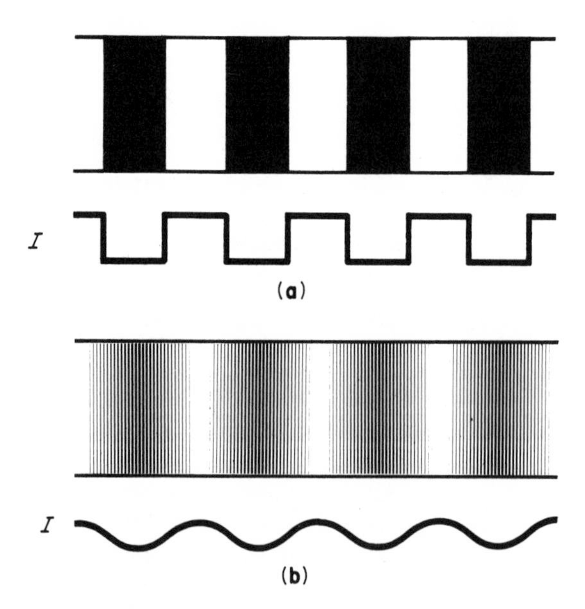

Fig. 7.29 Grating patterns, with their intensity profiles shown below. (**a**) is a square-wave grating and (**b**) is a sine-wave grating

function of distance in a direction perpendicular to the stripes, it would have a square-wave profile as shown. In the same way, sinusoidal gratings have an intensity profile that is sinusoidal (Fig. 7.29**b**). In each case, one can describe the grating in terms of its *spatial frequency* (i.e. the number of cycles per degree) and its contrast (defined as the difference in intensity between the peaks and the troughs, divided by twice the mean intensity). Thus a pattern of alternate pure black and pure white strips, each 1° across, could be described as a square-wave grating of 100 per cent contrast and spatial frequency 0.5 cycles per degree. A simple experiment is to ask a subject to view a sinusoidal

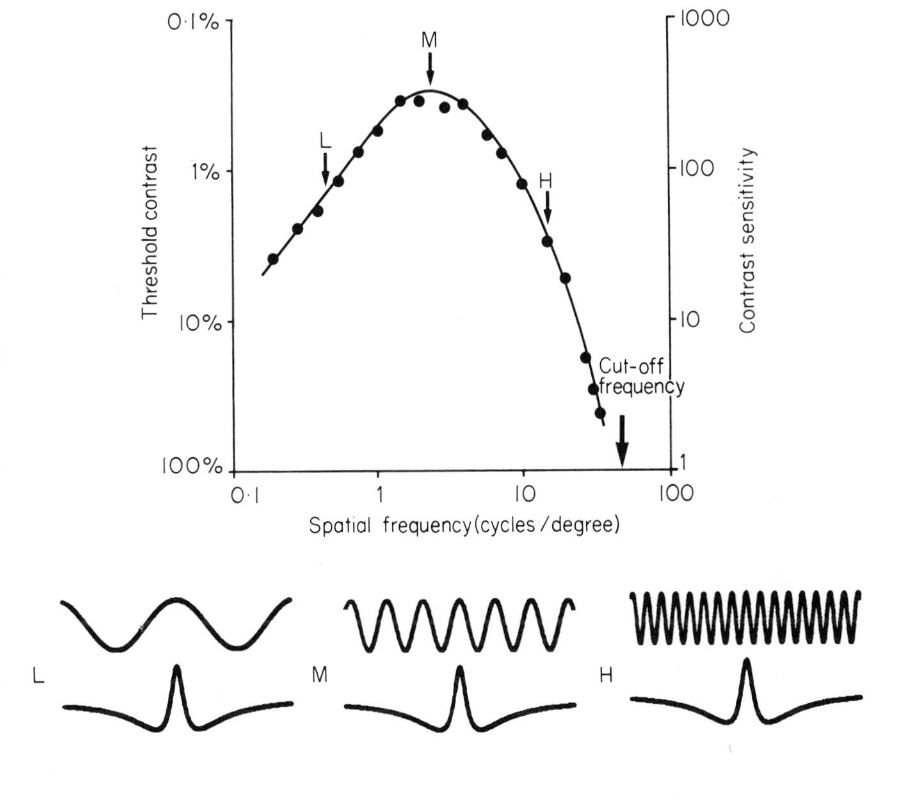

Fig. 7.30 Visibility of sine-wave gratings of different spatial frequency. Above, frequency-sensitivity function for sine-wave gratings, showing a peak (M) around 2–3 cycles per degree and a fall-off in sensitivity both at higher (H) and lower (L) frequencies. Below, schematic representation of a sinusoidally modulated image falling on a receptive field having excitatory centre and inhibitory surround. When the centre is roughly matched in size to the peak of the sine-wave (M), the response is maximum. With lower or higher frequencies (L,H) there is a relative increase in the degree of stimulation of the inhibitory surround. (Data from Campbell and Robson, 1968).

grating of a particular spatial frequency, and then reduce its contrast until he reports that he can no longer see it. If we plot this threshold contrast as a function of spatial frequency, we typically obtain a curve such as Fig. 7.30. Because a blurred pointspread function affects high spatial frequencies much more than low, the contrast required to see the grating increases sharply as its frequency is increased, until at about 40 cycles per degree (the *cut-off frequency*) the subject cannot even see a grating of 100 per cent contrast. Because of the steepness of this cut-off, a small amount of extra blur causes a large increase in the contrast needed, and so the method provides a sensitive measure of acuity. However, it is not an easy one to use in ordinary ophthalmological testing, and simpler tests of resolution are preferred. One such test is the Snellen chart (Fig. 7.31), in which rows of letters of diminishing size are to be read: by discovering the row at which the subject finally stops, and knowing the size of letters and the subject's viewing distance, one can estimate his minimum resolvable angle. In practice, each row is marked with the distance at which a standard observer should just be able to read it, i.e. the distance at which the details of the letters subtend a minute of arc; a subject who could only read the 8-metre row at 6 metres would be described as having an acuity of 6/8. The difficulty of the Snellen chart for scientific work is that the test is only partly one of resolution, some letters being recognized more easily than others

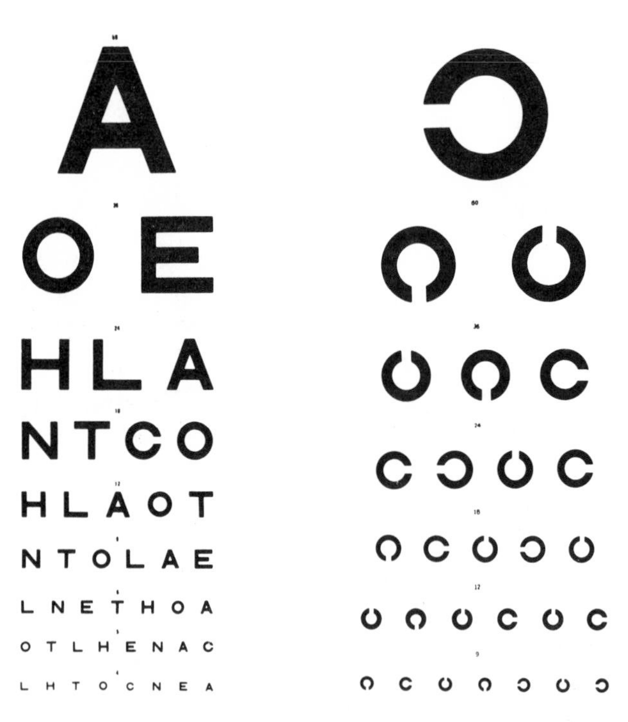

Fig. 7.31 Charts commonly used for routine testing of visual acuity. Left, Snellen chart; right, a similar chart using Landoldt C's: the subject must name the position of the gap in the circle

because of their overall shape; and for some purposes the Landoldt C chart (Fig. 7.31), used in the same way, is preferable because it provides no extraneous clues to the subject.

Factors affecting visual acuity

What contributions do the various parts of the visual system make to visual acuity? Clearly the quality of the image on the retina is an important factor, and this in turn depends on the quality of the optics of the eye. But even if the eye is fully corrected for errors of refraction and astigmatism, the cornea and lens will still introduce a certain amount of blur. There are three sources of this degradation. The first of these is the existence of the various *aberrations*, already discussed. Chromatic aberration can be reduced by limiting the range of wavelengths in the image: in monochromatic yellow light, visual acuity may be improved by some 25 per cent. The only practical way to reduce spherical aberration is to limit the area of lens and cornea that contributes to refraction of the incoming rays: other things being equal, the smaller the pupil, the more nearly the optical surfaces will approximate to their ideal forms, and the less noticeable the aberrations will be. But reducing pupil size cannot improve acuity indefinitely (quite apart from its undesirable effect of reducing the light-catching power of the eye) because of the second of the optical factors to be considered, namely *diffraction*. When light is imaged by a lens or other optical device of finite diameter, the edges of the aperture cause a diffraction blur of the resultant image. The width of the resultant pointspread function is of the order of λ/d radians, where λ is the wavelength, and d the aperture of the system. For a pupil of diameter 2.5 mm, and with green light, this amounts to a little less than one minute of arc. In other words, under these conditions acuity is effectively limited by diffraction at the pupil. In the dark, with a pupil of some 8 mm diameter, the corresponding figure is about 17 sec arc, but the actual pointspread is very much wider than this because of the increased contribution of the aberrations when the lens is widely exposed: in fact the pointspread function actually gets wider with increasing pupil diameter beyond 3 mm or so. Thus if acuity were the sole consideration determining pupil size, we would expect it to remain fixed at some value near 3 mm. But as we shall see, under conditions of dark adaptation the intrinsic acuity of the neural processing of the retinal image is so low that the poor optics contribute little to the overall blur, and the advantage of being able to increase retinal sensitivity by catching more light with a dilated pupil outweighs the disadvantage of slightly decreased acuity. The third source of optical blur is *glare*, caused by the diffuse scatter of light from the optical surfaces and media of the eye. Glare is scattered rather uniformly over the retina, so its effect on acuity is due not so much to spatial effects as to the reduction in the contrast of the retinal image that it produces, by superimposing on the image a more or less uniform background whose illuminance is of the order of 10 per cent of the mean illuminance of the retina. This means that if we look at a target whose actual contrast is 100 per cent, the effect of this scatter is to reduce the contrast of the retinal image to something nearer 90 per cent.

The relative importance of optical as opposed to retinal and neural factors in determining acuity can be determined directly by arranging to project a grating on the retina in such a way that its contrast is unaffected by the quality of the optics. One way of doing this is to generate interference fringes on the retina by means of two point sources of coherent light from a laser: the resulting interference pattern is in effect a sinusoidal grating, whose frequency depends on the separation of the two sources, and whose contrast is substantially unaffected by the quality of the optics. One can then measure the subject's threshold contrast as a function of frequency, in the manner already described, and compare the result with what is found when viewing a 'real' sinusoidal grating. Although there is some improvement when the optics are bypassed in this way, even when the eye is fully corrected (as would be expected from our calculation of the limiting effect of diffraction when the pupil is small), it is not a very great one. This suggests that the retina is in a sense 'matched' to the eye's optical properties. Clearly it would make little sense to have a retina with tiny receptors capable of resolving detail that could never be found in the retinal image in practice, because of optical blur. Nor for that matter would it be very sensible to have optics producing beautifully precise images, if the receptors were so gross that they were incapable of appreciating them. Under photopic conditions, aquity is thus essentially limited by both optics and receptor size together: it will be recalled that under these conditions it is the central fovea that is used to examine the fine detail, and the cones here are spaced some 30 sec arc apart (i.e. about the same size as the pointspread function) and connected in a one-to-one manner with the optic

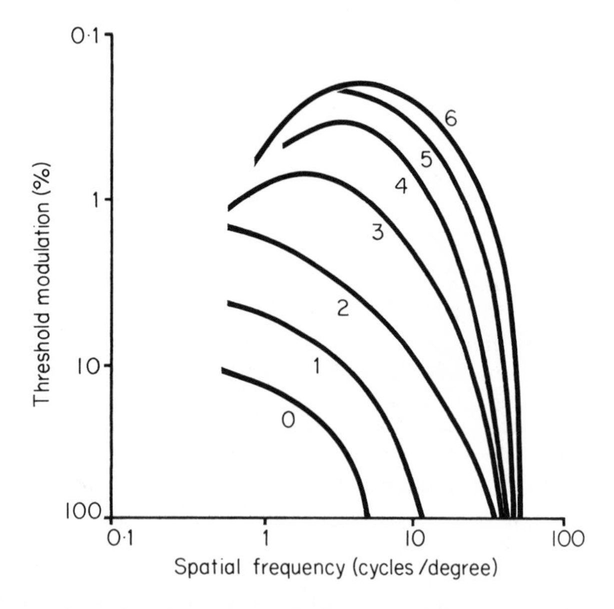

Fig. 7.32 Effect of adaptation level on the visibility of sinusoidal gratings. Each curve is a contrast sensitivity function of the kind shown in Fig. 7.30, measured at a different background intensity level ranging over 6 log units (the number by each curve indicates background intensity in relative log units). (After van Nes and Bouman, 1967)

nerve, so that no further blur is introduced by sideways diffusion of information through neural collaterals. But if the contrast threshold as a function of spatial frequency is measured with a fixed pupil during progressive stages of dark adaptation, one finds a steady decrease in the cut-off frequency (Fig. 7.32), the result of changes in the neural organization of the retina. We have already seen that one of the adaptational responses to reduced light levels is an increase in the effective size of the ganglion cells' summation pools, so that they can catch more light. But this obviously has the effect of increasing the degree of neural blur, and hence of reducing the overall acuity.

To summarize the factors that contribute to visual acuity, we have, first, factors that are properties of the *target*: its contrast, its colour (shorter wavelengths reduce diffraction, and monochromatic targets reduce chromatic aberration) and its luminance (bright targets allow smaller retinal fields and a smaller pupil; very dim targets near the absolute threshold show a further reduction in acuity because of quantum fluctuations). Secondly, there are *optical* factors: errors of refraction, aberrations, and the effect of pupil size on both of these and on diffraction, and glare from scatter. And finally there are factors related to the *retina* itself: the spacing of the receptors (closer in the fovea, closer for cones than for rods and for red and green cones than for blue, resulting in poor acuity in blue light) despite better diffraction, and the size of the receptive fields, being larger in the dark-adapted state and in the periphery of the visual field. In a good light, acuity is limited by diffraction, and thus about as good as could be expected from an eye of the size we have been given.

Lateral inhibition

The general properties of lateral inhibition, and its desirability in sensory systems, have already been discussed in Chapter 6. It is a very important feature of visual processing, and is indeed one of the first things that the retina does—at the horizontal cell level—to the signals that come from the receptors. It is strikingly demonstrated in the Hermann grid (Fig. 7.33**b**), and certainly

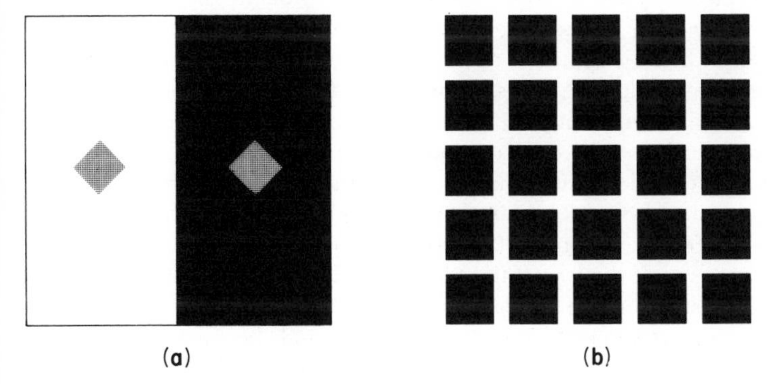

Fig. 7.33 Two illusions caused by lateral inhibition. (**a**), simultaneous contrast: the two grey squares are of equal luminance, but appear different in brightness because of the backgrounds they are seen against. (**b**), the Hermann grid: illusory dark spots are seen at the intersections of the white bars. The effect is much less striking near the fovea than in the periphery

accounts for some of the phenomena of simultaneous contrast (Fig. 7.33**a**) which, like adaptation, helps to ensure that the subjective sensation of brightness is on a sliding scale and thus more closely related to albedo than to luminance. It can be estimated quantitatively by using sinusoidal gratings to measure contrast threshold as a function of spatial frequency, as already described on p.169. We have already seen how the presence of large excitatory receptive fields has the effect of reducing the effective contrast of high-frequency gratings. Lateral inhibition has in a sense the opposite effect: it reduces the contrast of low-frequency gratings. To see why this is so, consider the response of a unit with a receptive field consisting of an excitatory centre and inhibitory surround, as it views sinusoidal gratings of various spatial frequencies (Fig. 7.30). If the frequency is in effect matched to the dimensions of the field, in the sense that if a peak in the grating corresponds to the peak of the excitatory centre, then the troughs on each side correspond to the inhibitory troughs of the field, and the response of the unit will be a maximum. Increasing the spatial frequency will reduce the response, because of the smudging effect of the central area, already mentioned. But reducing the frequency will *also* decrease the response, because the bright part of the stimulus will begin to invade the inhibitory surround and lead to inhibition. In the extreme case, at zero spatial frequency when the unit is viewing a field of uniform luminance, the excitatory and inhibitory areas will both be maximally stimulated, and thus exhibit their maximal antagonism. Consequently the shape of the contrast-threshold curve as a function of frequency for such a unit will be as shown in Fig. 7.30. Just as the high-frequency cut-off tells us about the size and shape of the excitatory centre, so the low-frequency part of the curve tells us about the inhibitory surround. The reader can demonstrate the low-frequency cut quite easily for himself by means of Fig. 7.34 which is a sine-wave grating of low contrast and low spatial frequency. Seen at a distance of a metre or so, the grating is easily perceived, but paradoxically the more closely it is examined the more difficult it is to see: eventually it disappears altogether because its spatial frequency is reduced to below the cut-off for the contrast concerned. By measuring a number of contrast-threshold curves at different levels of light adaptation, one can follow the changes in effective field configuration as a result of changes in the retina: under bright conditions, the excitatory centre is small and the surround prominent, so that there is a marked low-frequency cut but a good high-frequency response. As the illumination is reduced, the increase in size of the excitatory area brings the high-frequency cut down to lower frequencies, as already described on p.173, and the simultaneous reduction in lateral inhibition gradually flattens the low-frequency response. At the lowest light levels the low-frequency cut cannot be seen at all, corresponding to the fact that under dark adaptation, inhibitory surrounds of the ganglion cells in experimental animals become progressively less and less prominent, and finally may not be found at all.

Flicker sensitivity

A sinusoidal grating can be thought of as an area whose luminance is modulated sinusoidally as a function of distance, but is constant in time. A

Fig. 7.34 Sine-wave grating of low spatial frequency and contrast, more easily visible at distances of a metre or so than close-to, when its spatial frequency is too small

related stimulus is one that is spatially uniform, but whose luminance is altered sinusoidally as a function of time. With a sinusoidally flickering stimulus of this kind, we can perform an experiment that is analogous to determining the threshold contrast of a sinusoidal grating as a function of its frequency, previously described; this time we ask the subject to reduce the contrast of the flicker until he can only just see it, and determine how this threshold contrast varies with the temporal frequency of the flicker. The resultant curve shows many points of similarity with the spatial one. At the high-frequency end, there is a cut-off frequency (the *flicker fusion* frequency) at which the flicker cannot quite be seen even with 100 per cent contrast; and at the low-frequency end, one again finds that sensitivity begins to fall off as the rate of change of luminance declines, reflecting the inability of the eye to perceive slow changes because of its adaptational mechanisms. A further striking parallel between the two experiments is that if the flicker sensitivity curve is measured at progressively lower light levels, it undergoes very similar changes to those observed with sinusoidal gratings. The increasing sluggishness of the system when luminance is low is reflected in a progressive lowering of the flicker fusion frequency, while the gradual loss of the fast adaptational component causea a flattening of the low-frequency end. The close parallelism of the two effects, spatial and temporal, tempts one to speculate that the same fundamental mechanism might be responsible for both.

Colour vision

The visual equivalent of a pure auditory tone—a sinusoidal sound wave—is a light that is *monochromatic*, whose photons have all the same wavelength: colour is then the sensory correlate of wavelength, just as pitch is of auditory frequency. But natural light sources generally produce photons of many different wavelengths, and the eye responds to them in a way that is very different from the response of the ear to complex sounds. Whereas the ear is capable of *analysing* mixtures of pure tones into their components—so that a competent musician can listen to a chord played by an orchestra and say which instruments are playing which notes—not even the most perceptive and experienced observer can say by looking at a coloured light exactly what wavelengths are present in it. The reason for this difference lies in what might be called the number of degrees of freedom in the two systems. In the case of the ear, there are hundreds of thousands of different units in the auditory pathway, each responding best to a frequency that is slightly different from the preferred frequency of any of its neighbours. Thus a particular mixture of pure tones will excite a corresponding pattern of receptors; and because there are so many of them, the details of this pattern will reflect the detailed structure of the original sound spectrum. But in the case of the eye, although it is true that there are many millions of colour receptors—cones—in each retina, it turns out that as far as sensitivity to different wavelengths is concerned they fall into one of only three classes. Information about the spectral composition of a light can only be conveyed by the relative degree of activity in each of these channels: the system has only *three degrees of freedom*, a condition known as *trichromacy*. It is rather as if we tried to classify people using only the three parameters of weight, height, and size in shoes: although this would enable us to discriminate quite well between particular pairs of people, it clearly could not reflect the rich variety of human forms that actually exists.

One result of this fact is that it is possible to 'fool' the colour system in a way that cannot be done with the ear. If we sound two pure tones simultaneously, they sound like what they are: a mixture of two frequencies. If we shine two monochromatic lights on a screen, the result is a new colour that often bears no obvious relationship to its components—red and green for example making yellow—and does not even look like a mixture. It is this fact that enables artists to paint with only a small number of pigments on their palette, and makes colour printing and colour television a practical possibility. And in fact a good deal may be learnt about how the colour system works by investigating experimentally the rules that determine the results of such mixtures.

Colour mixing

Imagine a creature with only two types of colour receptor, red and blue, with sensitivity curves as shown in Fig. 7.35**a**. It is clear that any single

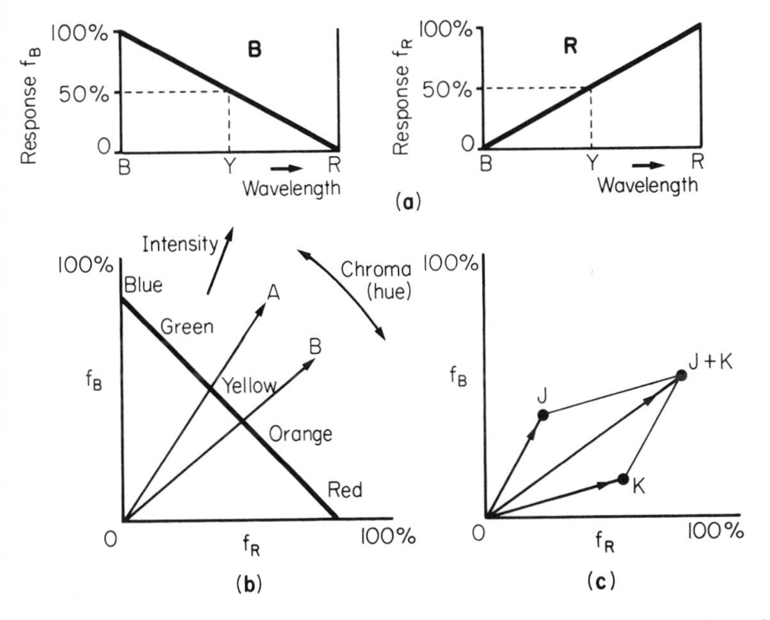

Fig. 7.35 Colour vision in a hypothetical dichromat. (**a**) The spectral sensitivities of the two types of receptor, B (blue) and R (red). (**b**) The colour response space: the axes show the degree of activity of each of the two channels, and any particular colour will result in a particular point in the diagram. The thick line indicates how the B and R responses vary as a light of constant intensity is varied in wavelength, while lines such as OA, OB represent stimuli of constant wavelength but variably intensity: in each case the *ratio* of B and R activity is constant, and they are thus of constant hue or chroma. (**c**) J and K represent two stimuli of different intensity and wavelength. If they are added together, the resultant response is given by the vector sum (J + K) of each separate response

wavelength within the total range will always produce a characteristic *ratio* of the rates of firing (f_R/f_B) of the two types of unit. Altering the *intensity* of the light will not alter this ratio, which is precisely why it is a useful measure of colour: experience tells us that altering the intensities of coloured lights does not—within limits—affect their perceived colours. Such a creature certainly enjoys colour vision, since it can distinguish wavelengths independently of their intensities. But what will happen if two wavelengths are presented simultaneously? Suppose we shine on this eye an extreme red and extreme blue light (R,B), each of unit intensity: the ratio (f_R/f_B) will then be unity. But a precisely identical response could have been achieved by shining a light of single wavelength (Y) on the eye, such that the two types of receptor were equally stimulated. In other words, R and B together here produce the same neural response as (and therefore presumably 'look like') another wavelength, Y; and in general, *any* pair of wavelengths whatever will look the same as some other single wavelength. We can derive rules to predict *what* wavelength they will look like quite easily from the two spectral sensitivity curves, by means of a diagram showing the 'response space' of the two channels (Fig. 7.35**b**). Here the two axes show frequencies of firing in the R and B channels, and so every point in the figure corresponds to a particular pair of frequencies and hence to a unique sensation. Lines like OA, OB join points for which the ratios of these frequencies are constant, and therefore correspond to lights having the same perceived colour (*chroma* or hue). The distance of a point along such a line from the origin corresponds to its intensity. The thick lines show the way in which f_R and f_B vary if a light of fixed intensity is varied in wavelength. If more than one light is present, the result of the mixture can be deduced at once from the diagram; if the channels are linear, the total activity in each will be the sum of the activities generated by each component separately. Thus if we simultaneously present two lights J and K of different intensities and wavelengths (Fig. 7.35**c**), the result will be the vector sum of the two (which may be found by constructing the familiar 'parallelogram of forces' as shown), and the position of the resultant point will indicate the perceived colour of the mixture. It is clear that any colour can be matched exactly by using appropriate proportions of any *two* other given wavelengths (with the proviso that the apparent wavelength of the result must lie between the wavelengths of the two components) and this is what is meant by *dichromatic* vision. It follows that if we want to be able to produce *any* colour by mixing two others, we must choose an extreme red and an extreme blue in order to encompass the entire range of possible wavelengths. These may then be called *primary* colours, stimuli from which all other hues can be made by mixture.

Now although there are many dichromatic animals, and some colour-blind humans, for whom the above provides an adequate description of their colour sense (except that the shapes of the sensitivity curves in Fig. 7.35 are over-idealized), it is the normal human eye with which we are concerned, and it has not two but *three* classes of receptor with respect to wavelength sensitivity, and hence three degrees of freedom. Plate I (top) shows the spectral sensitivities of the three types of human cone. It is clear that we cannot match a colour such as Y by using only two others together (such as B and R):

although we may get f_R and f_B right, the value of f_G will in general be wrong; and we cannot correct f_G without messing up one of the other channels. In fact we now need *three* colours to match any other, to take care of the three degrees of freedom involved: a pair of colours will only match if f_R for one is equal to f_R' for the other, f_G equal to f_G', and f_B equal to f_B'. Because of this, the diagram of the response space corresponding to Fig. 7.35 now has to be three-dimensional rather than flat; and because the individual sensitivity curves do not have the simple form of Fig. 7.35**a**, the locus of a light of fixed intensity whose wavelength is altered is no longer the simple straight line of Fig. 7.35**b** but one having a twisted three-dimensional shape (Plate I, middle).

This looks awkward to use: but a simplification can be made. As before, the distance from the origin to a particular point represents brightness, whereas its direction represents colour. If it is only the latter quantity that interests us, we can reduce the whole diagram to a two-dimensional form by the following simple expedient. Suppose we set up an equilateral triangle whose vertices are at equal distances along the three axes: then the point where the line joining the origin to a particular light intersects the plane of this triangle will depend only on the light's colour and not on its intensity. This triangle is called the *colour triangle* (Plate I, bottom) and each point on it correspond to a different colour. Its centre, W, corresponds to white light, that stimulates each type of receptor equally, and lines radiating from this point are lines of equal chroma. Along such a line, the nearer a point is to W the more *unsaturated* it is, i.e. the more it is diluted with white. Thus pink and red lie on the same line of chroma, but pink is nearer the centre. The thick line represents the locus of a light whose wavelength is varied over the visible range: it therefore represents colours that are of maximum saturation. The fact that this locus does not reach the G vertex reflects the fact that there is no wavelength that can stimulate G alone without the same time stimulating R or B or both, as can be verified in the spectral sensitivity curves of Plate I. The edge RB of the triangle represents saturated violets and purples that are not in the spectrum, but can be formed by mixtures of deep red and deep blue. The rules for colour mixture can now be stated simply, and were described in this form by Newton in 1704. To find the result of mixing two colours J and K (Fig. 7.36, left), join JK with a straight line: all the colours on that line can be made by mixing J and K in different proportions, and the more J is used, the nearer the resultant will lie to J. In general it can be seen that the effect of mixing two colours is to produce a new one of intermediate chroma, but less saturated, i.e. nearer to W: such mixtures are therefore paler than pure spectral colours. If JK happens to pass through W, J and K are said to be *complementary*: by mixing them in suitable proportions, pure white can be made. By mixing three colours (J,K,L) we can produce any colour lying within the triangle JKL; so long as this includes W, this means that any chroma can be mixed from *any* three colours (Fig. 7.36, right). Obviously, the bigger the triangle JKL is, the more saturated are the colours that can be mixed: but even if twe use pure spectral wavelengths as our primaries, there will be certain colours (notably saturated blue-greens or yellow) that cannot be matched. In practical colour mixing, as with colour film or colour television, one tries to choose primaries that make the triangle JKL as large as possible: but one's choice is limited by

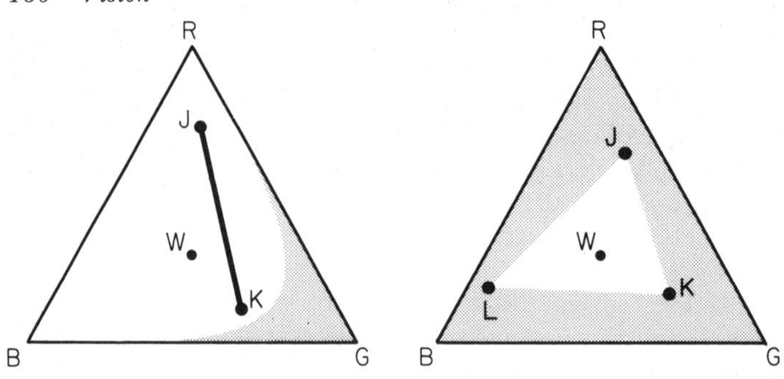

Fig. 7.36 The laws of colour mixing. Left: by mixing two colours (J, K) in different proportions we can form any of the colours lying on the line JK. Right: with three colours (J,K,L) we can form any colour lying within the triangle JKL. If this triangle encloses the white point (W), then any hue can be formed by mixing them in suitable proportions, though not with full *saturation*

the dyes or phosphors actually available. The reason why colour television in particular is often unsatisfactory can be seen by looking at the position of the primaries used—they are in fact close to the J,K,L of Fig. 7.36: the result is that blue-greens and purples are rather desaturated, though grass-greens and flesh tints are relatively good. The only practical solution to this sort of problem problem is to use more than three primaries, and high-quality colour printing—for example of reproductions of paintings—may use six or more to get as close as possible to fully saturated mixtures.

The essential point, then, about colour vision is that although there are three degrees of freedom, the *sensation* of colour itself is two-dimensional—it can be described by the two variables of chroma and saturation—and the third quality, intensity, is not essentially a colour attribute at all. That is not to say that it doesn't contribute to the popular idea of 'colour': the colour brown, for example, is only an orange of low intensity. The way in which the two dimensions of colour are abstracted from the signals in the three receptor channels is by making a comparison of the red and green activity, and of the blue and yellow (Fig. 7.37); and we have already seen that there is good neurophysiological evidence for just such a colour-opponent system at the level of the thalamus and of the cortex. What one might call psychological colour space thus has two perpendicular axes, one being blue/yellow and the other red/green. The fact that yellow, on such a representation, is on a par with red, green and blue although it has no corresponding cone type is probably the explanation for the fact that yellow behaves in many ways like a subjective primary colour: it doesn't, for example, look in the least like a mixture of red and green in the way that corresponding mixtures of red and blue or blue and green do. Evolutionarily, it seems that the blue/yellow axis is the more fundamental one: it is essentially a division of the spectrum into

Plate I Top, visible spectrum, showing the approximate average spectral sensitivities (on a log scale) of primate 'blue', 'green' and 'red' cones (data from Marks *et al.*, 1964). Middle, the three-dimensional locus traced out by a light of fixed intensity and varying wavelength; at any particular wavelength, the R-, G- and B- coordinates correspond to the degree of activation of red, green and blue receptors, as above. The locus may be projected on to the colour triangle, as shown. Bottom, an approximate representation of the colour triangle: the 'real' triangle is of course continuous rather than discrete. (The colours on this plate are not exact, because of the unavoidable problems associated with colour printing—discussed in the text!)

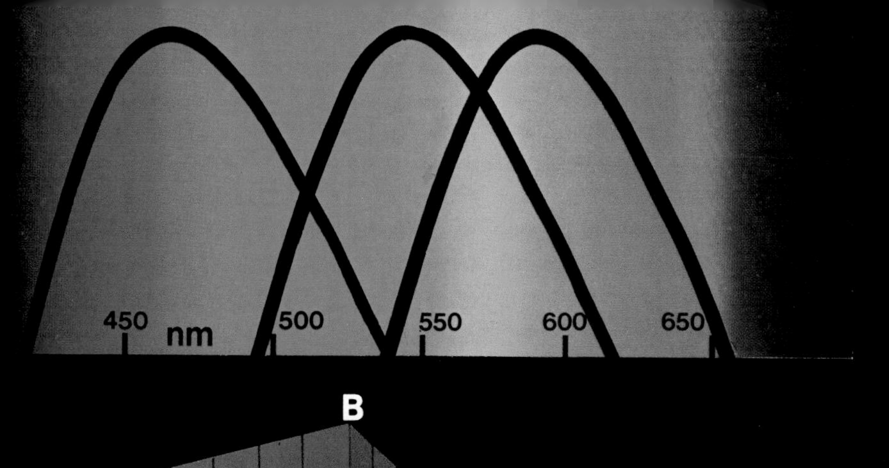

450
nm
500
550
600
650

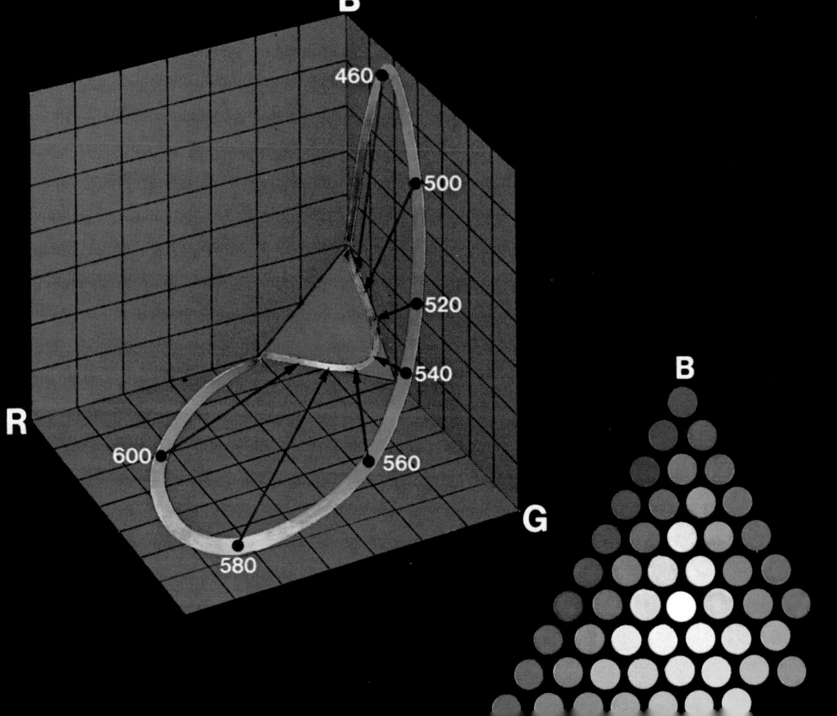

B
460
500
520
540
R
600
560
580
G
B

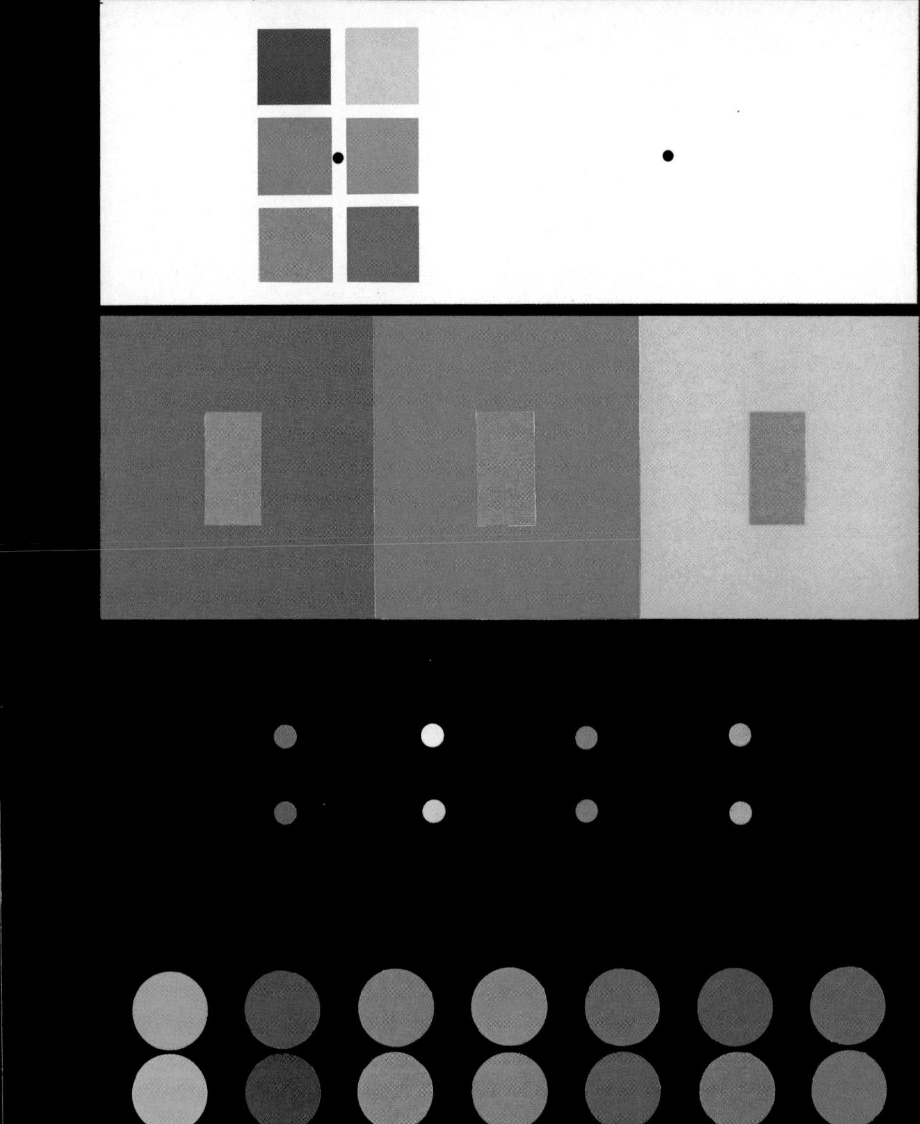

short wavelengths versus long. The red/green axis seems to have evolved later (though the evolutionary history of colour vision is complex and shows many anomalies), and it has been suggested that it was the need for primates to distinguish between the reds and greens of ripe and unripe fruit that led to the split of the original yellow cone into red and green varieties.

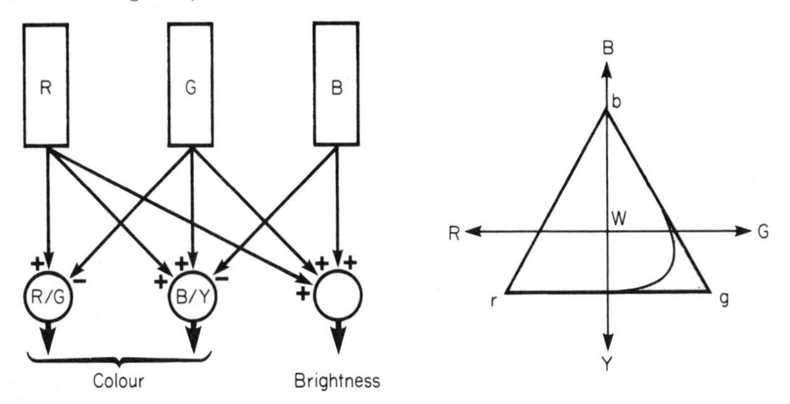

Fig. 7.37 Above, how information from the three receptor channels (R,G,B) might be neurally recoded into a red-versus-green channel (R/G), a yellow-versus-blue channel (B/Y) and a brightness channel. The resultant form of colour space, based on R/G and B/Y axes, is shown below with the colour triangle superimposed

Chromatic adaptation

Just as adaptation is important in helping the visual system to register the albedo of an object independently of the intensity of the light that falls upon it, so *chromatic adaptation*—the independent adaptation of the individual colour channels—can help the visual system make allowance for the *colour* of the illumination. Suppose we have a painting whose full range of colours falls within the triangle JKL of Fig. 7.38**a** under white light. If we now illuminate it with blue light, the effect will be to reduce the signals in the red and green channels relative to those in the blue, and the result will be a distorted triangle (jkl) and hence misperception of hues. But the red and green channels will respond to their reduced stimulation by increasing their sensitivities, which will have the effect of restoring the total range of colours perceived to something like its original extent. In fact the eye is surprisingly tolerant of changes in the spectral composition of the illuminating light, and the sensation

Plate II Peculiarities of colour vision. From the top: **(a)** After-images and complementary colours. If the black spot on the left is fixated for half a minute, and the gaze shifted to the spot on the right, coloured after-images of the six squares will be seen. The colours in each of the three pairs have been chosen to be approximately complementary to one another so that in the after-image each pair appears to swap over. **(b)** Simultaneous colour contrast. The central rectangle is in fact the same colour in each case, though it tends to take on the complementary colour to its surround. **(c)** Tritanopia with small targets. In each of the four pairs, the colours differ only in the degree to which they stimulate the blue channel. When viewed from a distance (at least 5 metres) they appear indistinguishable because of the failure of the blue mechanism (tritanopia) with small targets. **(d)** Matching by colour-blind subjects. Each pair of colours in the row was accepted as a good match by a fully red-green blind subject; some but not others were acceptable by a subject with partial red-green deficiency

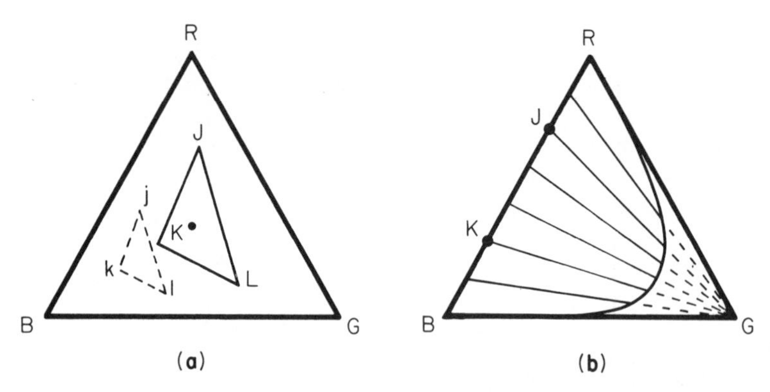

Fig. 7.38 (**a**) Effect of coloured illumination. A picture whose range of colours falls within the triangle JKL under white light is illuminated instead with blue. The effect is to shift the triangle as shown, to jkl; however, the *perceived* changes of colour will generally be very much smaller than this, because of the effect of chromatic adaptation: the sensitivities of the R and G channels increase, and that of B decreases. (**b**) Effect of lack of one channel on colour perception. Colours lying on a line such as GJ or GK differ only in the amount by which they stimulate the G channel. If, in a colour-blind subject, this channel were inoperative, such colours would all look identical: lines like GJ, GK can then be called lines of confusion. Consequently such a subject could match any colour by using only mixtures of deep red and deep blue light, in appropriate proportions

of colour is much more closely related to an object's relative albedo for short, long and medium wavelengths than it is to the actual spectral composition of the light reaching the retina. These adaptational changes can easily be demonstrated by means of successive contrast, or coloured after-images. If the coloured spots of Plate II**a** are fixated for about 20 seconds in a good light, and the gaze then transferred to the blank area next to them, striking after-images of the complementary colour will be seen that are due to distortion of one's colour space because of adaptation: what was originally a white that fell in the centre of the colour triangle now falls to one side, in a direction opposite to the adapting colour. The same explanation probably underlies simultaneous contrast (Plate II(**b**)): an area of pale colour lying next to a strong one tends to take on the complementary tinge.

Clinical disturbances of colour vision

If one of the three channels were inoperative, vision would become dichromatic, and the colour triangle would collapse into a single dimension (Fig. 7.38**b**). Such a subject would be able to match any colour in the spectrum by mixing blue and red in suitable proportions. Defects of this kind are often seen: loss of the red mechanism is called *protanopia*, of the green, *deuteranopia*, and of the blue, *tritanopia*. The first two are much commoner than the last and give rise to what is commonly called red-green blindness: the resulting confusion of red with green forms the basis of clinical tests (Plate II**d**). Some 8 per cent of males are thus affected, but a much smaller percentage of females: the gene in question is sex-linked and recessive. Tritanopia is much less common, though often in normal subjects the central fovea is tritanopic as can be seen from the demonstration in Plate II**c**. More severe defects, involving the functional loss of more than one channel—for example, the rod monochromat, whose retina contains only rods—are also found. Colour deficiencies might be due to a loss of one or more types of pigment, to lack of development of adequate neural connections, or possibly to a mixing of pigments or connections, so that discrimination is lost. Studies using retinal densitometry (see p.152) have shown that in some dichromatic subjects, at least, one of the cone pigments appears to be missing. One might wonder if it was possible to know what the world actually *looks* like to a colour-blind person. Thanks to a very rare condition indeed, in which only one eye is colour-blind, the answer is probably yes: such a subject seems to see everything in terms of blue and yellow with the affected eye, so that the spectrum appears deep blue at one end, fading to white in the middle, and then passing through progressively deeper shades of yellow to the long-wavelength end.

Less severe than than actual colour-blindness are the various colour *anomalies*. A protanomalous subject, for example, is trichromatic, but if asked to match a particular yellow by means of red and green will tend to use more red than a normal subject; deuteranomolous subjects use more green. It may be that such defects are due to imbalance in the amounts or spectral sensitivities of the cone pigments.

The use of vision

The visual system provides two distinct types of information: it tells us *what* objects are in our vicinity (recognition); and *where* they are (localization). From the point of view of the motor system, each kind of information is used at a different hierarchical level. The first essentially helps us decide what to do, while the second assists the detailed planning and execution of the resultant actions. Localization is essentially a simpler function than recognition, and will be dealt with first.

Localization

There are two components to localization, *direction* and *distance*. Clearly the spatial organization of the retina provides immediate information about the *relative* visual angles between different visual objects, since the retinal distance between two receptors that are stimulated is directly proportional to the angle between the corresponding stimuli in the outside world: but equally clearly, the motor system needs to know about the position of objects relative not to the eye but to the body as a whole. We shall see in Chapter 11 how knowledge of the position of objects relative to the eye is combined with information about eye position derived from the commands that are sent to the eye muscles, in order to compute the position of objects relative to the head; and how in turn this information, combined with signals from the vestibular system and from the neck, enables the motor system to calculate the position of objects both relative to the body as a whole, and also to absolute frames of reference such as the direction of gravity. Similar considerations apply to the visual appreciation of motion: again, the retina can only signal the relative motions of objects within the visual field. When two areas of the field move relative to one another, it is often the larger area that is assumed to be stationary: on a cloudy night, the moon appears to sail through the clouds. Of course it is the moon that is stationary and the clouds that move, but the latter occupy so much more of the visual field that the visual system assumes that they are stationary. The fundamental source of information about the movement of retinal images comes presumably from movement detectors in the cortex and elsewhere (p.159), and the fact that they adapt under prolonged stimulation gives rise to the striking *waterfall illusion*. If one stares for a minute or so at a surface that fills a substantial part of the field and is in continuous motion in one direction—like a waterfall—and then turns away to look at a stationary scene, one has the strong and persistent notion that it is moving in the opposite direction. Presumably the sense of visual motion depends on the balance between the rates of firing of movement detectors having opposite preferred directions: after adaptation an imbalance is caused by the depression of one set of these, leading to the illusion of movement in the opposite direction.

The sense of distance is a little more complex, relying more heavily than the sense of direction on what might be called 'high-level' cues. Some of this information derives from information about differences in the retinal images of the two eyes (binocular cues), while some is essentially monocular. The use of one eye rather than two substantially reduces the accuracy with which judgements of distance can be made, but does not abolish it altogether. It is convenient to consider the monocular cues first. The simplest, though probably the least important, is *accommodation*. For objects within a metre or so of the eye, the amount of effort of accommodation needed to focus an object certainly contributes to our sense of its distance, though in isolation this source of information is rather inaccurate. Much more important is information derived by moving the head: objects that are close to us then move more rapidly relative to the horizon than those that are far away (as can be seen on looking out of a train window) and this *movement parallax* can provide very

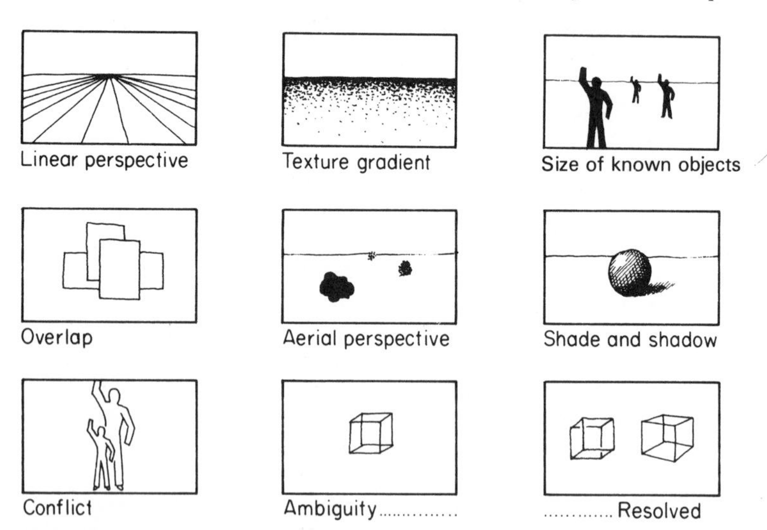

Fig. 7.39 Illustrating schematically some of the monocular depth cues. The third row illustrates first how conflict may arise, in this case between size and overlap; and secondly, how it is possible to construct figures that may be interpreted in more than one way (in this case, as a cube viewed either from above or below). Addition of extra depth information resolves the ambiguity

accurate information about distance. It can often be seen when a cat is preparing to leap onto a ledge and needs to know its distance: it pauses first, then moves its head up and down to generate motion parallax. The other monocular cues are those that require some prior knowledge of the real world, and are the ones used by artists in portraying depth: painters have a particularly difficult job in trying to do this, because both accommodation, movement parallax, and the binocular cues combine together to tell the viewer that the picture is really flat. Some of these higher-level cues are illustrated in Fig. 7.39. They include *overlap* (nearer objects tend to obscure further ones), *size of known objects* (if we know the actual size of an object and the angle it subtends at the retina we can deduce its distance), and various miscellaneous cues that are really special cases of 'size of known objects'. These include *linear perspective* (for example, the apparent convergence of parallel lines) and *texture gradient* (that the spatial frequency of a pattern or texture gets higher the further away it is). Finally, *shadows* can give useful information about three-dimensional shape, and at great distances *aerial perspective*—the fact that distant objects are fuzzier and bluer than near ones—may be used. It is possible to create artificial situations in the laboratory in which these cues are contradictory, and work out from subjects' response to them which cues are given more weight by the visual system than others. Many well-known illusions occur because assumptions about the distance of an object affect one's estimate of its size—demonstrating that of all the cues to depth, linear perspective is probably the strongest (Fig. 7.40).

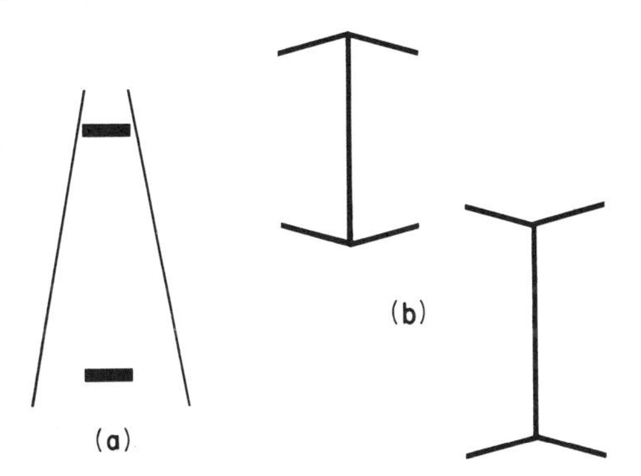

Fig. 7.40 Illusions that probably result from false depth interpretations. **(a)** The Ponzo illusion: the lower bar looks smaller than the upper, probably because it is perceived to be nearer (the sloping lines being perceived as parallel but receding into the distance). **(b)** Müller–Lyer illusion: the upright in the right-hand figure looks larger than the other, perhaps because the figure is interpreted as the far corner of a room or box, whereas the left-hand figure is taken as the near corner

What extra information is available if we use two eyes rather than one? Because we now see the visual world simultaneously from two different points of view, there are bound to be differences in the images that fall on each retina, that will be more marked the closer an object is to the eyes. Imagine the two eyes fixating a point A in the middle distance (Fig. 7.41). The image of A will fall on the fovea of each eye, and points like B that are at the same distance as A will be at the same angular distance from A as seen by each eye, and their

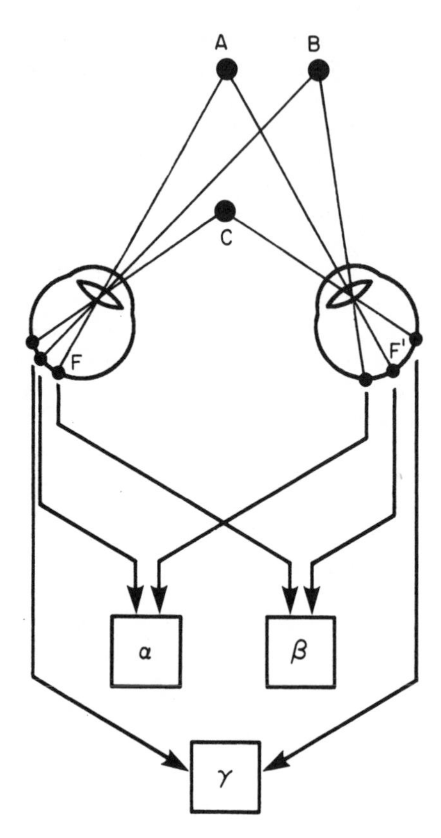

Fig. 7.41 Retinal disparity and disparity detectors. When binocularly fixating A, targets such as C that are at a different distance give rise to disparate images on the two retinae. Of neurones in the visual system that respond to signals from both eyes, many are stimulated by corresponding points in each retina (α, β); but some are connected to disparate points (γ)

images will therefore fall on points that are at the same distance and direction from the fovea in each retina: such points are called *corresponding* points. (Strictly speaking, as may easily be verified geometrically, points like B must lie on a circle that passes through the centres of the two eyes and also through A. Such a circle is called a *horopter*.) However, a point like C that is at a *different* distance from the eye will form images in the two eyes that are at different positions relative to the fovea: these are called *disparate* images. Now it turns out that if we examine the visual fields of cortical cells that have a binocular input (p.160), although many of them are connected to corresponding areas in the two eyes, others are connected to disparate regions of the retina, and may be called *disparity detectors* (Fig. 7.41). So with the eyes both fixating A, a unit like α will be responding to the image of A, a unit like β with zero disparity will be responding to the image of B, and a unit like γ with marked disparity will be looking at the image of C. Thus even with the gaze fixed on one point in space, some cortical units will be 'looking' at points lying in planes either in front of or behind the target, and will thus provide immediate information about depth. Subjectively, there is indeed an area around the horopter, called Panum's fusional area, in which one is not aware of the double image normally perceived when an object is out of the plane of fixation; the disparity units seem in some way to have fused the two images back together again in one's perception. It is important to emphasize that though this mechanism of disparity-detection is a very sensitive one indeed, it can only provide information about the distance of an object *relative* to the plane of fixation. As in the case of direction perception, we need further information about the positions of the eyes, about their angle of *convergence*, before it can be used to compute absolute depth. Since it turns out that knowledge of the convergence of one's eyes is rather imprecise, it follows that disparity, though highly accurate for determining relative distances, is not of great use for absolute estimates. It is not altogether clear what is in fact used to sense absolute depth: it may well be that the visual size of very well known objects, particularly of parts of the body such as the hand, whose distance can be checked by direct proprioception, provides the ultimate measure by which the rest of visual space—beyond one's own reach—is calibrated.

Recognition

Conceptually, recognition is perhaps the most difficult of all the visual functions, and one whose neural mechanisms are largely a complete mystery. Our lack of understanding in this area is reflected in our inability, at the time of writing, to build machines that will carry out more than the crudest kinds of pattern recognition. The difficulty lies in the fact that objects that look quite different from one another—for example the set of letter As in Fig. 7.42—must

Fig. 7.42 What is 'A-ness'?

be recognised as belonging in the same category. If the objects to be recognized are highly stereotyped, like £1 notes, it is not difficult to build a machine that looks for a match between a stored 'ideal' bank note in the machine's memory, and the actual specimen that is presented. But it is hard to define an 'ideal' letter A, or say what essentially is the 'A-ness' that all the examples of Fig. 7.42 share. This problem is exacerbated by the fact that objects in the real world are seen at different times under lighting of different intensity and colour, and from different distances and directions. A particular retinal image of a cube under particular conditions is as much a coded version of the cube, that has to be deciphered, as are the four letters CUBE: in many ways the latter is actually the easier task! In other words, the visual system must make a distinction between those aspects of an object's image which are its *essential* attributes, and those which are merely *accidental* and the result of temporary circumstances. The point has already been emphasized with respect to luminance and illumination: it is the albedo of an object which is essential in this sense, and not its luminance. A pattern-recognition system must therefore do two things: first, it must filter out accidental features of a stimulus; and then it must relate the features that remain to those that in the past have been found to be associated together.

The first stage, that of filtering out irrelevant information, is one that we have encountered at many levels of the visual system. It appears in the mechanism of adaptation, which allows the eye to ignore the intensity or colour of illumination, and in lateral inhibition which filters out information about areas of constant luminance and thus emphasizes the edges that define an object's form. If it were not for this filtering process, the job of the artist in representing the visual world would be all but impossible: it is after all rather remarkable that a few strokes of a brush can convey something as subtle as the emotion expressed by a human face. What the artist is doing, in effect, is to filter out irrelevant information before it even reaches the eye. The astonishing thing is not so much that this works, but that one accepts a representation so far removed from what the retinal image of a real face would look like without even noticing that it *is* remarkable.

The second stage of recognition is that of forming associations in the brain between elements of objects that occur together, and recognizing that they form a whole. We recognize a figure 3 because it has certain topological features that are generally found together: a single continuous line with a cusp in the middle to the left and a couple of bulges to the right. It is not difficult to imagine neural mechanisms by which units that often fire together might tend to strengthen their mutual connections and hence form a functional cluster corresponding to a particular object in the outside world: they are discussed when considering memory in Chapter 14. For example, it is easy to see how a cortical line detector might be 'built' in this way. Imagine that the projection of thalamic units to cortical neurones is initially rather random, with a good deal of convergence and divergence (Fig. 7.43). At first the receptive fields of these 'naive' cortical units will be chaotic and disorganized. Let us suppose also that the synaptic connections from thalamus to cortex have the property that when the cortical unit fires it strengthens those synapses that are active at the same time, and weakens those that are not. On looking at a straight line,

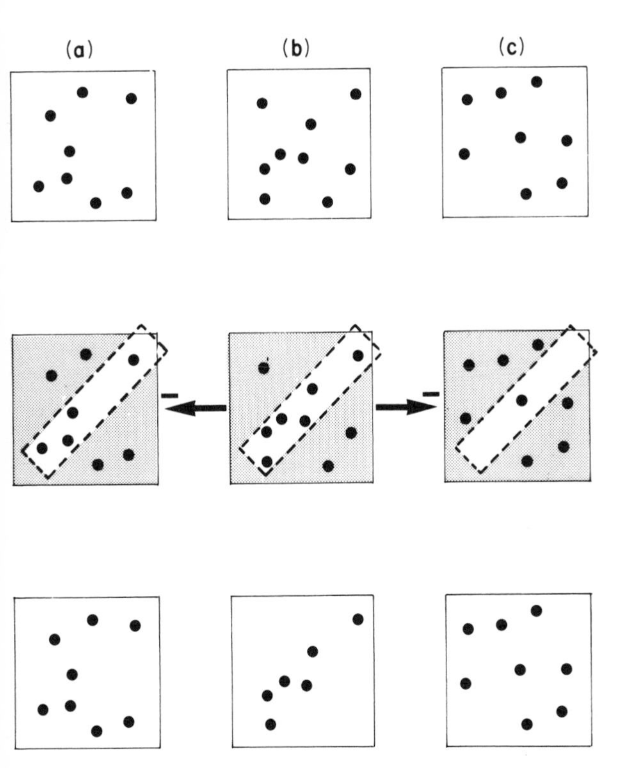

Fig. 7.43 Hypothetical mechanism by which the specificity of central visual neurones might grow from experience. Top row: receptive fields of three 'naive' neurones, indicated by dots. Middle row: on stimulation with a slit of light at a particular orientation, only (**b**) fires: its active afferent fibres grow stronger, while the others decay. Bottom row: after a sufficient number of presentations of this type, (**b**)'s receptive field is closely matched to the slit, and (**a**) and (**c**) are still available to learn some other stimulus.

although no line detectors as such will yet exist, it is clear that some cortical units will fire and others not. Of the afferents going to the cells that fire, the ones corresponding to retinal units lying on the line will be strengthened, while the others will weaken. In time, when a sufficient number of straight lines of the same orientation have been experienced, it is clear that the inputs from the line will have been reinforced, and the other, irrelevant ones, will have ceased to function: the receptive field of the cortical unit will then be that of an ordinary line detector. (Such a model can be extended to cover the generation of inhibitory surrounds as well; but it must be said that recent work has indicated that the true mechanism, though not understood in full, is not quite as simple as the one presented here). It is not difficult to extend such a notion to yet higher stages of cortical processing, and imagine units that could learn in exactly the same way to respond to the more complex sets of essential features that make up things like teacups and human faces. It is certain that some such mechanism of learned connections must exist, for we know that young kittens brought up in a visual environment consisting entirely of lines having a single orientation are found on subsequent testing to have cortical units that respond only to lines of that same orientation. Once such a set of features have been associated together in this way, the detector may not mind very much if some of its inputs are missing on a particular occasion: so long as it fires more actively than any of its neighbours in response to a particular object, then it will in effect form a hypothesis about what is present, and the subject may then think he sees features of the object that he expects but are not in fact there (Fig. 7.44).

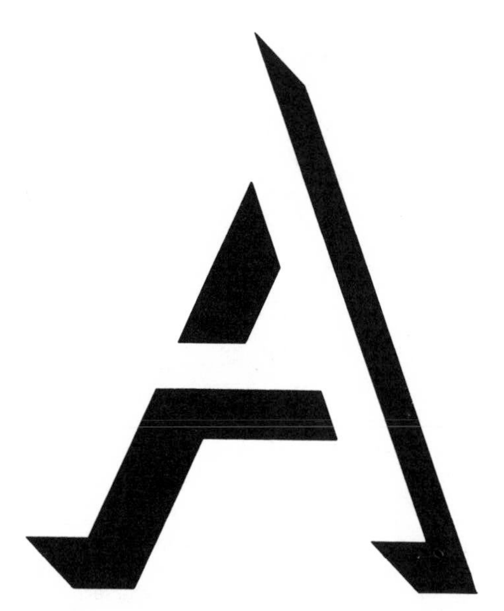

Fig. 7.44 Phantom contours

The final stage of visual recognition is not of course simply the realization that a particular retinal image has elements that have often been found together in association in the past, and therefore correspond to some real object in the outside world; it is necessary not only to recognize objects, but to attach *meaning* to them. This implies, in effect, associating them not just with other visual stimuli, but with the other senses: with words, actions, and above all with emotional states and with the satisfaction of physiological needs. The visual system is not, after all, merely an arrangement for providing a kind of in-flight entertainment for the soul: it is there to be *used* by the motor system both in the planning and execution of actions, and for this its information has to be translated into language the motor system and motivational systems understand. It is not difficult to see how this might be done, by an extension of the mechanism of forming associations by synaptic strengthening outlined above, and this topic will be pursued further in Chapter 14.

8

Smell and taste

All neurones in the brain respond to chemical transmitters, so chemosensitivity is hardly a specialization of function at all. We shall be concerned here only with chemical stimuli that originate outside the body, with *olfaction* (smell) and *gustation* (taste). The chemoreceptors that monitor the composition of the blood, and are used in the regulation of autonomic and hormonal functions, are discussed in other volumes in this series.

Olfaction: the receptors and their connections

Those readers who have had the privilege of dissecting a human head will be aware that the nose has an internal complexity that is quite startling in comparison with its rather drab exterior. The surface area of the nasal cavity is enormously inflated by the presence of three *conchae* on each side, highly vascular organs covered with erectile tissue whose function is primarily to moisten and warm the incoming air, and conversely to limit the loss of heat and water in the air that is expired (Fig. 8.1). The olfactory receptors form part of the *olfactory epithelium*, tucked away in the olfactory cleft right at the top of the cavity, and one might well wonder whether the inspired air ever gets anywhere near it in normal breathing. In 1882, the Viennese physiologist Paulsen performed an elegant but somewhat macabre experiment to find out. Having cut a human head down the middle, he placed tiny squares of red litmus all over the nasal cavities; then, sticking the two halves together again, he drew air laden with ammonia from a bottle held under the nose by appropriate manipulation of a pair of bellows attached to the trachea. On opening the head he could see by which pieces of litmus had turned blue what course the air had taken, and it turned out that in normal quiet breathing the air hardly reaches the olfactory epithelium at all. However, in sniffing, turbulences are set up round the conchae, and an appreciable fraction of the air gets to the olfactory receptors. This fraction is critically dependent on the state of the conchae: if you have a cold, they tend to become engorged with blood, hindering the passage of air to the higher regions and causing the familiar partial loss of smell.

Man is a *microsmatic* animal: smell plays a far smaller part in his sensory

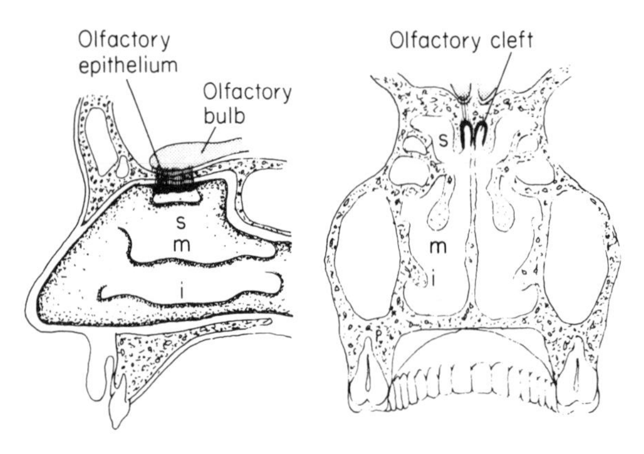

Fig. 8.1 The nasal cavity in Man; longitudinal and transverse section, showing the superior, medial and inferior conchae (s,m, and i) and the olfactory cleft

world and in the regulation of his actions than in the case of *macrosmatic* animals such as the dog, and his olfactory sensitivity is in general correspondingly reduced. To some extent this is reflected in the small area of his olfactory epithelium: about 5 cm^2 in all, compared with about twice that in the cat, despite its very much smaller head. This epithelium has a number of easily recognizable features. It contains *Bowman's glands* of tubuloalveolar form producing a lipid-rich secretion that bathes the surface of the receptors; an important consequence of this fact is that to have an odour, a substance must to some extent be lipid-soluble. Another characteristic is that one of the types of epithelial cells contains granules of *pigment*: the depth of colour in different species is often correlated with olfactory sensitivity, being light yellow in Man and dark yellow or brown in dogs. It has sometimes been suggested, as we shall see, that this pigment may play some part in the mechanism of olfactory transduction; albino animals, for example, are said to have abnormally reduced sensitivity to smell. Finally there are the receptors themselves, distinguished by a terminal enlargement above the surface of the epithelium, from which project some 8–20 *olfactory cilia* (Fig. 8.2). These cilia show the usual 9 + 2 fibril arrangement at the base, but are not thought to be motile: they form a dense and tangled mat that covers the olfactory area. Vacuoles can also be seen in the terminal enlargement, and experiments with colloidal gold have shown that they are actively pinocytotic: fluid is being continually drunk in by the receptors and passed down the olfactory nerves into the brain. The significance of this surprising feature is unclear.

Unlike many receptor cells, the olfactory receptors send their own axons to the CNS without an intervening synapse. These fibres, the *fila olfactaria*, make up the first cranial nerve: they are exceedingly fine and difficult to see with the light microscope. They pass through the cribriform plate in individual holes ('cribriform' = 'sieve-like') and enter the *olfactory bulb* which lies just above the olfactory epithelium (Fig. 8.3). Here they synapse with dendrites of the large *mitral cells* (they are supposed to look like bishops' mitres) in specialized nexuses

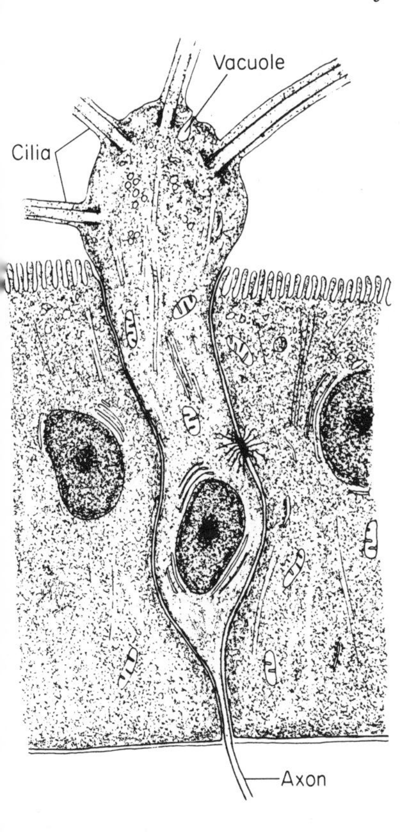

Fig. 8.2 Typical olfactory receptor, showing terminal enlargement with cilia and vacuoles projecting above the level of the surrounding epithelium. Note that the cilia are truncated in this picture: in practice they vary considerably in length, some being shorter than the receptor cell body, and some several times its length

called *glomeruli*. In rabbits, which have been particularly well studied, each glomerulus receives information from some 26 000 receptors, and has an output to about 24 mitral cells: there are probably only some 2000 glomeruli in all. The fibres entering any one glomerulus come from a wide area of the epithelium, so that detailed information about any spatial patterns of activity must be largely thrown away. The final output of the bulb consists of the axones of the mitral cells, forming the *olfactory tract:* this has two divisions, lateral and medial, of which only the lateral one appears to be important in Man.

However, the olfactory bulb is not just a simple relay, but has two other properties that we have already seen to be common to all the sensory systems examined so far, namely lateral inhibition, and negative feedback control of afferent information. The most prominent feedback path is formed by collaterals of the mitral cell axones, that turn back into the bulb to synapse excitatorily with *granule cells* (Fig. 8.3) which in turn inhibit neighbouring mitral cells. Other collaterals of the mitral cell axons synapse in the anterior olfactory nucleus with interneurones which also return and synapse with granule cells; but in this case some of the interneurones project *contralaterally* in the anterior commissure to influence the opposite bulb in the same way. The significance of this mutual inhibition between the two bulbs is not clear: one

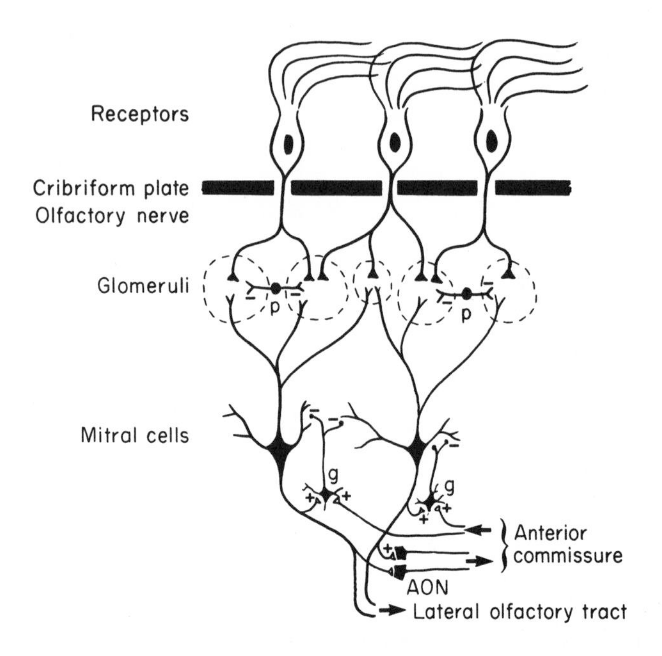

Fig. 8.3 Simplified representation of the cell types of the olfactory bulb, and their connections. p, periglomerular cells; g, granule cells; AON, anterior olfactory nucleus

possibility is that it may serve to enhance differences between the activities of the two bulbs—another kind of lateral inhibition—in a way that might perhaps be useful for localizing smells. Careful experiments have shown that even humans are capable of localizing odorous objects in a rather approximate way, presumably through slight differences in the timing or intensity of the stimuli in each nostril. Finally there are the *periglomerular cells*, which appear to subserve lateral inhibition at the level of the mitral cells and glomeruli respectively. The periglomerular cells make two-way synaptic contacts with the mitral cells, in a manner strikingly reminiscent of the relationship between horizontal cells and receptors in the retina.

Central olfactory projections

The central projections of the olfactory system provide something of an *embarras de richesses*, very different from the orderliness with which for example the optic tract projects to the lateral geniculate nucleus and thence to the cerebral cortex: indeed olfaction seems to be unique in projecting straight on to cortical areas without relaying in the thalamus or any equivalent structure. One must bear in mind that the olfactory system is very much older than such senses as vision and hearing, and in more primitive animals a very much larger proportion of the brain is directly or indirectly concerned with olfaction (Fig. 8.4). The reason for this is not hard to see: simple animals depend much more immediately than we do on knowing directly from their senses whether

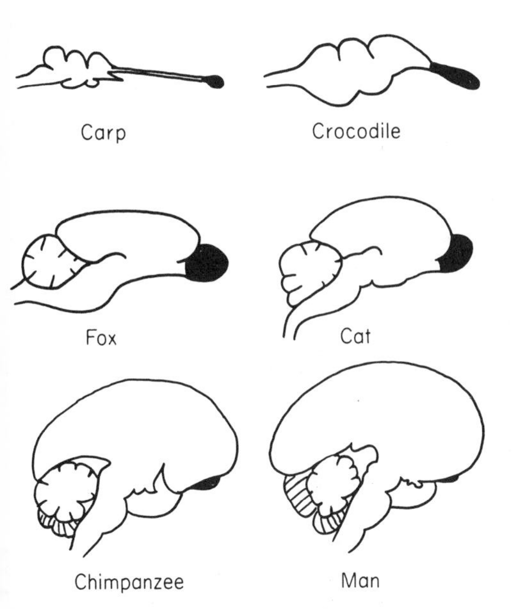

Fig. 8.4 Reduction in relative size of olfactory bulb (black) with phylogenentic development of the brain. (Partly after Carpenter, 1976)

food is in the vicinity, and their motor systems are likely to be more pressingly governed by the need to move towards nutrients and avoid poisons, and to seek out mates by recognizing the chemical attractants they release. Their *motivation*—the drive that tells their motor systems what to do—is essentially olfactory. Even in Man, the remains of this very basic system for chemical motivation and emotion (emotion being the sensory correlate of motivation) can still be seen in the central olfactory projections (Fig. 8.5). Many of these structures form part of the *limbic system*, a group of nuclei, cortical regions and connecting tracts of great evolutionary antiquity that appear to be concerned with precisely those kinds of function that one would expect in a primitive animal to be closely related to chemical stimulation: motivation, emotion, and certain kinds of memory. The limbic system and its functions are discussed more fully in Chapters 13 and 14; for the moment we may note for example that the septal nuclei and amygdala contain regions known as 'pleasure centres', in the sense that when electrically stimulated they seem to provide a kind of direct positive motivation. The hippocampus seems to be concerned with motivational memory, the ability to associate a previously uninteresting stimulus (like the bell in Pavlov's well known conditioning experiments) with the promise of food or pleasure signalled more directly by olfactory stimulation, and to recognize such a stimulus in the future as a source of motivation in its own right. What seems to have happened in the course of evolution is that this kind of *secondary motivation* by stimuli that only acquire their meaning through experience and learning has steadily grown in importance relative to that of primary olfactory motivation (Fig. 8.6). For

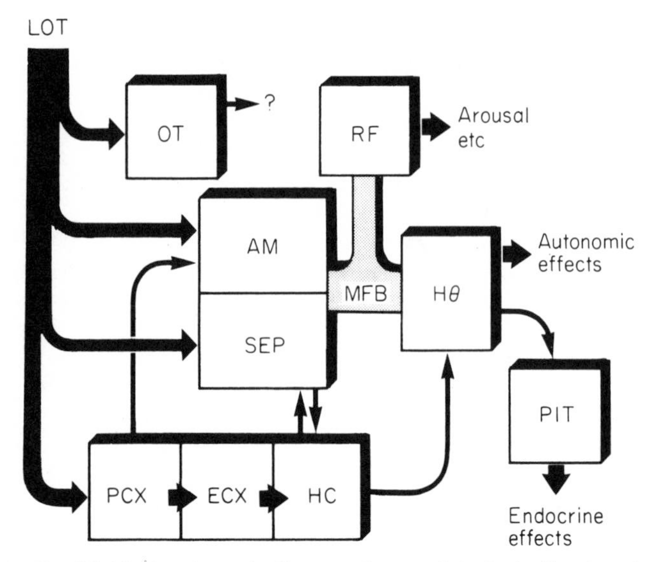

Fig. 8.5 Highly schematic diagram of areas of the brain directly or indirectly driven by olfactory stimulation from the lateral olfactory tract (LOT). OT, olfactory tubercle; RF, tegmental reticular formation; AM, amygdala; SEP, septal nuclei; Hθ, hypothalamus, especially preoptic and lateral nuclei; PIT, pituitary; PCX, periamygdoloid and prepyriform cortex; ECX, entorhinal cortex; HC, hippocampus and subiculum; MFB, medial forebrain bundle

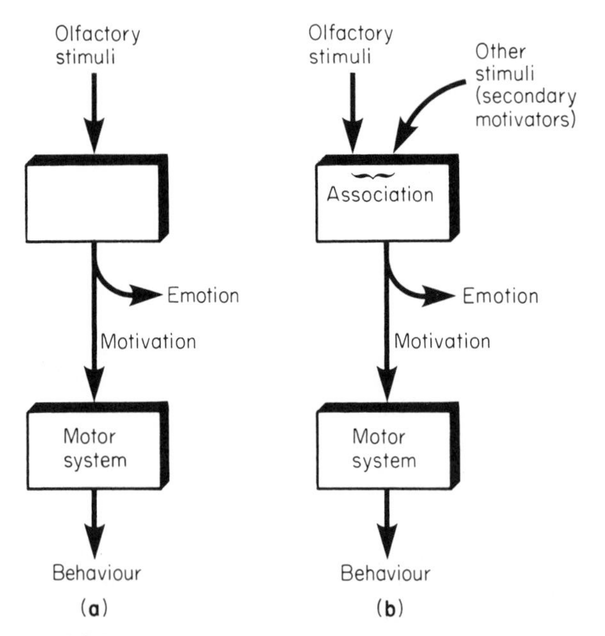

Fig. 8.6 Schematic representation of motivation through primarily olfactory stimuli in primitive organism (**a**), and in a more developed animal (**b**) through other stimuli (secondary motivators) that become potent through association

Man, money is perhaps the most powerful secondary motivator of all, and given the choice between a plate of fish and chips and a plate of £10 notes, there is no doubt which would cause the greater motivational drive! For this reason, limbic structures that were originally subservient to olfaction are now not primarily olfactory at all; and the name *rhinencephalon* ('nose-brain') which is sometimes given to the limbic system is an inappropriate one in higher animals. Finally, there are important connections between the limbic system and the hypothalamus, providing routes by which olfactory stimuli can cause such obvious autonomic and hormonal effects as sexual arousal and modification of reproductive cycles (even abortion), salivation and other secretory responses to food smells, and in many species instinctive fear reactions to threatening odours from predators. Many creatures emit specific *pheromones* (air- or water-borne hormones) that act as sexual attractants, often over large distances. A curious feature of some of these lures is that they often resemble, in species as different as the civet and the moth, the steroid reproductive hormones themselves (at least in overall shape: Fig. 8.7).

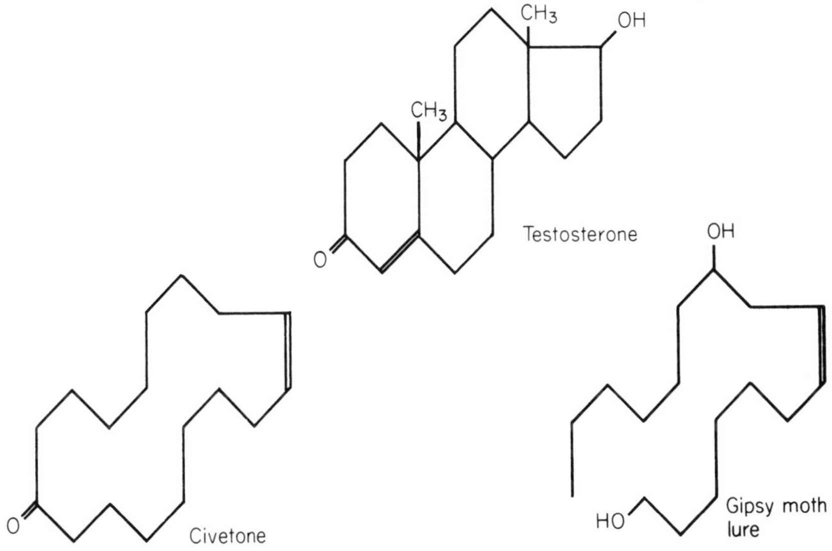

Fig. 8.7 The structure of testosterone (above) and of two olfactory sexual attractants (below)

Furthermore, some of them—such as civet oil and musk—are used in perfumery and presumably act as lures for human males as well! There are said to be marked sex differences in the olfactory threshold for some of these macrocyclic compounds, which may also vary with the phase of the menstrual cycle.

Lastly, another indication of the psychological links between olfactory and limbic functions in Man is the very striking way in which odours may call up—often with surprising intensity—recollections of past experience; and it is interesting how often such evocations are not just of the objective circumstances of a particular event, but also of the mood or emotion that was felt at the time, in a way that is seldom experienced with purely auditory or visual stimulation.

Recordings from olfactory cells

Perhaps because smell does not seem as vivid or important to us as say vision or hearing, and also because of certain difficulties of experimental technique, our knowledge of the electrophysiology of olfaction is still somewhat rudimentary. One of the difficulties is that whereas it is comparatively easy to stick an electrode into the optic nerve or auditory nerve and record the way in which single fibres respond to particular stimuli, it is not easy to do the same thing in the case of smell: the olfactory fibres are exceedingly fine, rather short, and buried for much of their length in the cribriform plate. An electrode in the olfactory epithelium tends to pick up not spike responses from individual cells, but an averaged slow potential from many of them together, the *electro-olfactogram* or EOG. Fig. 8.8 shows some examples from the frog. If one experiments with puffs of air laden with different chemical substances, one may get all kinds of different sizes and shapes of response, with little obvious correlation between the nature of the response and the kind of substance that is applied. Fig. 8.8**a** shows the responses to two simple alcohols, methanol and butanol: despite the similarity of the stimuli (and they smell quite similar to us), the responses could hardly be more different. Most substances give responses like the lower one which are partly excitatory and partly inhibitory, and to make matters worse, if one examines the interactions between stimuli—for example a puff of methanol followed by a puff of butanol—the result is in general not one of simple summation.

If the electrode is driven further into the epithelium, one is occasionally

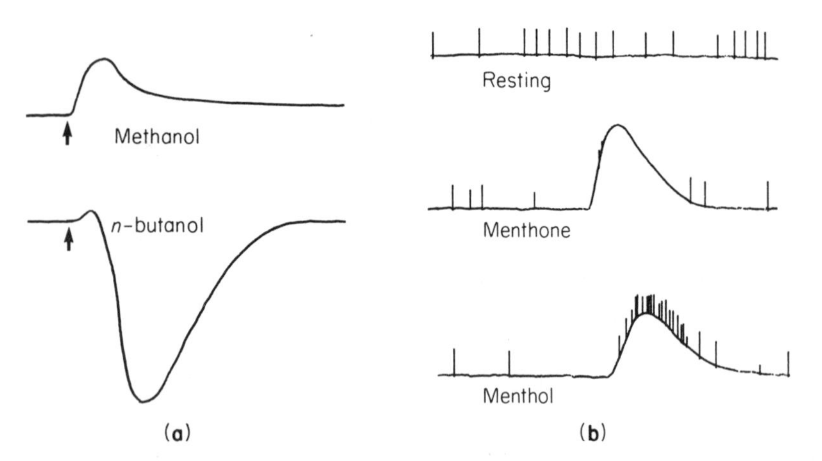

Fig. 8.8 Responses from frog olfactory receptors. (**a**) Slow potentials (electro-olfactogram) in response to a puff of air (at arrow) laden with methanol (above) or *n*-butanol. (**b**) Action potentials at rest, and in response to menthone and menthol. (Gestland *et al*., 1965)

lucky enough to record spikes from individual fibres, usually with the EOG superimposed on top (Fig. 8.8). Fig. 8.8**b** here shows the resting spike activity in the absence of stimulation and the responses to menthone and menthol, which again have very similar odours. It can be seen that although the EOGs are roughly comparable, in one case the fibre is inhibited, firing more slowly during stimulation, and in the other it is excited. In fact with a single-unit preparation of this kind one can draw up a list in two columns, showing for a particular cell which substances excite it and which inhibit it. Such lists turn out to be quite chaotic, with apparently similar substances like menthol and menthone often on opposite sides. Even more perplexingly, if one moves the electrode to record from a different cell, one finds in general an entirely *different* list, with substances that were excitatory for one cell being now inhibitory for the other, and substances that were on the same side of one list being on opposite sides of the other. In fact there seems no system whatever in the way in which chemical stimuli are coded into patterns of firing of the olfactory nerve: each unit has its own idiosyncratic view of the olfactory world, like a spoilt child who likes baked beans but not bananas, fudge but not fish fingers, in a wholly arbitrary manner. The situation could hardly be more different from a sensory organ like the retina, with each of its units closely specified in terms of position, intensity and colour. Randomness of this kind is

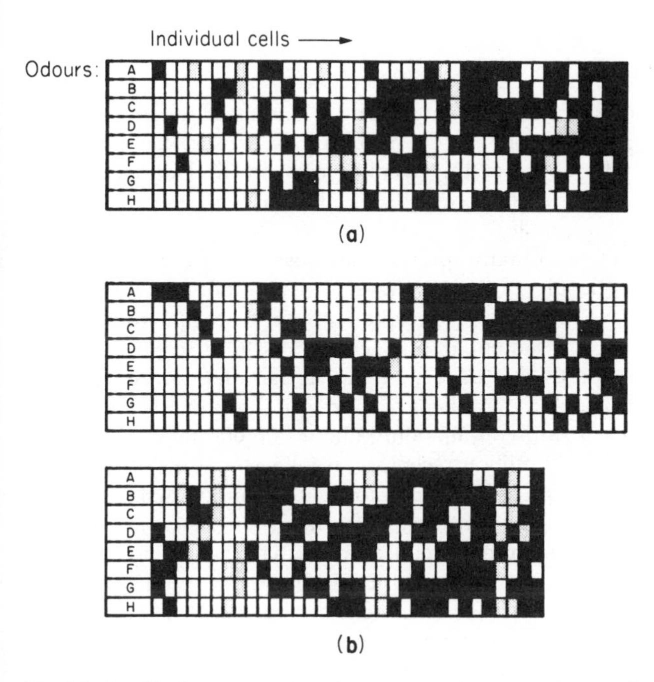

Fig. 8.9 Graphical representation of responses of neurones in the olfactory system of a monkey to each of eight different odours (A–H). Black = excitation, grey = inhibition. (**a**) 40 different units in the olfactory bulb; (**b**) 73 units in prepyriform cortex and amygdala: a larger proportion of the more central units respond to a smaller set of the stimuli applied. (Data from Tanabe *et al.*, 1975)

not necessarily a weakness in a sensory system; no information need be lost, since by looking at the *pattern* of response over the fibres as a whole the nature of the original stimulus can still be reconstructed. Imagine a nursery of spoilt children seated at a dinner table and provided with pushbuttons with which they can register approval or disapproval of what is set in front of them: if these buttons were connected to an array of lights on a screen, it is clear that although any individual child's reaction may be quite idiosyncratic, nevertheless any particular dish will result in a perfectly characteristic and reproducible pattern of lights by which it may be recognized.

Recordings from the olfactory bulb show a similar degree of chaos: Fig. 8.9**a** shows in graphical form the responses of a large number of units in the olfactory bulb of a monkey, and it can be seen that only a small proportion of the cells responded to just one of the eight chemicals used as stimuli, and most showed excitation or inhibition to at least three. Of more central areas little is certain: responses to olfactory stimulation can be recorded from wide areas of the brain, not just in the limbic system but as far afield as the basal ganglia as well. One region that has been studied in more detail is the prepyriform cortex: here units show properties that suggest that the chaos characteristic of preceding levels of the olfactory system is beginning to be sorted out, and a significantly larger proportion of units respond to just *one* of a series of chemicals (Fig. 8.9(**b**)).

Psychophysics of smell

The sense of smell shows a number of interesting and unusual features, which as well as shedding light on one's understanding of sensory processes in general, also account for some of the peculiar experimental difficulties of studying olfaction, and set certain bounds on the kinds of theories of its mechanism which are acceptable.

The *sensitivity* of olfaction in many species is astonishing: just as the rods in the eye respond to single photons, and the ear to vibrations of the air of subatomic dimensions, so the olfactory receptors are very near the theoretical limit of their sensitivity, and can apparently respond to the absorption of one or two molecules. Tests on tracker dogs have shown that they can respond to one millilitre of butyric acid (an ingredient of stale sweat) in some 10^{11} litres of air. This means that each sniff contains only about 200 000 molecules, and since the dog has roughly 200 million receptors, it follows that the absorption of two molecules at the most must be enough to excite an individual receptor. Sensitivity of this kind makes for considerable difficulty in experimentation, for when one measures an apparent response to one particular substance A, unless A is quite exceptionally pure, one can never be quite certain that what one is measuring is not the response to some other substance B present in exceedingly small amounts as a contaminant.

Another characteristic of olfactory sensitivity is that *adaptation*, though not particularly rapid, is usually absolutely complete. Thus men working in uncommonly smelly environments such as sewers or gas works soon become quite insensitive to the smells around them, and people are in general unaware of their own body odours. Professional food evaluators have to take special precautions to avoid adaptation of this kind: Scottish cheese-tasters, aware of this danger, take a nip of whisky after each sample to restore the keenness of their palates (this at least is the ostensible reason). This may perhaps be why the olfactory epithelium does not lie on the path of the incoming air in normal breathing: in sniffing, a sufficient *change* in odour concentration may be set up that overcomes unwanted adaptation.

The most fundamental way in which smell differs from the other special senses is in the lack of a systematic method of classifying and analysing different types of odour. To some extent this is because we ordinarily pay little conscious attention to smell, having enough to do in coping with the flood of more interesting information pouring in from our eyes and ears. Helen Keller, the well known writer who was blind and deaf from an early age, was able to develop her olfactory discrimination to an extraordinary degree through not being distracted by her other senses; she could for example recognize people she met and places she visited solely through their characteristic odours. Part of the trouble is undoubtedly our lack of a proper *vocabulary* for describing smells: if we want to convey to someone what a eucalyptus smells like, we are literally at a loss for words. The difficulty is that there exists no objective, physical, way of classifying smells systematically in the way that we can, for example, order colours into a spectrum or tones into a scale. In the case of vision our system of classification leads us to formulate simple rules using the colour triangle that enable us to predict the result of mixing colours in certain proportions to produce other colours: but in the case of smell this is quite impossible. We can never predict in advance what the result of mixing two odours together will be, and the results of doing so are frequently quite paradoxical. For example, the smells of iodoform and of coffee, individually strong and characteristic, are said to cancel each other out if appropriately mixed; other examples of cancellation of this kind have sometimes been exploited commercially to produce specific deodorants. Worse still, many odours smell quite different at different concentrations. Indole, which is a major component of dog excreta, and smells like it when concentrated, has a pleasant floral smell when very dilute and has actually been used in cheap perfumes! Many of the organic sulphides smell appalling at close range, but in small quantities turn out to be mainly responsible for the appetizing smell of foods such as roast beef and onion. Conversely, the nose is usually unaware without special training whether a particular smell is pure, in the sense that only one kind of molecule is present, or a mixture. Many natural odours that seem perfectly unitary and pure, like that of raspberries, are in fact composed of dozens of components, many of which taken by themselves are rather unpleasant, and cannot be detected for what they are in the whole ensemble.

For all these reasons, although in the past strenuous efforts have been made to try to classify smells into primary odour classes, like the nineteenth-century six-fold classification shown in Fig. 8.10, no classification is ever really satisfactory because it does not enable one to predict the results of making mixtures, in the way that the colour triangle does for colours. There is no such thing, in fact, as a *primary* odour. The explanation for this unsatisfying state of affairs lies in the chaotic way in which individual

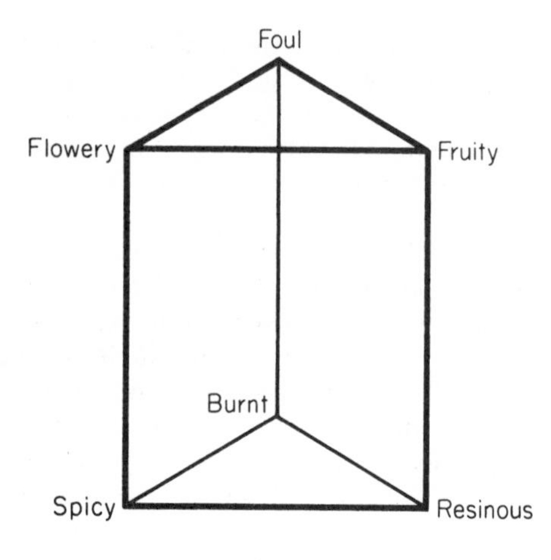

Fig. 8.10 Henning's prism: an attempt to divide olfactory stimuli into six 'primary' classes

receptors respond to particular chemicals. If we have two substances A and B which each produce characteristic patterns of activity in the olfactory nerve as a whole, then the response to A and B together will not be simply the sum of the responses to each separately: it will be a *new* pattern altogether. Thus the number of 'primary odours' will be of the order of the number of types of receptor, which seems to be indefinitely large.

Theories of the olfactory mechanism

There are currently three very different theories of olfactory transduction: it has to be said that no one of them is very compelling. The simplest in concept is perhaps Amoore's (1963). He suggests that either on the receptor membrane, or possibly inside, there are hollow receptacles of molecular proportions, which accept or reject odorant molecules according to how well they fit the site (Fig. 8.11). In his theory he chooses just seven types of site, and each site has its corresponding odour quality—floral, minty, and so on:

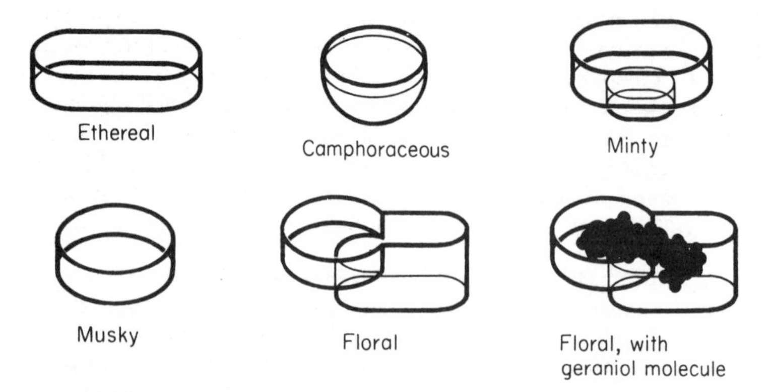

Fig. 8.11 Five of the receptor sites proposed by Amoore, and (bottom left) the floral site occupied by a molecule of geraniol, a constituent of the smell of roses. (After Amoore, 1963; copyright Macmillan Journals Ltd.)

two of the sites are for electrophilic and electrophobic molecules. In general terms the theory is plausible enough, except for the very small number of primary classes that is envisaged: we have already seen that a classification with only a few primary odours is quite insufficient to describe the richness and complexity of the real olfactory world. Another problem is that in practice there is often a striking lack of the expected correlation between a molecule's overall shape and what it smells like. Camphor and hexachlorethane smell practically identical to us, yet one could hardly imagine two molecules more different in size and structure (Fig. 8.12). Again, optical isomers—molecules having the same structure but mirror images of one another—invariably have identical smell: it is difficult to see how the same site could fit both forms.

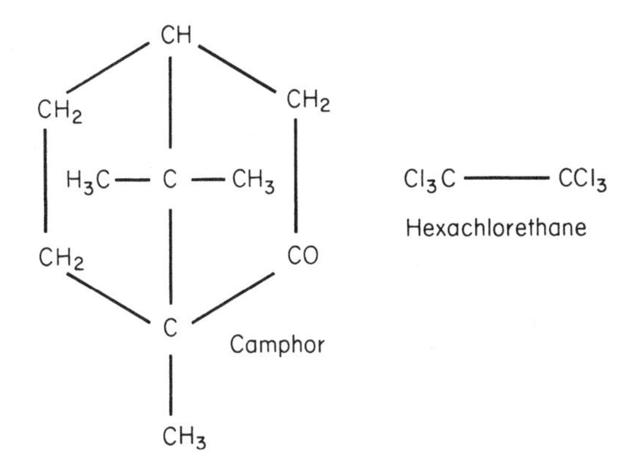

Fig. 8.12 Two substances whose smell is extremely similar though their shapes and chemical properties are entirely different

A second theory that was devised partly to get round this problem of lack of consistency between molecular shape and odour-quality is the infra-red vibrational theory developed by Wright (1964). The basic notion here is that while odour must ultimately be determined by chemical structure—the structure after all defines what the chemical is—overall shape is by no means the *only* property of a molecule that is determined by its structure. All molecules undergo mechanical vibrations, and the frequencies of these vibrations lie mostly in the infra-red region, and depend in a rather complex way on the molecule's structure. In principle one could certainly imagine a sensory system that analysed the spectrum of these frequencies of vibration, perhaps through receptors tuned to different wavelengths. In such a case one might well find two substances with similar shape but different smells because their frequencies of vibration were different; or conversely, substances sharing particular vibration frequencies and thus smelling similar, but of very different overall shape. There are a number of examples of the latter phenomenon that lend some support to the theory: nitrobenzene, benzonitrile and alpha-nitrothiophen, all of which smell of bitter almonds, happen to have many of their vibrational frequencies in common, but have widely differing shapes. Optical isomers necessarily have identical vibrational frequencies, and smell identical as well. Recently, further evidence supporting this idea has come from a study of the curious way in which male moths are attracted by candles. It turns out that candles have characteristically spiky infra-red spectra, and that a number of the emission lines coincide with the vibrational frequencies of the female moth's pheromone: other sources such as hurricane lamps that emit as much infra-red but lack the spikes are much less powerful attractants. So it seems that the suicidal fascination of the candle for the moth is a sexual one: the candle is the *flamme fatale* of mothdom! But while the theory has a certain plausibility as far as creatures like moths are concerned, in the case of warm-blooded animals there is a grave physical objection: it is very difficult to see how an olfactory system working with infra-red radiation could function with such exquisite sensitivity—responding to single odorant molecules—in the presence of the inevitable background 'noise' generated by the body's own heat. It might conceivably be that the pigmentation of the olfactory epithelium could play some role in the absorption of radiant energy, and its presence is otherwise somewhat puzzling. But few physiologists would care to accept the infra-red theory as the explanation of olfaction in higher animals.

The third theory is due to Davies and Taylor (1959), and is less exotic. The notion here is that an olfactory molecule may be adsorbed onto the receptor surface membrane and then pass through it, leaving in its wake a temporary hole through which ions may flow and produce a local depolarization (Fig. 8.13). Different sizes of molecule will produce different-sized holes, and the length of time the holes remain open will depend on such factors as the diffusion rate across the membrane and the affinity of the molecule for the membrane itself. Indeed it is possible on the basis of the theory to calculate what the expected relative thresholds ought to be for different classes of

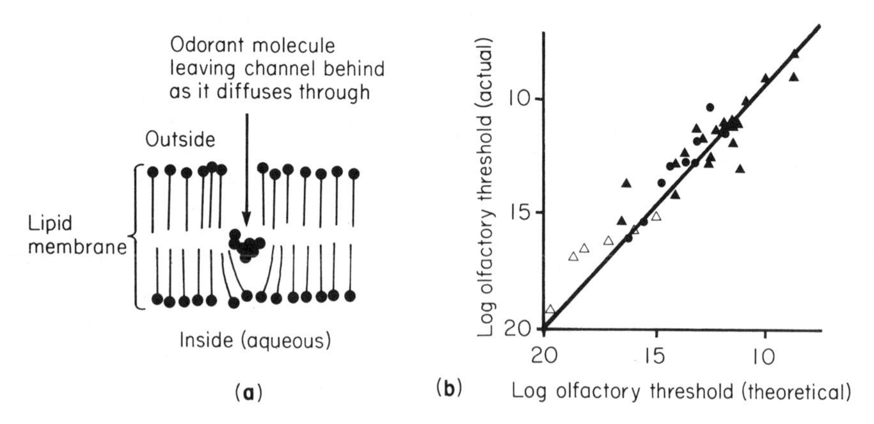

Fig. 8.13 Davies and Taylor's theory of olfactory transduction. (**a**) An odorant molecule crossing a cell membrane and leaving in its wake a temporary hole through which ions may flow. (**b**) Comparison of olfactory thresholds for a number of classes of organic compounds with the prediction of the theory. (After Davies and Taylor, 1959 and Davies, 1969)

chemical substance: these predictions turn out to be remarkably good in a large number of cases (Fig. 8.13), providing the main support for the theory. The fundamental difficulty with it is that there is no hint as to how different molecules give rise to different *qualities* of sensation, and in that respect the theory is not a very satisfying one. Penetration of the cell membrane by the molecule may simply be a preliminary to the transduction process itself, of an entirely different nature, which might explain to some extent the accuracy of the predicted thresholds. It may very well be that Amoore's theory is basically correct, but with many hundreds of different receptor sites rather than just seven, whose affinity for different molecules is determined by something rather more sophisticated than simply their overall shape. There might for example be receptor proteins in or on the cell, responding to odorant molecules in much the same way that other receptors respond to hormones or synaptic transmitters. If these proteins were distributed in a haphazard way amongst the receptor cells, it would explain very well their chaotic electrophysiology. Finally, it is interesting in this context that many otherwise normal people show specific anosmias—'blindness' to particular smells (that of freesias being a common example)—that are inherited as single recessive genes, suggesting perhaps the loss of a single receptor protein.

Gustation: the receptors

What the man in the street means by 'taste' is actually very largely *smell*, with purely somatosensory contributions such as texture, temperature and even pain (as in pepper) playing a part as well. People with anosmia, perhaps as a result of a cold, find their sense of 'taste' profoundly disturbed: apples taste like onions, vintage port like blackcurrant syrup. In fact the human tongue appears to have only four modalities of taste apart from ordinary cutaneous sensation: salt, sour, bitter and sweet. These four qualities have obvious physiological significance: sweet things are on the whole sources of metabolic energy; bitterness is usually associated with poisons; sourness is simply a measure of acidity, and salt of [NaCl]. To some extent taste thresholds and preferences are under the influence of the body's state of physiological need: salt-deprived rats show a preference for drinking salt solutions instead of water, and will tolerate strong saline solutions that other rats will refuse to drink. Coal miners, who sweat a lot, often used to put salt in their beer, though to others it tasted revolting.

If one stimulates the tongue with solutions applied locally through small pipettes, it is evident that certain areas are more sensitive than others to particular stimulus modalities (Fig. 8.14), and furthermore that the sense of taste is confined to special structures on the tongue called the *papillae.* In Man, three types of papillae have been described: one of these, the *filiform* papilla, is not concerned with taste at all but is specialized for rasping and particularly well developed in meat-eaters like the cat; *circumvallate* papillae, associated with sour and bitter taste, consist of a sort of dome surrounded by a moat (Fig. 8.15), and in the walls of this moat one can find sensory cells with microvillae arranged in pit-like invaginations, forming *taste-buds*. Similar buds are found in the *fungiform* papillae, which respond to salt and sweet. Unlike the olfactory

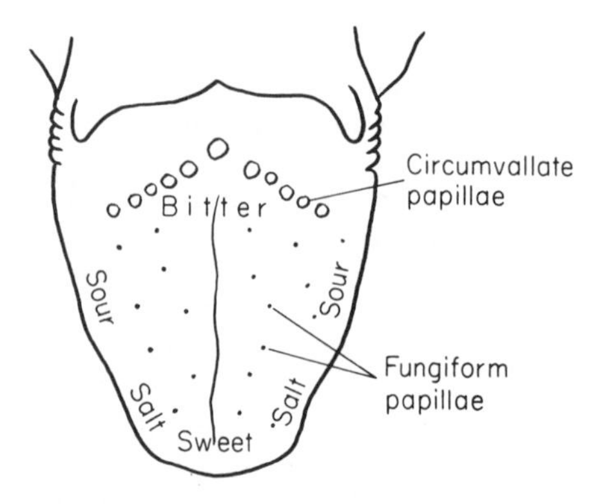

Fig. 8.14 Human tongue, showing regions where the four components of taste are most readily evoked, and the approximate distribution of two kinds of papillae. (After Moncrieff, 1967)

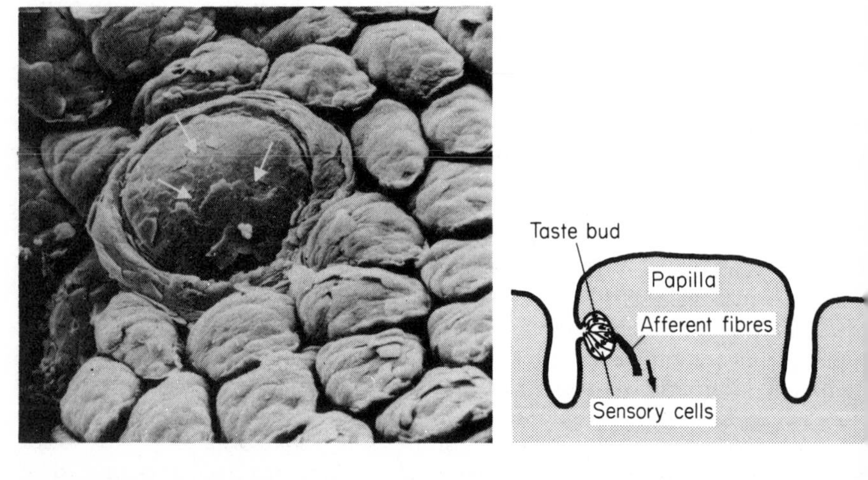

Fig. 8.15 (**a**) Scanning electron micrograph of circumvallate papilla; the arrows show the openings of taste pits or buds on the surface of the papilla (Kessel and Kardon, 1979; copyright W. H. Freeman and Co.). (**b**) Schematic section of such a papilla, showing the sensory cells lying within the taste bud

receptors, these do not send their own axones to the CNS, but are innervated by fibres of cranial nerves VII and IX, whose cell bodies are in the geniculate ganglion and glossopharyngeal ganglion respectively; the tongue is also innervated by the trigeminal nerve (V), providing ordinary somatic sensibility. The afferent taste fibres relay in the rostral part of the *nucleus solitarius*, and again in the medial part of the ventral posterolateral area of the thalamus, whence there is a projection to a small area of the cerebral cortex (Fig. 8.16); thus the whole system is much more like that of ordinary cutaneous sense than is the case for olfaction, though some gustatory projections to the periamygdaloid cortex and other limbic areas certainly exist.

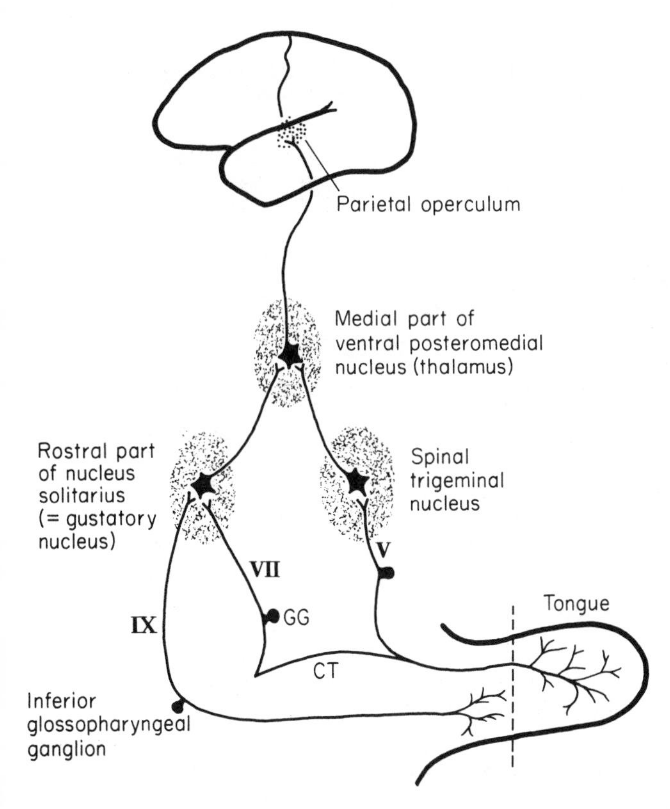

Fig. 8.16 Simplified scheme of the main afferent gustatory pathways (V,VII,IX, cranial nerves; CT, chorda tympani; GG, geniculate ganglion)

Electrophysiology

One branch of the lingual nerve passes rather conveniently in the *chorda tympani*, where it is relatively accessible for recording. Single units show almost the same chaotic properties seen in the olfactory fibres (Fig. 8.17); instead of responding strictly to just one of the four modalities, they tend to respond to a random assortment of them (and often to water as well, which in many species should really be thought of as a fifth gustatory modality). Once again, each unit seems to have its own viewpoint of the gustatory world, and stimuli are encoded in the spatial pattern with which the ensemble of afferent fibres discharges. At the thalamic level (Fig. 8.17) the situation seems hardly less chaotic, although there may be more of a tendency for units to be selective for one type of stimulus.

Theories of gustation

For two of the modalities of taste, salt and sour, the complexities so characteristic of olfaction seem to be absent: sourness depends in a simple way

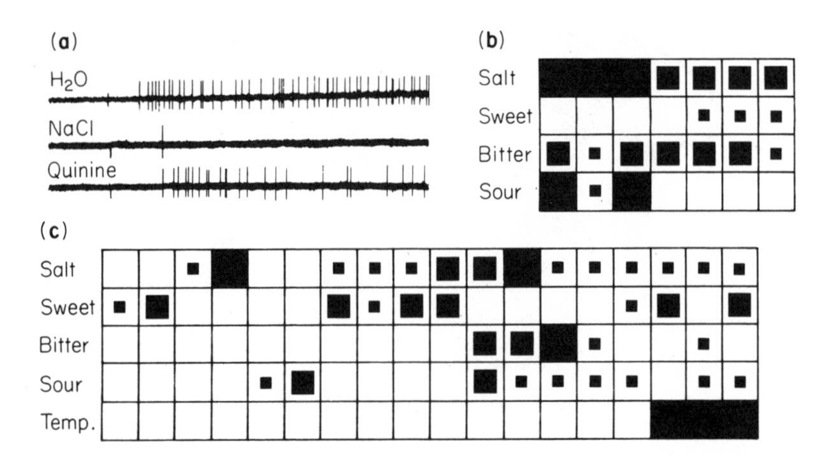

Fig. 8.17 Responses of gustatory neurones. (**a**) Records of activity of a single fibre in the chorda tympani of a cat, showing discharges in response to water and quinine (bitter), but not to salt. (**b**) Diagrammatic representation of the responses of seven individual taste receptors in the rat to four different stimuli, showing the varieties of stimulus preference. (**c**) Similar representation for 18 taste units in monkey thalamus. (Size of square indicates size of response: data from Cohen *et al.*, 1955; Kimura and Beidler, 1961; Benjamin, 1963)

on pH (though not all solutions of equal pH are equally sour), and saltiness is a function mostly of the sodium and chloride ion concentrations, though to some extent of potassium, bromide, iodide and so on as well. It is not particularly difficult to think up plausible receptor mechanisms for such responses. But in the case of sweet and bitter, the situation is much more like that in the nose: once again we find a distinct lack of correlation between overall molecular shape and taste, as for example in the well known artificial sweeteners, whose thresholds are vastly lower than the actual sugars for which the receptors were presumably intended (Fig. 8.18). But the fact that we are

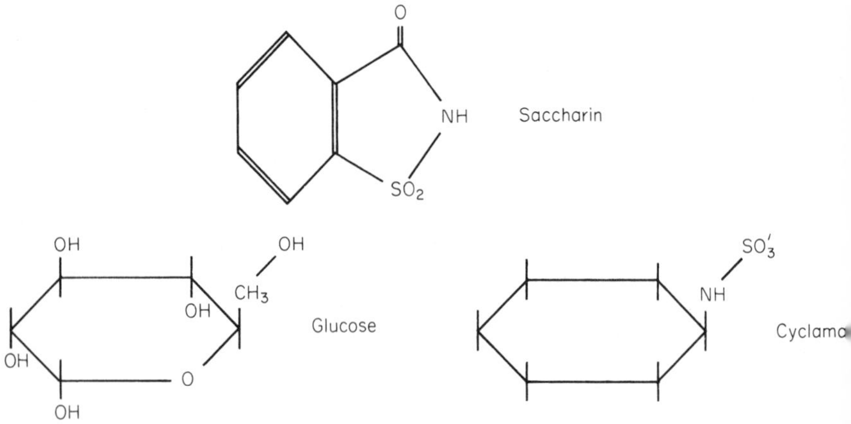

O
NH
SO_2
Saccharin

OH
OH
CH_3
OH
Glucose
OH
O
OH

SO_3'
NH
Cyclama

Fig. 8.18 Three substances that taste sweet to humans

dealing here with only two classes instead of indefinitely many simplifies things considerably, and it is now becoming clear that there are specific receptor proteins which in some cases have been extracted and found to bind to sweet or bitter substances, and are 'fooled' by false stimuli like saccharin in the same way that we are ourselves. These findings give some confidence to the idea that olfactory transduction might also be mediated by specific receptor proteins, although necessarily with an enormously greater repertoire of types of binding site.

9

Types of motor control

Motor systems—unfortunately—are intrinsically rather more complex than sensory ones, and our knowledge of them is correspondingly more rudimentary: it may be helpful to begin by considering why this should be so.

Difficulties in studying motor systems

Now a useful way to picture the nervous system as a whole is as a series of neuronal levels (Fig. 9.1), in which incoming sensory information S is transmitted from level to level, and modified at each stage until it becomes a response R at the output. What the network does as a whole is to generate specific patterns of response when particular patterns of input are present in S: this is achieved because, on a smaller scale, each of the neurones that makes up a particular level is itself only responsive to a particular pattern of activity amongst the neurones of the level immediately in front of it. Though any one

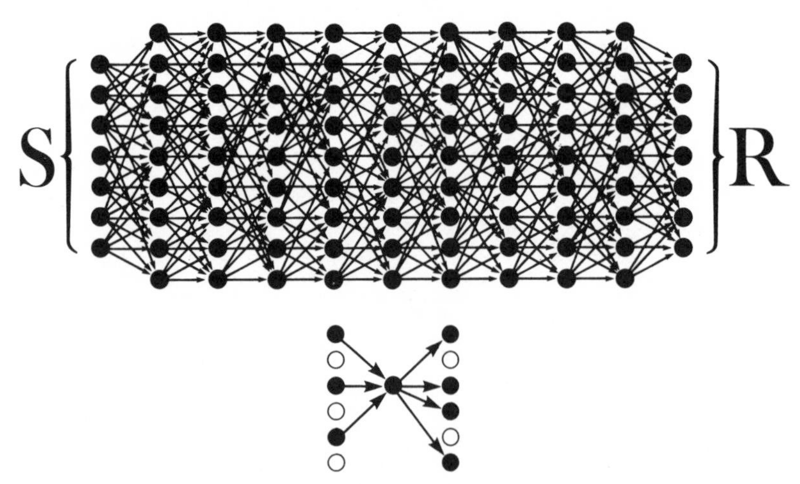

Fig. 9.1 Above, representation of the nervous system as a sequence of neuronal levels, though which sensory patterns (S) are converted into patterns of response (R). Each neurone itself responds to a particular pattern amongst the neurones of the preceding layer (below)

neurone may only be influenced by a tiny fraction of all the activity taking place in the previous level, nevertheless, because of the convergence implicit in the transfer of excitation from level to level, as one penetrates further and further into the model one finds neurones that are capable of responding to more and more complex and wide-ranging aspects of the sensory world. We have already seen this happening in the case of the visual system. Receptors in the eye convey information about only a minute part of the retinal image; but after a few levels have been passed, in the visual cortex, we find units that are able to respond to a specific type of stimulus, such as a moving edge, over wide areas of the visual field. We shall see later that as one gets to the deepest levels of the brain, there seems to be an increasing tendency for neurones to be *multi-modal*, responding not just to one kind of sense such as vision, but to others, such as sounds and to stimulation of the skin, as well. And of course if we consider the final level of all, the muscles that move our bodies, it is clear that they are utterly multi-modal, in the sense that we can learn to activate any particular muscle in response to *any* pattern of sensation, involving any or all of our sensory modalities.

To understand the details of the way in which such a system carries out this conversion of input patterns into patterns of response, we need to know how individual neurones at any particular level respond to different patterns of activity in the preceding level. Now in the case of sensory systems—that is, of the left-hand half of the figure—this presents few problems of experimental procedure. All we need to do is to record from a selection of single units at the level we are interested in, perhaps the visual cortex, whilst applying a variety of visual patterns. We will then discover which kinds of patterns the neurones do or do not respond to, and can then hope to deduce the nature of their functional connections. This task is made easier for us because we have a pretty good idea in advance of what sort of patterns to try, namely those that actually occur commonly in real life. Thus in the visual system it is not unreasonable to start with things like straight lines and edges: there would be little point in searching for units in a cat's cortex that specifically responded to letters of the Greek alphabet! But when it comes to the right-hand side of Fig. 9.1, the motor system, the problem is turned inside out. We now have to stimulate the neurones individually, and see what patterns of movement are generated as a result. The difficulty now is that to get any response at all, we have to provide a pattern of stimulation that will be recognized at the next level down the chain and result in activation of the muscles. Unlike the situation in sensory systems, we do *not* have any idea in advance of what is or is not likely to be a 'physiological' pattern of neuronal discharge to try to simulate. (This is quite apart from the extreme technical difficulty of actually generating such patterns in the first place). Consequently, what is found when trying to stimulate the motor system, particularly in its more central regions, is that single-unit stimulation seldom produces any effects at all, that many areas that are certainly 'motor' are apparently incapable of responding to electrical stimulation with motor responses, and that when responses *are* achieved, they are usually the result of using such large currents spread over such large areas that the specificity of the neurones is swamped, resulting in diffuse and unphysiological responses. Electrical stimulation has not therefore proved to be a very helpful way of studying motor systems; a more fruitful approach is an extension of the sensory method, applying 'real' stimuli to the senses and tracing the resultant activity deeper and deeper through the levels of the nervous system until they emerge again at the motor end. This is not yet technically feasible in systems as complex as those controlling for example the human hand; but where the number of levels is much smaller, as in more primitive brains like those of insects, or in simpler subsystems of the mammalian brain (like those controlling eye movements, which may have as few as three neuronal levels between input and output) the problem is more tractable.

A second source of difficulty in studying the motor system is that whereas sensory systems by and large form a straightforward progression from level to level, a characteristic of motor control is that every action necessarily results in sensory *feedback*. If we raise our hand, there is an immediate influx of sensory activity from the skin, from muscle and joint receptors, from vision and the other special senses as well. So the effects of trying to stimulate a particular region of the motor system are additionally complex: any movement that may result from it also generates new patterns of activity coming as it were up from behind the level at which we are stimulating, that alter the pattern of stimulation we are trying to apply. This feedback from the effects of motor responses is absolutely fundamental to an understanding of motor control, and a good way to begin a study of the motor system is to consider in what ways sensory information may be used to improve motor performance.

The use of feedback

One can of course imagine a motor system that made no use of feedback of any kind. A spermatozoon, for example, progresses by a flagellar movement that is under no immediate sensory influence at all: if it is pointing the wrong way, that is just hard luck! Blind behaviour of this kind can sometimes be seen in animals with much more sophisticated motor systems; a classical example is the nest-building behaviour of the brown rat, described by Lorenz. When a brown rat decides to build a nest, it performs a characteristic series of actions: it runs out to get nesting material, drags it back to the centre of the nest, sits down and forms it into a sort of circular rampart, pats it down and smooths it, and then runs out to get more material; and so on until the nest is finished. This certainly looks like intelligent and purposive behaviour: yet a simple experiment shows it to be nothing of the kind. If a naive rat is not given enough to make a nest, it *still* runs out to grab the (non-existent) material, goes through the motions of dragging it back, forming it into a rampart and patting and smoothing it, even though in reality there is nothing there! And of course we ourselves have all experienced occasions on which we have absent-mindedly carried out some equally complex series of actions—perhaps putting tea in the kettle instead of the teapot—in wholly inappropriate circumstances. Besides, there are many circumstances when our motor systems are *forced* to act blindly because for one reason or another they are deprived of normal sensory feedback; any throwing action comes in this category. When I nonchalantly toss something into a waste-paper basket, I have actually been obliged to work out *beforehand* the precise sequence of motor commands necessary in order to produce the correct pattern of muscular contractions that I need to achieve my goal, and it is clear that once the object has left my hand, no amount of sensory feedback about its trajectory is going to enable me to modify its flight. Motor acts of this type are called *ballistic*—a word meaning 'thrown'—and their control can be represented schematically by a diagram like that of Fig. 9.2. Here one begins with a *desired result*, in this case the presence of the object in the basket. This is translated into an appropriate pattern of *commands*, drawing on a 'library' of *motor programs* suitable for different acts; and these commands produce the *actual result* through their effect on the body's muscles. If the programs that were selected were good ones, then the actual result will equal the desired result. A well known example of such a system is the control of ballistic missiles. Here the operator decides which particular portion of the globe he wishes to destroy, and makes the necessary computations to determine where to point the missile and how large a thrust is required at take-off: but once it is launched he takes no further action—the missile is in any case no longer under his control—beyond hoping that his calculations were in fact correct.

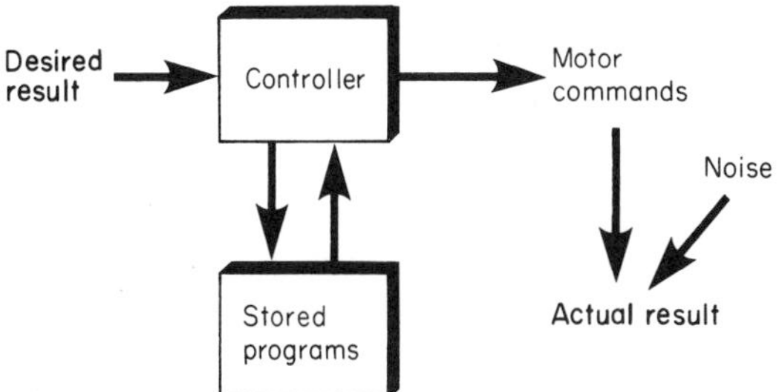

Fig. 9.2 A ballistic control system

Though ballistic control is in a sense conceptually simple, there are two kinds of problem associated with it. In the first place, the calculations that are needed before the action takes place are in general extremely complex—even throwing something in a waste-paper basket requires, in effect, the solution of a set of partial differential equations with countless variables—and it is not altogether plausible that the brain could actually have at its disposal a library of such routines so vast as to be able to deal with *all* the possible motor tasks it might ever encounter during its lifetime. Secondly, and perhaps more fundamentally, such a system is extremely vulnerable to what systems engineers call *noise*. Noise is any kind of unpredictable disturbance that prevents objects or systems from behaving in real life in the satisfyingly regular way that mathematical models and other abstractions do. In this case, the inevitable existence of noise in the outside world means that a particular set of motor commands will never produce quite the same result twice in succession. A given pattern of innervation will produce different movements of a limb on different occasions, depending on the nature of the load to be moved and also on a host of internal factors such as the body temperature, the degree of fatigue, the amount of energy available, and so on. Or again, our ballistic missile, if the wind happens to be blowing the wrong way, may land on Aberdeen instead of Moscow. The problem of coping with noise, and also of having to make complex computations in advance, demands an entirely different mode of control called the *guided system*, which is in a sense the exact opposite of the ballistic system.

Here we start as before with a *desired result* (Fig. 9.3); but this time the first

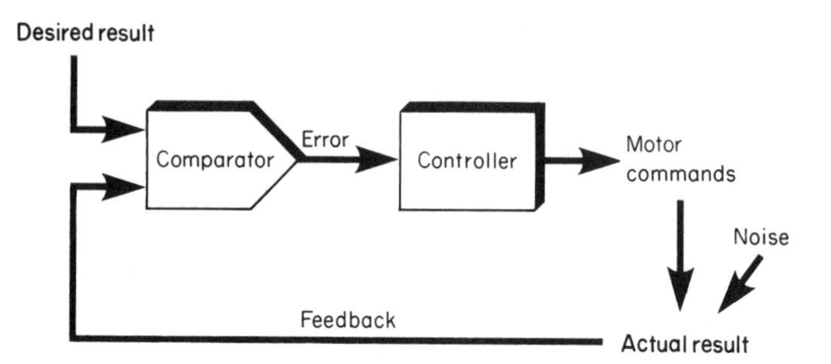

Fig. 9.3 A feedback control system

thing we do is to compare at every moment the desired result with the actual result (as known by feedback sensory information); the difference between the two is a measure of the *error*, and the smaller this difference is, the more nearly we are succeeding in what we are trying to do. Guided missiles are controlled by systems of this type: here the error signal might be something like the angle between the direction in which the missile is pointing and the direction of its target. This error signal is then used to generate motor commands whose function is essentially to reduce the difference between the desired and actual result; the computation of these correcting commands can in general be a very much simpler process than the calculations needed in a ballistic system. Thus in another familiar example, a domestic central heating system, the thermostat acts as the *comparator* of Fig. 9.3, and generates an error signal which consists very simply of one of just two possible messages: either that the actual temperature is below what is wanted, or alternatively that it is above it. The subsequent computations of the motor commands could hardly be simpler: in the former case the boiler is switched on, in the latter case it is switched off. Another, physiological, example also illustrates this essential simplicity of guided systems. If we instruct a subject to look at a small light such as A, Fig. 9.4, and then suddenly move it closer to his nose as at B, we find that his eyes converge smoothly and quite quickly in such a way that in the end the image of the light still falls exactly on the fovea of each retina. The velocity of the convergence movement is quite high at first, but declines exponentially as the eye gets closer and closer to its target. This is rather what would be expected of a guidance system, in which the eyes are essentially driven by an error signal indicating how far the retinal image of the light was from the two foveas. It turns out that this is indeed precisely what is happening, and that it is *disparity* between the two retinal images, presumably

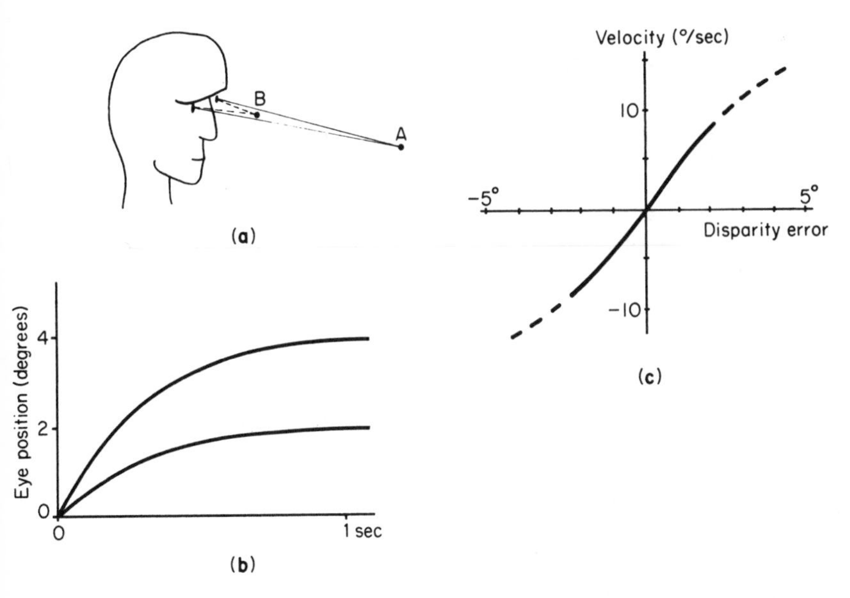

Fig. 9.4 An example of feedback control. (**a**) Movement of a target from A to B elicits a convergence movement of the eyes. (**b**) Time-course of the eye movement in such an experiment, for two different values of the distance AB. The eyes move roughly exponentially towards their target. (**c**) Observed relationship in one subject between disparity error at any moment, and resultant eye velocity, showing nearly linear relationship. (Data from Rashbass and Westheimer, 1961)

sensed by disparity detectors of the kind described in Chapter 7, that provide an error signal which moves the eyes in such a way that their velocity is at all times proportional to the size of the error. As the eyes reach their goal, the error gets smaller, and so the rate of movement correspondingly declines to zero. Fig. 9.4**c** shows the observed relationship between degree of disparity and the resultant velocity of the eye in human subjects, and shows how simply the correcting commands may be derived from error signals in guided systems. The overwhelming advantage of guided systems, however, is perhaps not so much their simplicity, but rather the fact that they are almost immune to the effects of noise. For if something unexpected happens that upsets the normal relationship between command and performance—if someone leaves all the windows open in the house with the central heating system—this fact will be noticed at once (because it will result in a new error) and appropriate commands to achieve the desired result despite the existence of the disturbance will be automatically generated: the theromostat will sense the sudden drop in temperature, and the boiler will automatically be switched on until the temperature once more reaches the desired level. The power and elegance of such a system is thus that it will guarantee to achieve what it has been designed to achieve, even in the presence of types of interference that could not have been anticipated by its creator; it is capable of producing results that *look* intelligent even though—in sharp contrast to the ballistic system with its large library of programs for different occasions—it knows very little (only the size of the error) and remembers nothing! Its weakness is that its proper functioning depends critically on the fast and reliable transmission to the comparator of information about the progress of the action. In physiological systems, where both the sensory receptor processes and the transmission itself may be rather slow, this can be a serious problem; delay of this kind 'round the loop' will mean that instead of responding to the error as it actually is, the system will be responding to the error as it *was* however many milliseconds ago it takes for the information to get back to the brain. For example, we know that the visual receptors are rather slow, and that the shortest reaction times for any kind of vision task are of the order of 200 milliseconds. Consider a batsman in a game of cricket; one might think that he could use a system like that of Fig. 9.3 to bring his bat up to the ball under visual guidance, using error information about the distance between bat and ball. But the existence of this large visual delay means that any such information is hopelessly out of date: if the bowler is delivering at 90 m.p.h. the ball will travel half the length of the pitch in the 200 milliseconds it takes for any visual information about its position to be of use. Thus the last useful visual fix on the ball is when it is still more than 10 metres away; clearly the bat cannot in any sense be guided on to it.

The third type of control system to be considered is a sort of compromise between a guided and a ballistic system, and uses *parametric feedback* (Fig. 9.5). The main forward pathway here is exactly the same as in the ballistic system of Fig. 9.2: the desired result is again converted into a suitable pattern of output commands by means of a set of stored programs. What is new about this system is that it also incorporates feedback about the actual result which is compared with the desired result to provide an error signal: but this error signal, instead of being used to generate commands directly, in the course of the movement, is used instead to modify the parameters of the program itself. This is a system that in effect *learns* to improve its performance through

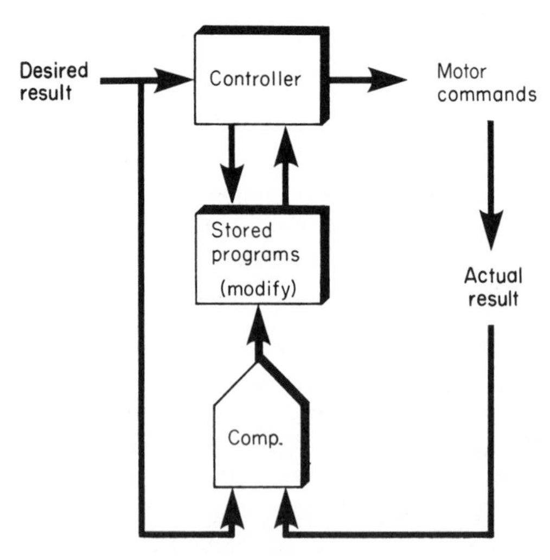

Fig. 9.5 A parametric feedback control system

experience: if our missile *keeps* on landing on Aberdeen instead of Moscow, then the faulty programs are altered in such a way as to improve things in future. There are two advantages of such a system over a simple ballistic one. In the first place, it does not have to have stored programs ready in advance for any conceivable kind of action: by starting with rather simple, all-purpose programs, one may refine, through trial and error, what is needed for the tasks that are actually encountered. Secondly, although it cannot deal with immediate errors introduced by noise, what it *can* do is use error information from one attempt to improve performance on the next. It goes without saying that this kind of behaviour is highly characteristic of the way in which our motor systems learn to execute complex actions. In playing darts, a novice may at first use pre-existing programs developed perhaps from his experience of throwing other objects such as cricket balls (and ultimately from throwing rattles out of his pram): but as he practises, the feedback from each throw, though obviously arriving too late to use immediately, *is* used to reduce future errors, and in the end he may gradually evolve very accurate programs specifically for dart-throwing. A great deal of the learning of motor skills can usefully be thought of as a parametric feedback of this kind, in which errors are used to modify one's stored motor programs.

The final type of control system to be considered is similar to parametric feedback in that it involves modification of behaviour through experience, but has closer affinities with the guided system of Fig. 9.3 than with a ballistic one, and uses *internal feedback*. The notion here is that if it is difficult to obtain feedback about actual results sufficiently quickly for them to be of use during an action, nevertheless it may be possible as the result of experience to *predict* what the result of a particular motor command is going to be, before the actual result is known. From a general knowledge of the mechanical properties of one's hand and arm, and information about the kinds of loads that are present, one can form an estimate in advance of what position the limb is going to adopt in response to any particular pattern of motor commands that is sent to it. Since this estimate is formed entirely within the brain, it may well be available long before any feedback from the actual movement has found its way back from the periphery. When things are happening fast, such an estimate—one may call it the *predicted result*—will at least be better than no information at all. Thus in an internal feedback control system (Fig. 9.6) the desired result is compared not with the actual result but with this predicted result, which is in turn derived by sending a copy of the motor commands (an *efference copy* signal) to a neural model of the mechanical properties of the body, which is used to predict the probable result. A final refinement of this basic system is that the actual result may also—later—be compared with the predicted result, and any errors used to correct the model itself: another variety of parametric feedback, that continually improves the accuracy of the predictions.

One excellent example of a physiological system that seems to work in this way is in the control of saccadic eye movements. Saccades are the eye movements made when a subject shifts his gaze from one target to another at the same distance from him, but in a different direction. Most saccades are made with virtually constant velocity—which may be as much as 900 degrees

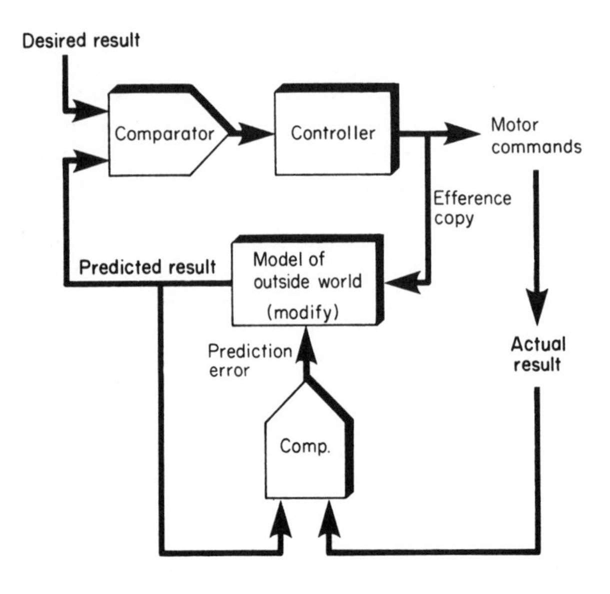

Fig. 9.6 Internal feedback control system, with automatic correction of the model of the outside world

per second—whatever the distance between the targets; thus the duration of a large saccadic movement is a nearly linear function of its size (Fig. 9.7). At first sight, one might think that the eye was simply moving off towards the new target at a constant rate, and that as soon as the visual system senses that the target has been reached, the brakes are applied and the eye comes to rest. But it is easy to calculate—as in the case of the batsman and cricket ball—that this cannot possibly be the true explanation. Because the movement is so extremely rapid, few saccades last more than 100 msec, and most are between 20 and 40; since visual processes in general take considerably longer than this, by the time the brain had recognized that the target had been reached, the eye would have grossly overshot. Thus a simple feedback loop like that of Fig. 9.3 cannot possibly be used. However, the control of eye movements is an almost ideal example of a case where internal feedback can be profitably employed: unlike our limbs, our eyes are not (unless carrying contact lenses!) subject to varying loads, and consequently we can form a very good idea of where the eye is pointing from knowledge of the commands we send it. Thus the internal model of Fig. 9.6 is not difficult to conceive in neural terms, and in fact we shall see later that there is good evidence that efference copy is indeed the means by which we normally sense the position of our eyes. In the saccade, efference copy information can be used in this way to deduce what angle the eye has moved through, and when this calculated eye position is equal to the desired new position previously requested by the visual system, the drive to the muscles is in effect switched off, and the eye comes to rest on the new target; a system, in fact, precisely of the type shown in Fig. 9.6.

On the face of it there is perhaps no obvious advantage is using internal

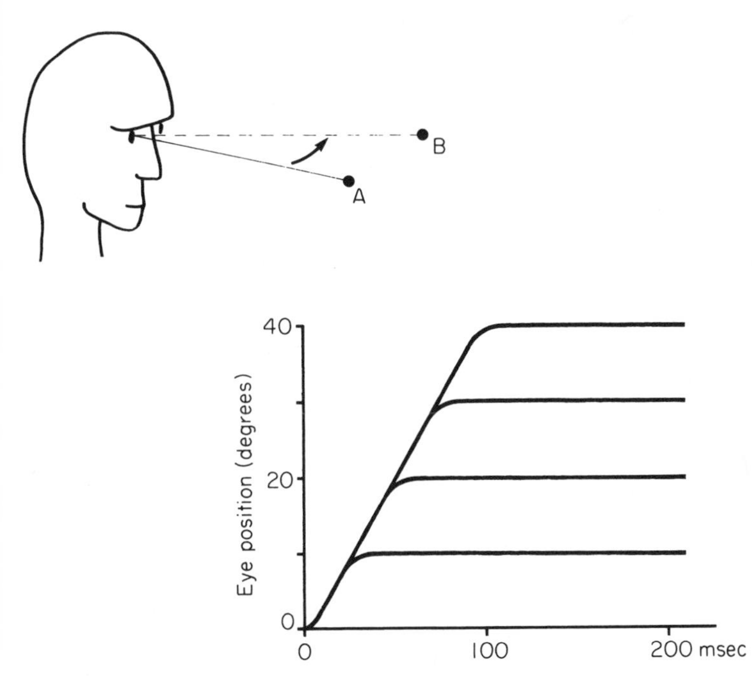

Fig. 9.7 Above, saccadic eye movements are produced when the subject looks from A to B. Below, time-courses of saccades of different size, showing the approximately constant velocity of the movement (somewhat idealized)

feedback rather than parametric feedback. But because the model of the body that is embodied in the former system is essentially a general one, not tied to any particular type of action, it means that experience in carrying out one kind of skilled motor function will benefit the performance of other ones in a rather more direct way than was the case for something like Fig. 9.5, particularly when, as in the case of the eye, the expected result can be computed relatively easily from the motor commands. Learning motor skills then becomes a matter of learning to predict the behaviour of one's own body.

The existence of parametric or internal feedback obviously makes the interpretation of the results of experimental stimulation or lesions no easy matter. It is not at all obvious what the effect would be of artificially stimulating the parametric feedback system of Fig. 9.5 at the 'modify' input. Equally, we would anticipate that lesions in such regions as the library of motor programs, or the neural model inside the internal feedback system of Fig. 9.6, would result in complex and subtle effects: not just simple paralysis, but perhaps loss of *quality* of performance, of the ability to *modify* responses through experience, and perhaps the appearance of rigidly *stereotyped* patterns of behaviour not properly adjusted to their objects. Defects of just these kinds are indeed characteristic of many types of clinical derangement of the higher levels of the motor system.

The hierarchy of control

Finally, another factor that makes both the anatomy and the physiology of motor systems alarmingly complicated is the way in which it has developed in the course of evolution. Whereas the behaviour of the very simplest organisms can be largely described in terms of simple local segmental mechanisms at a peripheral level—as for example the co-ordination of a centipede's legs when it walks—in ascending the evolutionary tree we find more and more domination of the special senses, and as a consequence of this, a corresponding degree of *encephalization*: control by higher centres grouped near these sense organs, in the head. It is important to appreciate that by and large this has been a process of *accretion*. Simpler mechanisms are not in general displaced by more recent ones: they are left essentially intact, but supplemented and controlled from above. They are, after all, carrying out useful functions. Man's walking movements are in essence not so very different from the centipede's, and associated with rather similarly stereotyped sequences of muscle actions mediated by spinal mechanisms of the same general character: such sequences can often be evoked from spinal preparations, animals in which the higher levels of control have been surgically disconnected. It would clearly be foolish for the brain to build its own neural circuits that merely duplicated what the spinal cord was already doing perfectly well, and it is important not to underestimate what the cord is capable of. Classical examples include the spinal dog wagging its tail after defaecation, or the wiping reflex in the frog: if a small piece of filter paper is moistened with acid and placed on its back, it will quite accurately use the nearest leg to wipe it off the skin; if that leg is held down, after a short delay another leg is used! It is clear that one should not think of the spinal cord merely as a sort of speaking-tube down which the brain shouts its orders to the muscles: rather, it provides a repertoire of fragments of action, 'party pieces' that can be called on when necessary by the higher levels. The main difference, in fact, between the spinal cord of a 'higher' and 'lower' animal is that the former in a sense expects to receive more in the way of commands from above; consequently, when isolated from the brain in a spinal preparation, it may *appear* less responsive. This phenomenon is known as *spinal shock*: immediately after making the cut that separates cord from brain, spinal reflexes are depressed or absent, because the usual 'permission' from above is not there. But after a period of time the cord becomes more lively, and may in the end actually show a greater degree of responsiveness than before the operation. This period of time depends markedly on the degree of encephalization: in Man it may take many months; in a dog, days; and in a frog perhaps only a few minutes, reflecting the differing degrees of control normally descending from the brain.

Experiments of this kind lead naturally to the idea of a *hierarchical* organization of the motor system into a series of functional *levels* (Fig. 9.8), the higher levels having more diverse kinds of sensory information at their disposal and therefore able to plan and anticipate more effectively than the lower. Because of their ability to store experience through memory, they can also be more flexible in their responses and learn to conform to the outside world in a way the spinal cord cannot. It follows therefore that as well as being

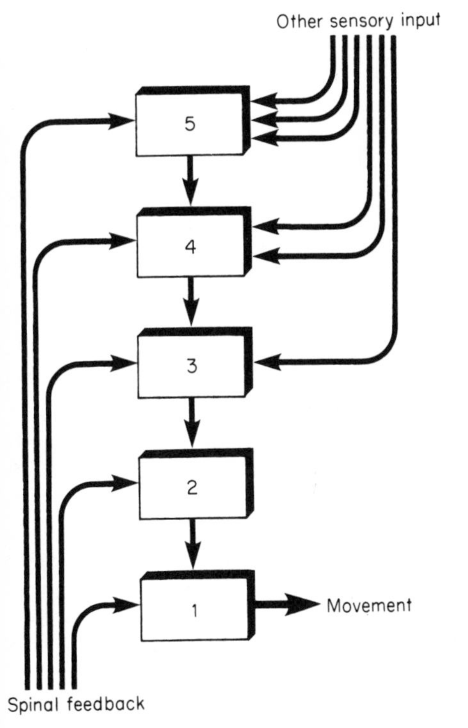

Fig. 9.8 Schematic representation of hierarchy of levels in the control of movement. Higher levels have more access to integrated information from diverse sources; levels on the whole act by controlling the levels immediately below them, rather than generating movements directly

able to stimulate the spinal cord to generate particular patterns of output, these higher levels must also exert a tonic inhibitory influence on lower levels. Brain and cord may well often have conflicting ideas about what is the right thing to do in a particular situation, and this conflict needs to be won by the brain. Consequently the effects of lesions in higher levels of the brain are usually two-fold: first a *loss* of function, particularly of the more flexible and integrated kinds; and secondly the *appearance* of abnormal and more primitive modes of response. The latter phenomenon is often described by neurologists as *release*: the lower centres are released from the restraining influences of the higher, like schoolboys when the master is called from their class.

An understanding of the concept of hierarchical control is essential if one is to make sense of the clinical effects of lesions of the motor system, and analogies with man-made hierarchies are often useful in thinking about them. In an army, for instance, if a general wants to invade Italy, he doesn't himself give detailed orders about who is going to do what, where the latrines are to be dug, and so forth. What he does is to indicate the outline of the general strategy that is required to his immediate subordinates, who elaborate them a little and pass them on to *their* subordinates, and so on until eventually detailed patterns of activity are carried out by the men themselves. The general has the advantage of having integrated information available to him from a wide variety of sources, which he can use to develop wide-ranging strategies: the men have the advantage of immediate and detailed experience of local conditions, with which they can modify their individual behaviour. A consequence of such a hierarchy is that the effects of blowing up a platoon ar very different from those of shooting a general. In the first case, the defect i obvious, immediate, and limited: very specific jobs no longer get done an there is a clear correlation between the 'lesion' and the 'symptoms'. In th second case, at first nothing may appear to be wrong at all: the army sti functions. But gradually more subtle defects may begin to show themselves such as a lack of long-term planning or co-ordination. At the same time, *new* patterns of activity may start to become apparent as the general's subordinate begin to put their own ideas into practice without restraint: symptoms, in other words, of 'release'. These are precisely the kinds of disorders commonly described after damage to higher motor regions of the central nervous system A classical example is the Babinski sign, or 'up-going big toe'. If the foot of a

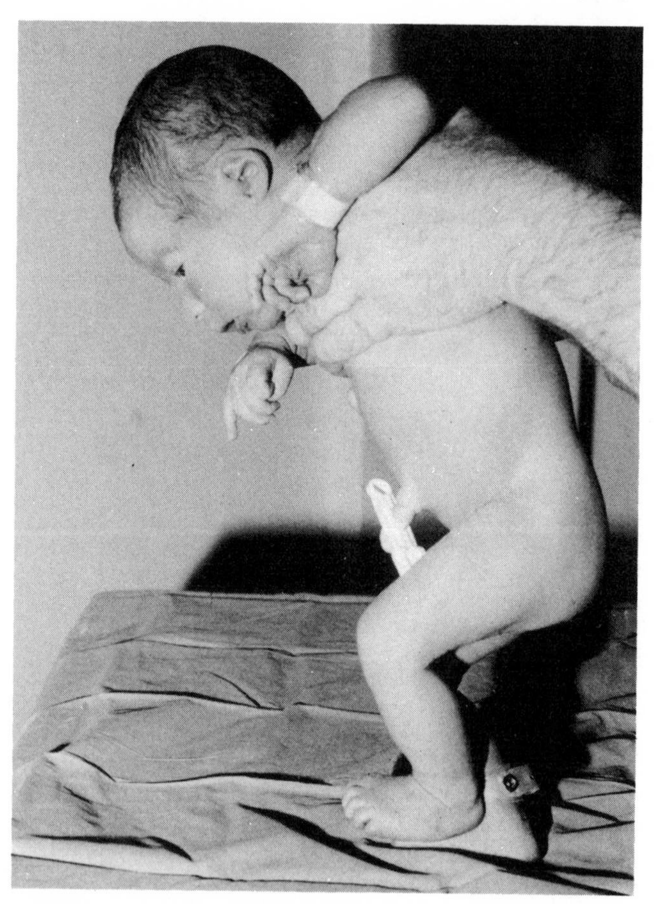

Fig. 9.9 A newborn child walking: this ability will be suppressed within a few days by the brain, even though adult walking patterns are not developed until a year later. (With kind permission of Dr N. R. C. Roberton, Department of Paediatrics, Addenbrooke's Hospital, Cambridge.)

normal adult is firmly stroked, the immediate response is an involuntary flexion of the foot and toes; but in certain kinds of brain damage, as also in newborn children, the reaction is the exact opposite: the toes curl upwards. It is clear in this case that cord and brain have different ideas about what to do when the foot is stroked; in the adult, the brain wins. Another example is the co-ordination of walking movements. A newborn infant is actually able to walk after a fashion (Fig. 9.9), so long as its weight is supported, but one of the first things that the developing brain does is to suppress this primitive response, many months before it develops its own much more sophisticated patterns of walking, that make better use of integrated sensory information.

With this notion of hierarchical control in mind, it is appropriate to begin the motor system by considering what the lowest level of all, the spinal cord, can and cannot do, and the ways in which descending pathways from the brain may control and modify its activity.

10

The spinal level

The final output from the central nervous system to the muscles of the body is from the *ventral horn cells* of the spinal cord: these motor neurones are large, on account of the length of their axons. Each controls a group of individual muscle fibres called a *motor unit*: a unit may comprise as few as five or six fibres, as in some of the muscles of the middle ear and in eye muscles, or as many as 1500 in large and crude muscles such as gluteus maximus. In the cat's leg, a single unit is capable of exerting a maximum force of some 10 g. Gradations of force are brought about partly by changes in the firing frequency of individual motor neurones, and partly by a process of *recruitment*, in which more and more units are brought into play as the required muscle tension increases. Although individual units may sometimes fire at very low frequencies, this does not normally cause discrete twitching of the muscles because different units fire out of synchrony with one another. One of the functions of the Renshaw cells mentioned in Chapter 3 is probably to provide a kind of lateral inhibition between motor neurones that discourages synchronization of this kind.

Ventral horn cells have an orderly and systematic arrangement within the cord, in clumps not quite well enough defined to be called nuclei, that reflects the topology of the muscles they serve (Fig. 10.1). Medial neurones innervate the muscles of the trunk, the most distal parts of limbs are governed by the most lateral neurones, and flexors and extensors tend to be under the control of the more dorsal and ventral groups respectively. On these cells terminate all the afferents from interneurones, and in some cases receptors, that serve the various spinal reflexes and responses, as well as certain of the descending paths from higher levels. Because these cells represent the ultimate gate through which all nervous excitation must pass whenever a motor act is made, whatever its source, together they form what is sometimes called the *final common path*. The quantity of information that thus converges on a single ventral horn cell is enormous, and is reflected in the huge size of its dendritic tree, which may often extend over a large part of the grey matter of the cord (Fig. 10.1).

It is those afferents whose origins, perhaps via one or more interneurones, are sensory fibres that enter the dorsal roots that give rise to *spinal reflexes*. It is never easy to define exactly what is meant by a reflex. Any response that can

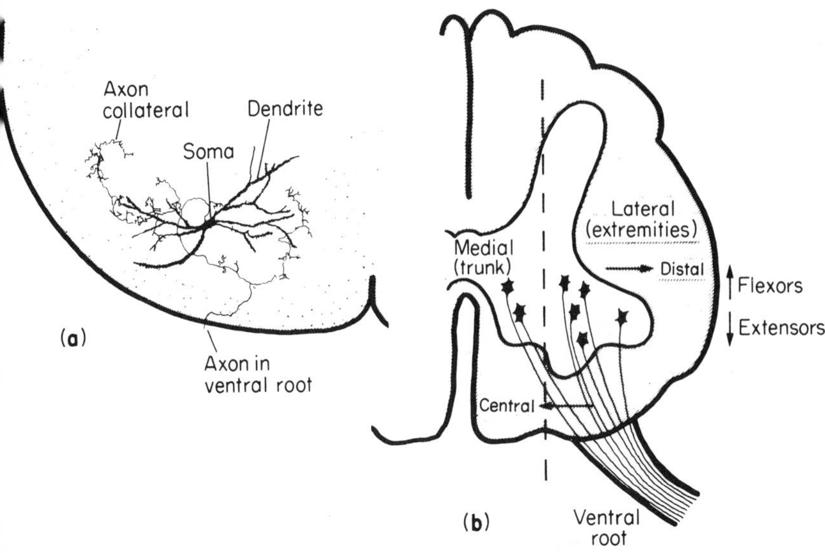

Fig. 10.1 (**a**) Golgi preparation of a single ventral horn cell, showing the enormous extent of its dendritic field (after Scheibel and Scheibel, 1960). (**b**) Schematic representation of localization of motor neurones corresponding to various groups of muscles

be elicited from a spinal animal is certainly a spinal reflex. But we have already seen that although a response may be essentially a spinal one, in the sense that its neural circuitry lies entirely within the spinal cord, yet one may not be able to elicit it in the spinal preparation because the usual facilitating, permissive, influences descending from the brain are absent. It is not quite enough to describe a reflex as an automatic, reproducible, response that is independent of the will, as one may readily influence one's own spinal reflexes by willed inhibition or facilitation from above. For example, there is a mechanism in the cord called the *withdrawal* or *flexion reflex* that causes a rapid flexion when the skin is touched by a hot object or other noxious stimulus: yet we all know from personal experience that in cases of necessity—when carrying a plate that proves to be hotter than we thought when we picked it up—we can (up to a point!) inhibit the reflex completely. We shall see later on that what at first sight appears the simplest and most automatic spinal reflex of all, the monosynaptic stretch reflex, is in fact under almost total control from other, mainly descending, influences. In fact when we look at the mode of termination of the tracts descending from the brain, we find that the great majority of them end, not on the motor neurones themselves, but rather on the interneurones that form part of these reflex arcs. Descending control is not so much of muscles as of *actions*, amounting to a selection from the cord's repertoire: the brain plays on the spinal cord not as one plays a piano, but rather as one selects a disc from a juke box.

Descending pathways

There are five important descending tracts; four of these come from closely neighbouring parts of the brain, in the brainstem and medulla. These are the

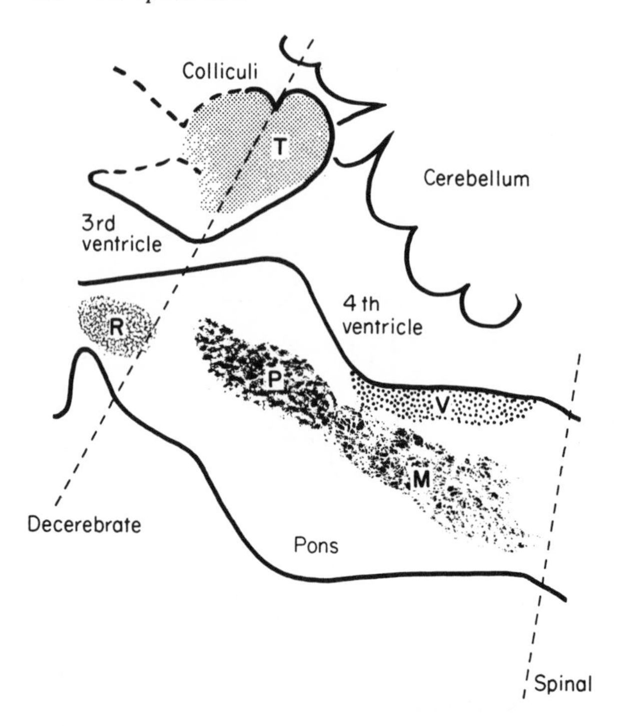

Fig. 10.2 Sagittal projection of structures in the brainstem of the cat, showing reticular nuclei (P, pontine; M, medullary, including gigantocellularis), and other areas that are sources of descending tracts: R, red nucleus; T, tectum; V, vestibular nuclei. The dashed lines show the approximate positions at which cuts are made in the spinal and decerebrate preparations

reticular formation, the vestibular nuclei, the red nucleus, and the tectum; the fifth origin of descending fibres is the cerebral cortex (Fig. 10.2). The *reticular formation* ('rete' = net) is one of the most important and oldest structures in the brain, a direct descendant of the nerve nets controlling creatures like the sea-anemone: in protochordates such as *Amphioxus* all descending influences from the brain have to be relayed via the reticular formation. Although it is a somewhat diffuse structure, comprising both cell bodies and connecting fibres, it is possible to make out certain condensations in it that are effectively nuclei. It stretches from the superior cervical spinal cord up to the intralaminar thalamic nuclei with which it merges at the top. Two areas in particular send important tracts down to the cord: they are the *medullary* reticular formation, especially the *nucleus gigantocellularis* (the largeness of its neurones reflecting of course the length of their axons) that gives rise to the *lateral reticulospinal tract*; and the *pontine* reticular formation, giving the *medial* reticulospinal tract. The positions of these tracts within the cord are shown diagrammatically in Fig. 10.3: both are mainly homolateral with only a little crossing of fibres. Most of the fibres terminate on interneurones rather than on the motor neurones themselves, except for certain fibres of the lateral tract. It is difficult to

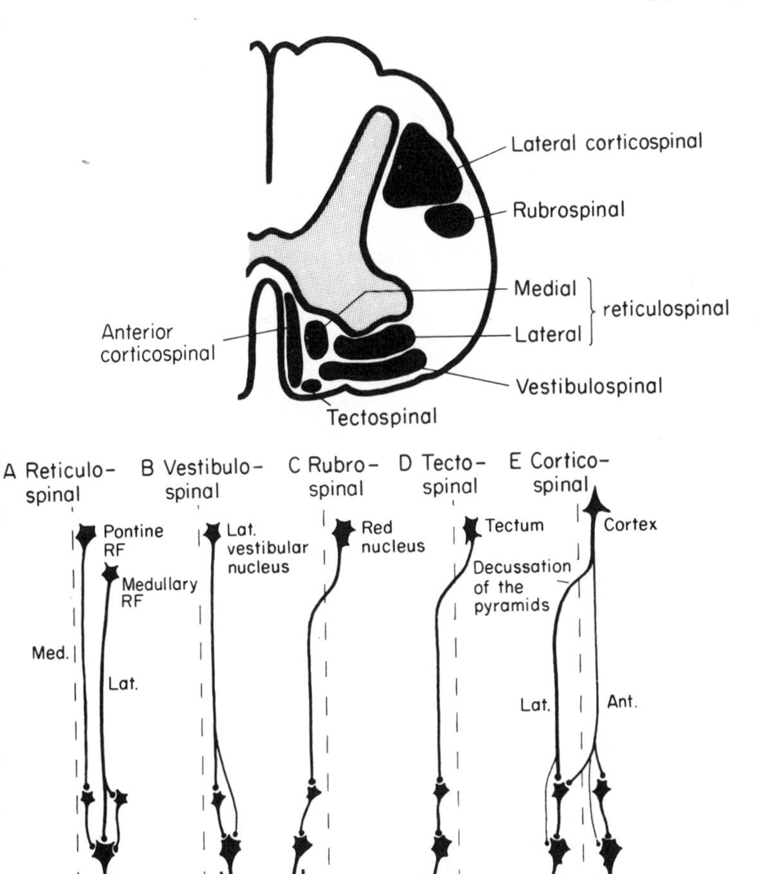

Fig. 10.3 Above, section of cord showing the approximate positions of the major descending tracts. Below, diagrammatic representation of arrangement of the descending fibres in the major motor tracts. RF, reticular formation

generalize about their function, except to say that on the whole they are concerned not with the fine or fast control of skilled movements under voluntary control, but rather with basic, instinctual reactions such as the startle reaction—the dramatic tensing of the whole body seen in response to a sudden stimulus like the sound of a gunshot—and also in some of the postural mechanisms that will be described in Chapter 11.

In the course of evolution, areas of the reticular formation concerned with particular sources of sensory stimulation have tended to condense together as nuclei, and migrate together towards their common source of excitation. The *vestibular nuclei* seem to have emerged in this way under the influence of afferent fibres from the vestibular apparatus, which as we saw in Chapter 5 are concerned with sensing movements of the head and the direction of gravity. At least four nuclei make up the vestibular complex: one of them, the *lateral* or *Deiter's* nucleus, is the origin of an important descending motor tract, the *lateral vestibulospinal tract* (Fig. 10.3B). As might be expected from the reticular origin of the vestibular nuclei, the course of this tract is similar to those of the reticulospinal tracts: virtually all the fibres are uncrossed, most end on interneurones, but some excite motor neurones monosynaptically. A medial vestibulospinal tract also exists, projecting bilaterally mainly to cervical and upper thoracic regions. The functions of these tracts are essentially to maintain posture and to support the body against the force of gravity: consequently they are mostly concerned with the control of extensors rather than flexors. The vestibular nuclei are also closely associated with the *cerebellum*, one of the largest and most important higher motor areas, and their descending tracts may provide one way in which the cerebellum may control the spinal cord: it has no direct projections of its own.

The third descending pathway is the *rubrospinal tract*, which is derived from the *red nucleus* (Latin 'ruber', red), a well defined region lying above the pons in the midbrain (Fig. 10.2). This area, being highly vascular, is pinkish in fresh specimens, giving it its name. Like the vestibular nuclei, it forms an important output relay from the cerebellum, but is not associated with any particular sensory modality; it also receives descending fibres from the cerebral cortex. In animals it is more prominent than in Man, and projects mainly to caudal rather than cervical parts of the cord; it produces flexion rather than extension when electrically stimulated. Its fibres cross at a high level and then descend laterally, as shown in Fig. 10.3C. Its functions are something of a mystery.

The fourth motor tract is the *tectospinal tract* (Fig. 10.3D). 'Tectum' is Latin for roof, and anatomically the tectum is simply the roof of the fourth ventricle, comprising the superior and inferior colliculi in mammals. These are integrating centres for vision and hearing respectively, although as one ascends the evolutionary tree one finds their functions increasingly taken over, or at least supplemented, by the cerebral cortex. They seem to be concerned in particular with orientating responses, as for example in turning to look at the source of a sudden sound. In the superior colliculus one finds neurones that are responsive to visual stimuli in precise locations in the visual field, and also respond to electrical stimulation by causing the eyes to execute a saccade to the very same part of the field. As one might expect from such orientating responses, the tectospinal tract projects no further than cervical segments. The fibres are crossed, and end on interneurones.

Finally we come to a tract that has been investigated to a degree perhaps somewhat out of proportion to its real importance: the *corticospinal* or *pyramidal* tract (Fig. 10.3E). These are some of the longest neurones in the body, since they run from the cerebral cortex in the top of the skull all the way down into the cord. In Man, some 80 per cent cross in the medulla (in other species the proportion is greater: 100 per cent in the dog), and as they lie on the extreme ventral surface (the pyramids of the medulla, hence 'pyramidal'), the decussation can generally be seen with the naked eye. The crossed fibres descend as the lateral corticospinal tract, and the uncrossed ones as the anterior corticospinal tract. Both are relatively recent pathways, their development following that of the cerebral cortex itself, and consequently there is a good deal of species variation as to their size and disposition. Only in Man and some primates do any fibres terminate on motor neurones; in many species they project no further than cervical segments, and in any case form a relatively small component of the total white matter of the cord. Their clinical importance is however very great indeed, since at the upper end, where in ascending from medulla to cortex the fibres fan out into a sheet called the *internal capsule*, in order to squeeze past the thalamus and basal ganglia, they are peculiarly susceptible to damage from vascular accidents, resulting in the well known form of paralysis called *stroke*. The functions of the corticospinal tract will be considered in more detail both later in this chapter and also in Chapter 12; by and large it is concerned with fine, skilled, voluntary movements, and particularly with manipulation.

Some of the tonic actions of these various centres, and their interrelations, can be deduced from classical experiments involving lesions in the brainstem that effectively disconnect them from the spinal cord, or from each other. Despite what might at first seem to be the crudity of the techniques, it is nevertheless possible to come to quite firm, if general, conclusions from them. The *spinal* preparation has already been mentioned; this is one in which all the tracts are cut, so that the cord is completely isolated from the brain (Fig. 10.2). The result of this is a floppy or *flaccid* paralysis, in which there is loss both of voluntary movement and of muscle tone. Whereas a normal person's muscles are tonically excited by a steady level of motor neurone activity, and offer resistance to any movement imposed on the limbs from outside, in flaccid paralysis they are relaxed and offer no resistance at all: thus one may be able to pick up such a patient's arm and fling it in his face, something that cannot be done to a normal conscious subject. Another frequently studied preparation is the *decerebrate* preparation, classically produced by a cut at the level of the colliculi (Fig. 10.2). The effect of such a transection, once the animal is allowed to recover, is utterly different from the floppiness of the spinal preparation: the animal now has muscles that far from being flaccid and relaxed are tonically hyperactive, and the general picture is one of stiffness—*decerebrate rigidity*. The increased tonic activity of the decerebrate animal as compared with the spinal animal must presumably be interpreted as a release phenomenon of the kind discussed in the previous chapter, and due to unopposed activity originating in some structure that lies between the levels of the two cuts. A feature of decerebrate rigidity is that the pattern of stiffness in the legs depends markedly on which way up the animal is (Fig. 10.4), and it seems therefore very likely that the tonic overactivity is essentially due to the influence of sensory stimuli from the vestibular apparatus, normally held in check by centres lying above the brainstem: the rigidity is abolished by lesions in the lateral vestibular nuclei. A kind of rigidity may also be produced by destruction of the cerebral cortex (giving a *decorticate* preparation) rather than decerebration: but in this case lesions of the vestibular nuclei have relatively little effect. This second kind of rigidity—it differs in other ways as well, and is often called *spasticity*—is thought to be due to a tonic excitatory influence from upper areas of the reticular formation, which are disconnected from the cord

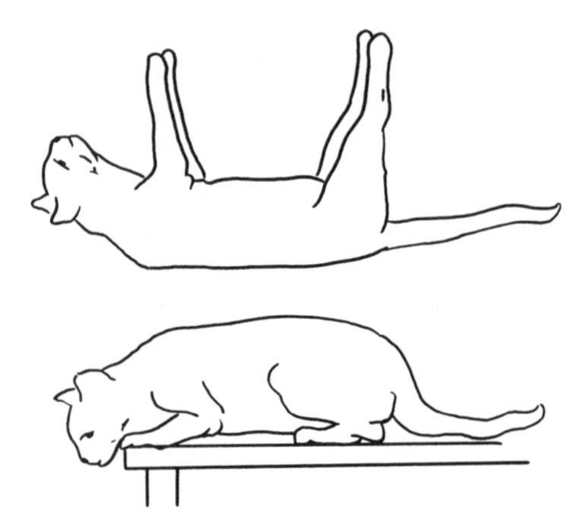

Fig. 10.4 Tone in the limbs as a function of head position in the decerebrate cat (Bell *et al.*, 1961)

in decerebration, and presumably normally inhibited by the cortex (Fig. 10.5). Direct confirmation of this has come from electrical stimulation of the upper reticular formation, which results in a general facilitation both of spinal reflexes and of the effects of electrical stimulation elsewhere in the brain. In this respect, the lower or bulbar reticular formation is exactly the opposite: electrical stimulation causes not facilitation but depression of reflexes and evoked movements. The tonic relationships between these structures that may be deduced from such experiments are summarized in Fig. 10.5; lesions and stimulation of the cerebellum and basal ganglia also influence rigidity, but these effects are more complex and will be described later on. A striking fact

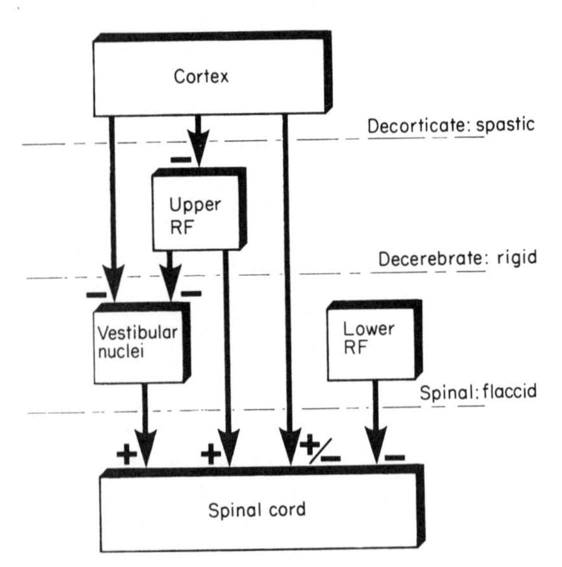

Fig. 10.5 Simplified scheme of apparent tonic facilitatory (+) and inhibitory (−) influences between various central regions, and their relationships to the levels of section in different experimental preparations. RF, reticular formation

about the relation between the higher motor levels and the cord is that the two largest areas of all, the basal ganglia and cerebellum, have absolutely no direct projections to spinal levels: all that they do *must* be achieved by indirect relay through one of the five tracts described above. Furthermore, these tracts themselves act for the most only indirectly, influencing spinal reflexes rather than motor neurones, underlining once again the essentially hierarchical nature of the motor system. The rest of this chapter will be concerned with one particular spinal reflex, the *stretch reflex*, which probably plays a more important part than any other in the control of movements, and illustrates something of the way in which the brain can make use of indirect control of this kind.

Sensory feedback from muscles

In the previous chapter, we saw that there is an intimate involvement of sensory feedback at every level of the nervous system, and examined some of the ways in which this feedback might be used to improve motor control. Here we shall be concerned with the very lowest level of this feedback, that from the muscles themselves. Figure 10.6 shows what an enormous quantity of this

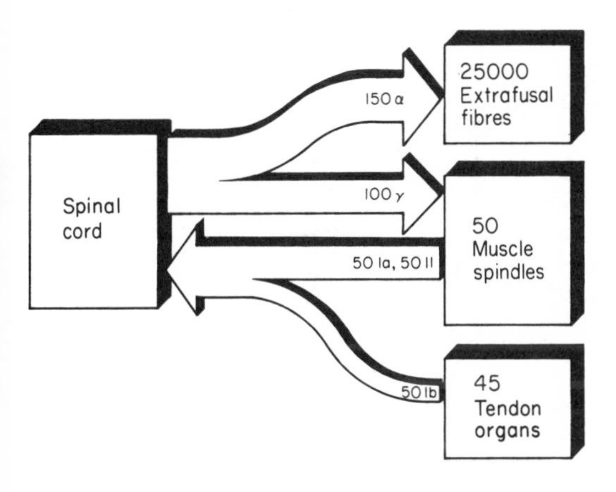

Fig. 10.6 Flow of information to and from a typical cat soleus muscle, showing some 250 efferent fibres and 150 afferents; of the total number of fibres, 250 are concerned with sensory information from the muscle, and only 150 are directly responsible for muscle tension. (Data from Matthews, 1972)

information there is. In this particular instance, compared with the 150 or so fibres that are truly motor, innervating extrafusal muscle fibres, there are some 150 sensory fibres, and another hundred or so γ-fibres, which as we saw in Chapter 5 modify sensory signals from the muscle spindles rather than directly causing contraction of the muscle itself. In other words, some 250 fibres are concerned with afferent information, and only 150 are strictly motor. Of the two types of sensory receptor within muscles, the spindle and

Golgi tendon organ, the latter is less well understood. We saw in Chapter 5 that being in series with the main contractile elements it acts as a *force* transducer; what is not altogether clear is how this information about muscle tension is actually put to use. One reflex for which it is probably responsible is the *clasp-knife reflex*: if you take hold of someone's hand and then push on it in such a way as to bend his elbow—having told him to push back as hard as he can to resist you—there will come a point (assuming that you are the stronger!) at which the force he exerts suddenly seems to give way, and the arm folds up like a clasp-knife. This reflex is thought to be brought about by some such neuronal circuit as in Fig. 10.7, in which the incoming tendon-organ fibres inhibit their parent motor neurone through an interneurone, with some kind of threshold. It is usually claimed that this reflex is protective in function, preventing damage to tendons by pulling on them too hard; if so, it is not very good at its job, since athletes do of course frequently 'pull' their tendons despite the existence of the reflex. It is difficult to believe in fact that this is all the tendon organs do, and a role for them in the control of muscle movement is suggested later in this chapter: it is the spindles with which we are mainly now concerned.

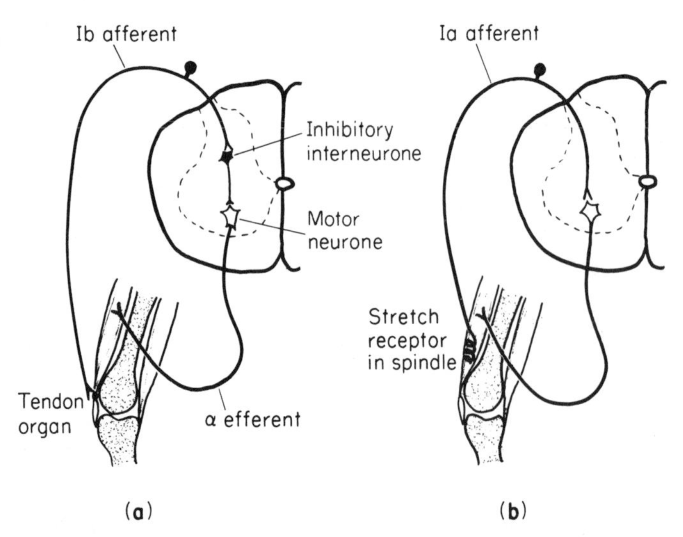

Fig. 10.7 Schematic representation of the neural circuits thought to underly two spinal reflexes. **(a)** The clasp-knife reflex: excessive muscle tension, sensed by the Golgi tendon organs, results in reflex inhibition of the muscle's motor neurones. **(b)** The monosynaptic phasic stretch reflex: rapid stretch activates the Ia fibres, which monosynaptically excite the motor neurones

Spindle reflexes

We saw in Chapter 5 that spindles essentially signal both muscle length and also—especially in the case of the Ia fibres—rate of change of length, and that the messages they convey are also modified by the activity of the γ-efferent fibres to the intrafusal fibres: to a first approximation, their response is a function of stretch plus degree of γ-activity. One may put it round the other way by saying that they signal muscle length, *minus* γ-activity. What do these signals actually do?

One of their best known actions comes about because the Ia afferents monosynaptically excite motor neurones of the same muscle (Fig. 10.7), forming the classical *monosynaptic reflex arc*, the simplest imaginable kind of neuronal circuit that could link a stimulus to a response. The result is that any stretch of the muscle, but particularly a brief, phasic one that will preferentially excite the rate-sensitive Ia fibres, will stimulate the motor neurone and cause a rapid contraction of the muscle. An easy way to elicit such a response is in the familiar *tendon jerk*: tapping a muscle's tendon produces just the right sort of fast-rising stretch to elicit a brisk reflex contraction. The patellar tendon is convenient, and gives an easily noticable response, but other tendons such as the achilles tendon will do just as well. Another effective way of stimulating the Ia fibres is by the use of massage vibrators; much of their 'exercising' effect is due to the fact that they induce tonic reflex contractions. In the normal person, these reflexes are rather feeble, but in certain experimental preparations—and pathological states—they are much increased, which is why they may often be valuable in clinical neurological diagnosis. In the decerebrate animal one may demonstrate not just a phasic reflex of this kind but also a tonic component, in which the muscle responds to *steady* stretch with a steady contraction—the *myotatic* or *tonic stretch reflex*. A record of this kind is shown in Fig. 10.8: in response to the 6 mm stretch shown by the upper line, the muscle responded with a steady tension of nearly 4 kg; some of this, but only a small part, was due simply to the muscle's intrinsic elasticity, which may be revealed if it is paralysed so as to suppress the reflex component. Since in this case a stretch of 6 mm generated about 3 kg of reflex tension, we can say that the *gain* of the reflex—the extra tension evoked per unit of stretch—was about 500 g/mm. The stretch reflex thus makes muscles appear stiffer—less elastic—than they naturally are. In fact it is

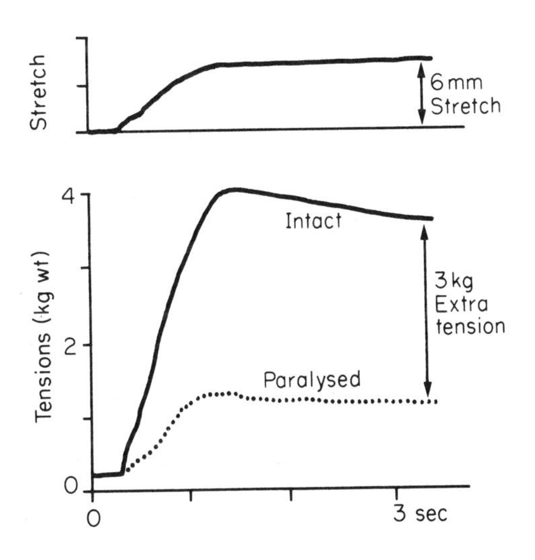

Fig. 10.8 Tonic stretch reflex in decerebrate cat. The muscle was stretched with the time-course shown in the top curve, and the resultant tension was measured (solid line below). The dashed line shows the result of the same experiment when the muscle was paralysed. (After Liddell and Sherrington, 1924)

easy to show that the stiffness of decerebrate rigidity is entirely due to overactivity of the stretch reflexes, probably through excitation of the γ-efferents, for if the dorsal roots are cut in such a preparation, preventing the Ia discharges from reaching the motor neurones, rigidity vanishes. Decerebrate rigidity might therefore be described as a sort of hyper-stretch-reflexia.

The servo hypothesis

Thus the first notion about the function of muscle spindles that came to be accepted was that by acting through the stretch reflex they were responsible for the generation of *muscle tone*, the constant muscular activity that is necessary as a background to actual movement in order to maintain the basic attitude of the body, particularly against the force of gravity. But tone is something that essentially opposes movement, that tends to keep muscles at preset lengths by making them resist any changes. Hence the idea arose that during movements one would have to alter the degree of tone in step with the movement if there was not to be a degree of conflict between the two, and that the γ-fibres were ideally suited to doing just this. If every time a command was sent via the α-motor fibres (the ones innervating the extrafusal fibres) to make the muscle contract, the γ-fibres were simultaneously activated in order to shift the operating region of the stretch reflex to a shorter muscle length, then all would be well. It then became apparent that one could carry this line of thought a stage further and envisage an even more active role for the γ-fibres than this. Imagine for a moment that the γ-fibres were stimulated *without* simultaneous direct activation of the α-fibres. What would happen? By shortening the intrafusal fibres of the spindle, such a stimulus would result in excitation of the sensory afferents just as if an actual stretch had occurred (Fig. 10.9); consequently there would be a reflex activation of the α-motor neurones, and the muscle would contract automatically until the stretch receptors found

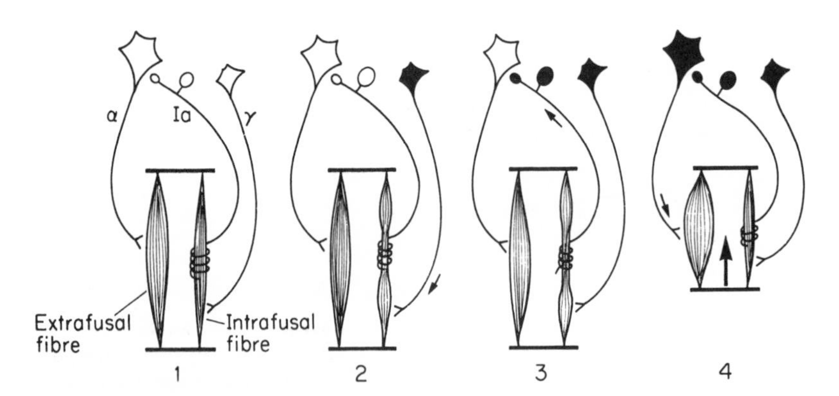

Fig. 10.9 Reflex contraction via γ-fibre stimulation. 1, muscle at rest; 2, activation of γ-fibres stretches sensory endings in spindle, leading (3) to excitation of spindle afferents which in turn (4) cause excitation of α-motor neurones and contraction of extrafusal muscle fibres

themselves back at their original resting degree of stretch. In other words, γ-stimulation could in principle *initiate* contraction, in exactly the same way as direct α-stimulation. But would there be any point in such a roundabout way of making muscles contract?

The answer is that there would. We would now have, in effect, a feedback or guided system in which the γ-fibres would signal 'desired length', the spindles would function as comparators and generate an error signal amounting to something like the difference between desired and actual length, and this error signal would be translated by the stretch reflex into tension, generated by the extrafusal fibres, that would bring the actual length into correspondence with what was desired (Fig. 10.10). The advantage of such an arrangement, like all feedback systems, is that it automatically allows for noise, in this case the existence of unpredictable loads that have to be moved.

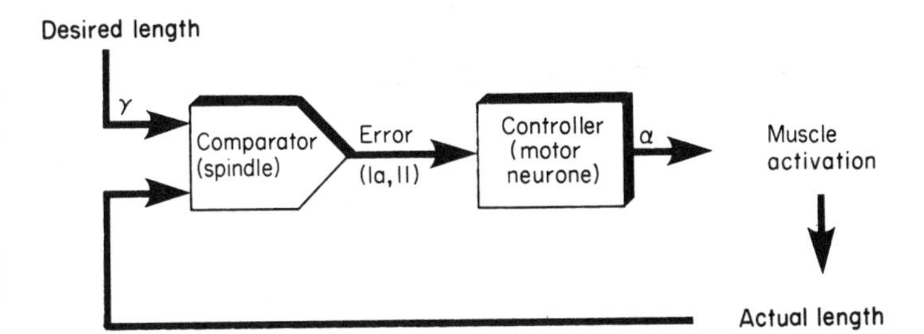

Fig. 10.10 How the spindle might function as a comparator in a simple servo-system in which a muscle's length is automatically made to conform to the desired length signalled by the γ-efferents (see Fig. 9.3)

Consider for example the problem posed by holding out a cup while someone is filling it with tea: clearly, one's task here is to keep the various muscles concerned at a constant length, despite the fact that the force required to do this is continually increasing as the load—the amount of tea in the cup—gets bigger. Now one could of course imagine the brain continually monitoring the situation, and deciding at every instant exactly how much direct α-excitation to send down into the cord to keep the hand steady. But how much simpler it would be just to send, once and for all, a message indicating not the force needed but the desired *position* of the hand, and leave the spinal cord to get on with the job of adjusting the force to the load automatically, by sensing the extent to which the actual position of the cup matches the brain's command. In other words, such a feedback system would provide load compensation, by acting as what is sometimes called a *follow-up servo*: the main muscles simply act as slaves that follow any length changes signalled to the intrafusal fibres.

Granted that the γ-fibres could in principle initiate movements of limbs on their own, is there any evidence that they in fact do so? For certain types of movement, the answer is a clear yes. Figure 10.11 shows one such instance: we noted earlier that moving the position of a decerebrate cat's head causes reflex changes in the tone of its limbs; (**a**) here shows the discharge from a spindle afferent from the leg during such a stimulus, and it can be seen that the

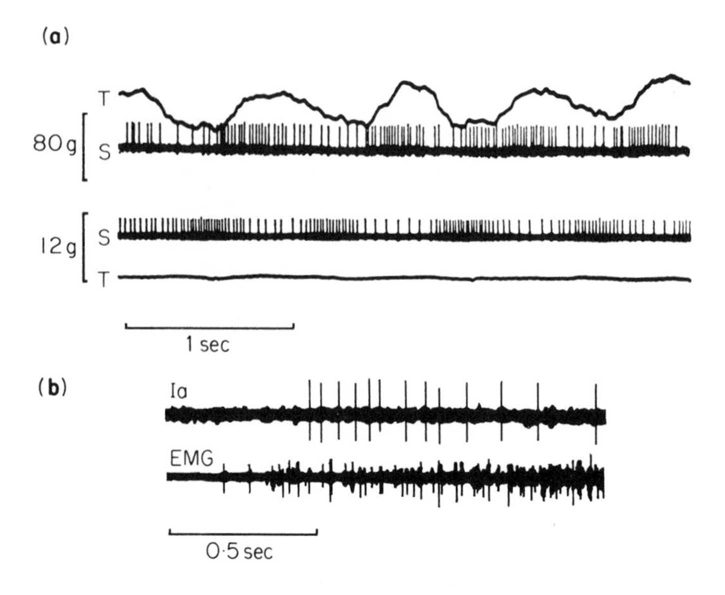

Fig. 10.11 (**a**) Demonstration that γ-fibres may be used to generate reflex movements. Above, tension (T) in decerebrate cat soleus muscle, and associated firing of a spindle afferent (S) during reflex contraction evoked by head movement. Below, after cutting the dorsal root, no contraction occurs, yet modulation of spindle discharge is much as before: this can only be through γ-activation (after Eldred *et al.*, 1953). (**b**) Records of Ia discharge and electromyogram (EMG) during voluntary human wrist movements; the Ia discharge clearly does *not* precede the muscle activity, as would be expected if the muscle were only driven by a servo-system like that of Fig. 10.10 (after Vallbo, 1971)

frequency of firing is modulated in time with the head movement. This in itself proves nothing, since it could merely be the result of changes in length of the muscle, rather than activation of γ-fibres. But if the dorsal roots are cut, preventing the stretch reflex from operating, one finds first of all that the limb movement is abolished—indicating that it must have been driven through the reflex rather than directly by descending pathways activating α-motor neurones—and secondly that the spindle discharge is *still* modulated by the stimulus. Since the muscle length is no longer changing, the only possible way in which this modulation can be taking place is by varying activation of the γ-fibres. In other words, it is quite certain in this case that the movement is indeed *initiated* by γ-activation of the stretch reflex servo. Of course, a decerebrate cat with its dorsal roots cut is hardly in a physiological condition: but other experiments in conscious human subjects have also demonstrated that γ-fibres may on occasion be used in voluntary movements. Figure 10.11 shows the results of an experiment in which recordings were made in Man of activity in Ia fibres during voluntary movement of the wrist, together with the electromyogram from the muscle itself; it can be seen that although it is shortening that is taking place—which by itself would of course reduce Ia activity—nevertheless there is an *increase* in spindle discharge associated with the movement, implying that stimulation of γ-fibres must be taking place as well. However, experiments of this kind, though demonstrating that γ-activation occurs during voluntary movements, also reveal some serious discrepancies with the notion that this activity is actually the *cause* of the movement. If the servo hypothesis were true, the sequence of events that we would expect to observe would be first activation of the γ-fibres, then the resultant Ia activity, and only *then* would we expect to see the discharge of the α-fibres and of the main muscle fibres themselves. But actual recordings of the relative timing of these events show that this is not what happens at all (Fig. 10.11): the Ia discharge actually occurs *after* the contraction has commenced, and therefore cannot possibly be its cause. Further, there is a theoretical consideration that casts doubt on the possibility of driving real movements solely by means of γ-activation, and that is the size of the *gain* of stretch reflexes; 'gain' here means how much extra tension is generated by a given error in length. Returning to the problem of holding out the teacup, it is a relatively simple matter to work out what gain this reflex would need to have in order to perform adequately (Fig. 10.12). From our definition of the gain,

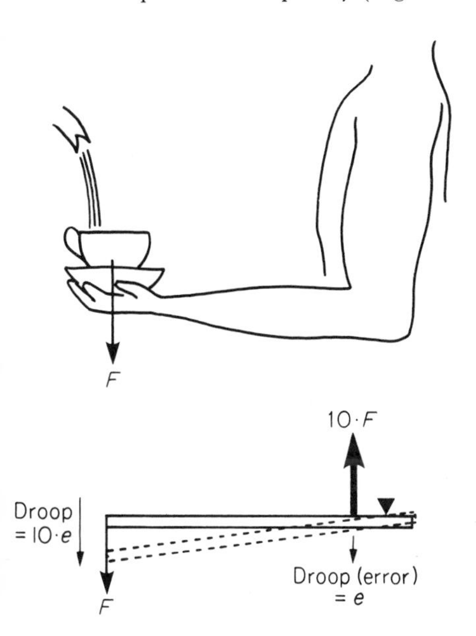

Fig. 10.12 The problem of holding a cup. Taking the mechanical disadvantage of the lever system of the forearm as 10, then a weight F in the hand must be balanced by a tension $10.F$ in the muscle. At the same time, an error e in muscle length will result in a descent $10.e$ by the cup. Thus if the static gain of the stretch reflex is G, the cup will fall by $100/G$ for every unit of weight

G, it follows that for every degree of error e (that is, for every value of the difference between actual and desired muscle length) there is a corresponding reflex force F that is developed by the muscle, where $F = e.G$. So when a muscle is subjected to a load F, its actual length will be slightly greater than what is desired, by an amount $e = F/G$. In the case of the muscle shown in Fig. 10.8, this means that a load of 500 g would cause an error of 1 mm, the stretch required to generate the 500 g needed to sustain the load. Now most muscles, because of the way they are attached to their bones, work under a considerable mechanical disadvantage: in the case of the human biceps, every kilogram of load on the hand results in some 10 kg of tension in the muscle; and conversely, a muscle movement of 1 mm moves the hand by some 10 mm. If for the sake of argument the gain of the stretch reflex being used to hold the teacup was also 500 g/mm, then this means that every 50 g of load in the *hand* will cause a muscle extension of 1 mm, and the hand will move some 10 mm. In other words, the extra 300 g or so produced by filling the cup would be expected to make the hand droop by no less than 6 cm! Somehow, the gain of the system must be automatically increased in step with the changed requirements as the cup is filled, perhaps as the result of pressure receptors in the hand, or by monitoring the tension in the muscle's tendons by means of the tendon organs. If for instance the gain was always increased in such a way as to be proportional to the load, then the ratio *F/G*—in other words the error or droop—would always be constant, regardless of the load applied. Recent experiments carried out in human subjects suggest that something of the sort does indeed occur. Here a subject was required to use his thumb to push a lever at a constant velocity against a fixed resistance: he was provided with visual feedback from the lever to tell him how well he was doing. Recordings of his electromyogram (Fig. 10.13) show a steadily rising averaged activity during the course of the movement, as the muscle shortens. If now, without the subject's knowledge, a stop is introduced into the apparatus that prevents the lever moving past a certain point, one finds that after a short latency the EMG quickly rises, reflecting the subject's effort to overcome the unexpected obstacle. In fact the latency is too short to be due to a conscious decision of this

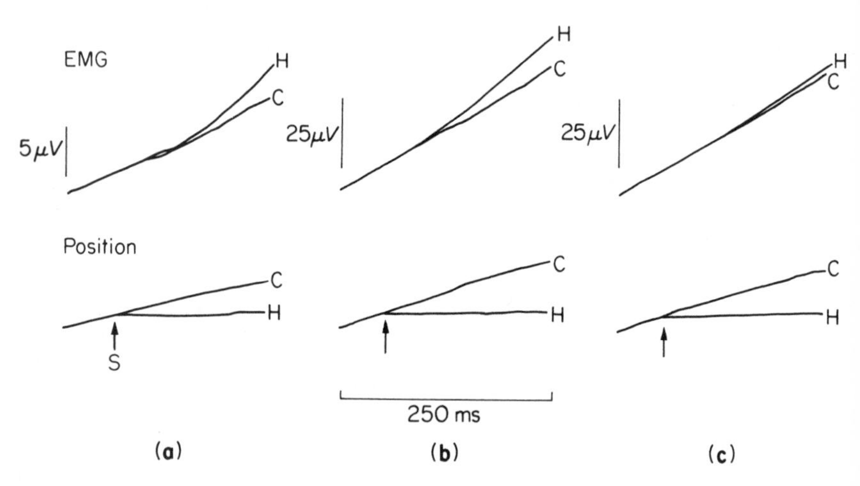

Fig. 10.13 Evidence for variable gain of stretch reflex. (**a**) A subject moves his thumb (lower trace shows its position) against a steady load so as to track a uniformly moving target; the resultant steady increase in EMG is shown above (C = control). If a stop is now introduced at the point *S*, after a latent period the EMG starts to rise more rapidly (H = halt trial). (**b**) If the load against which the thumb is pushing is increased by a factor of 10, the EMG in response to the error also increases by nearly the same factor. (**c**) as (**b**), but with the hand anaesthetized, resulting in almost complete abolition of the stretch reflex. (Note the change in scale of the EMG in (**b**) and (**c**); each trace is an average of 8 trials. After Marsden *et al.*, 1972; copyright Macmillan Journals Ltd.)

kind, and a simpler explanation is that because of the stop there develops an increasing error between desired and actual thumb position—the former increasing steadily, and the latter having stopped—and this causes a stretch reflex. This reflex cannot be of the ordinary tendon-jerk kind, however, because its latency is considerably longer. We can work out the gain of this response by seeing how much extra activity we get for a particular value of the error; what is found is that if the experiment is repeated with different degrees of resistance to the thumb movement—with different loads, in other words—the gain increases with increasing load in any particular trial (Fig. 10.13**b**. That this change in gain is caused by pressure receptors in the skin is suggested by the fact that if the thumb is anaesthetized the gain drops nearly to zero (Fig. 10.13**c**).

Servo-assistance

Consequently one is forced to conclude that in the control of movements there are two separate signals or commands that are sent to the spinal cord by the brain. One is a *position* command, that indicates, via the γ-fibres, what the desired length of a muscle is to be; the other is a *force* command, indicating (by control of gain) what load is to be encountered. In the case of the teacup, the latter information could be obtained from receptors in the skin, or for that matter from Golgi tendon organs: but many tasks are more ballistic in nature and require anticipation in advance of what the load is likely to be. In such cases past experience and the more sophisticated use of special senses like vision may be brought into play as well. When we go to pick up a sack of potatoes as opposed to a sack of waste paper, or when we fling open a swing door with which we are familiar—too little force, and we walk into it: too much, and we smash it—we have clearly estimated beforehand what the likely load will be, and thus what level of gain is required. This notion of simultaneous force and length command is sometimes called *alpha/gamma coactivation*; if the estimate of force is an accurate one, then the system behaves, in effect, ballistically, and there is no error for the spindles to have to correct. The stretch reflex then has only to deal with any residual errors left over after the estimated force is put into operation; it is not expected to provide the *whole* force necessary for the job, which we have seen it is too feeble to provide. A system of this kind is known as a *servo-assisted system*, and may be represented by an arrangement like that of Fig. 10.14: it is really a ballistic system with a safety net provided by a back-up guided system. In such a system the spindles play two distinct roles. In the first place, through the stretch reflex, they provide immediate correction of any errors in the estimate of force; in the second place, they may also supply *parametric* feedback that in the long term can make future corrections of the estimates themselves. We shall see later that there is reason to think that the origin of the force command may be the cerebral cortex, via the corticospinal tract, and that it may be the cerebellum, which is richly supplied with afferents from muscle spindles (unlike the cortex) that is the site of the motor program store. These ideas will be discussed later, in Chapter 12.

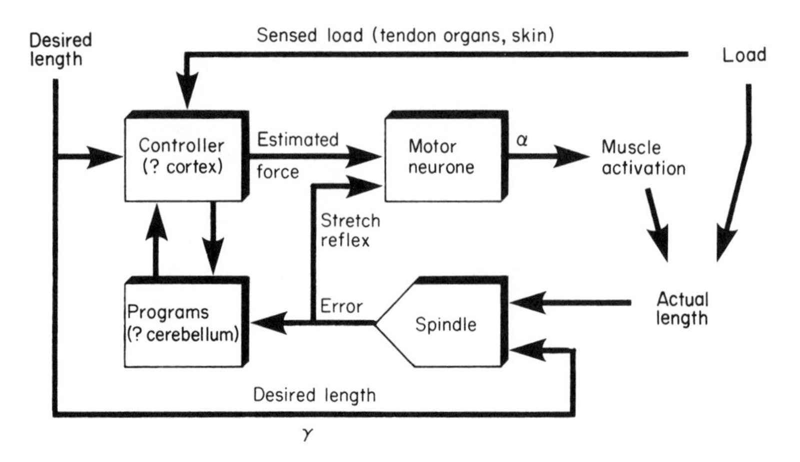

Fig. 10.14 Hypothetical scheme of a servo-*assisted* control system for muscle length. Spindles provide an error signal (discrepancy between actual and desired length) that is used directly in the stretch reflex to correct the response, and also indirectly to modify the ballistic programs for future action. The brain signals to the cord not only desired length but also an estimate of the force needed to achieve that length, derived partly from stored programs embodying previous experience, and also partly from immediate information from sensory receptors about the load that is present

Finally, it is perhaps worth mentioning that the muscles of the eye, though richly endowed with stretch receptors and γ-motor fibres, show no stretch reflexes whatever! It seems in this case that because of the predictable relation between force commands and resultant eye position, emphasized earlier, errors arise so seldom that no short-term correction mechanism is needed. It seems very likely that the sole function of these spindle afferents is to provide parametric feedback in order that the essentially ballistic control of such movements as saccades may in the long run be performed accurately.

11

The control of posture

Movement begins and ends in posture: and for most of the time, the motor system is not in fact concerned with moving the body at all, but rather with keeping it still. This is especially true in Man, where constant motor commands are needed to keep him upright on his precarious twin supports against the force of gravity. At first sight one might perhaps think that this was merely a matter of keeping sufficiently rigid, once a stable balancing position had been found. But a standing man's centre of gravity is so high off the ground that this kind of *passive* stability is not enough: one need only compare the ease with which one can push over a tailor's dummy with the near impossibility of doing the same thing to a living person to realize that stability must involve *active* processes as well, that use proprioceptive feedback information.

The importance of support

In physical terms, whether someone falls over or not is entirely a matter of the vertical projection of his centre of gravity relative to his supports (Fig. 11.1). If this line of projection lies within the area defined by the points of contact with the ground, then all is well—small disturbances will result in a turning couple

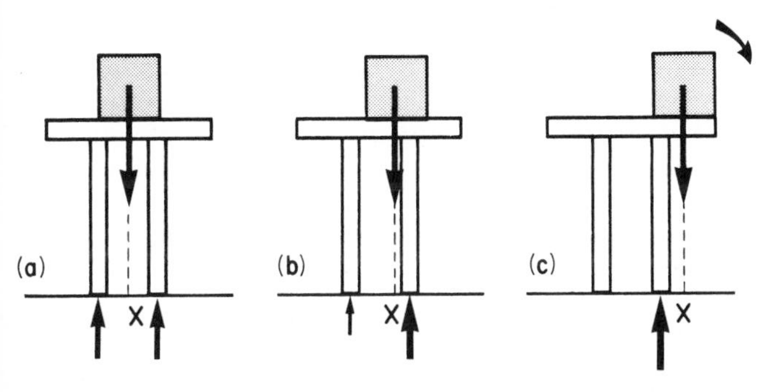

Fig. 11.1 A sufficient condition for postural stability is that the vertical projection of the centre of gravity on the floor (X) should lie within the area defined by the supports. The distribution of pressure between the supports (small arrows) provides sufficient information to determine the position of X

that will tend to restore the status quo. If it lies outside this critical area, then the system is unstable, and any further tilting will cause an ever-increasing couple that will make the person fall over. This critical area is much smaller in Man than in four-footed animals, and maintaining an upright posture is correspondingly more difficult: a tilt of only a few degrees is sufficient to cause instability. Thus proper standing is *not*, as is often implied, just a matter of keeping upright: the man in Fig. 11.2 clearly has an excellent upright posture, but is equally clearly about to experience a postural disaster, because the vertical projection of his centre of gravity lies outside his region of contact with his support. In other words, the control of posture is essentially a matter either of moving the centre of gravity relative to the feet, or of moving the feet relative to the centre of gravity.

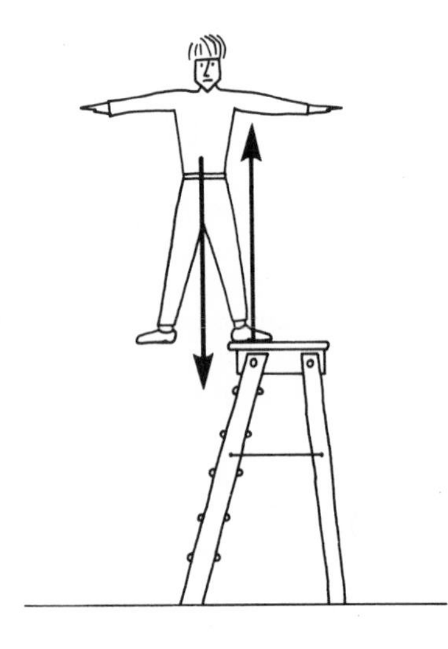

Fig. 11.2 A splendidly upright but nevertheless unsatisfactory posture

There are two sources of information that enable one to do this. The first is the existence of *pressure* receptors in the feet themselves, that provide information about the distribution of support. Knowledge of differences of pressure at different points of support tells us precisely what we need to know to determine our postural state, namely the position of the vertical projection of the centre of gravity relative to the body's supports (Fig. 11.1). The second source of information is the existence of special senses in the head that can tell us about the position and motion of the head relative to the outside world: these senses are the *vestibular* and *visual* systems. It is convenient to consider the use of information from the feet first, since this is a more direct process than the contribution of the special senses.

Imagine that we had to design an automatic system that would keep a lunar module standing upright on uneven ground. One simple solution to this problem would be to equip each leg with a pressure transducer, and arrange things so that any leg that experienced more pressure than the others was automatically extended, and any that experienced less than the others was shortened. The result would be a system that would always bring the vertical projection of the centre of gravity to the middle of the critical area defined by the points of support. It is not difficult to demonstrate a precisely analogous mechanism in four-footed animals: if we suspend a decerebrate animal in the air with its legs hanging down, and press the sole of one of its feet, the animal responds by extending the corresponding limb. This tonic response is called the *positive supporting reaction*; there is also a transient component of the response called *extensor thrust*, and it is often accompanied by stiffening of the limb and arching of the back (Fig. 11.3). It is not difficult to see how this

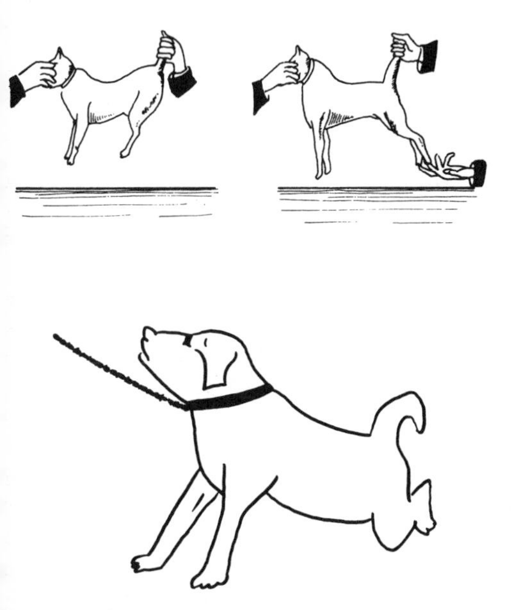

Fig. 11.3 Above, positive supporting demonstrated in decerebrate dog suspended in the air, on making contact with the back paws (after Walsh, 1964). Below, typical 'buttress reaction' on trying to pull a dog forward and thus increasing the differential pressure on the front foot; note obvious front-leg extension (from Rademaker, 1935)

mechanism would act to increase postural stability, as for example when the animal is standing on sloping ground (Fig. 11.4). On the level, there is a roughly even distribution of pressure amongst the four feet, and hence no tendency for one leg to lengthen more than another. But if the animal is facing up a slope (Fig. 11.4**b**) the pressure on the back feet is greater than on the front, and consequently the positive supporting reaction will result in rear-limb extension and front-limb flexion: the final result will be that the body adopts a more horizontal posture, and the projection of the centre of gravity is brought more nearly to the middle of the points of support. The same mechanism may be seen at work in the *postural sway reaction*: if an animal's body is pushed from the side, the shift in pressure on the feet results in marked extension of the limbs on the opposite side, and retraction of the others, so that

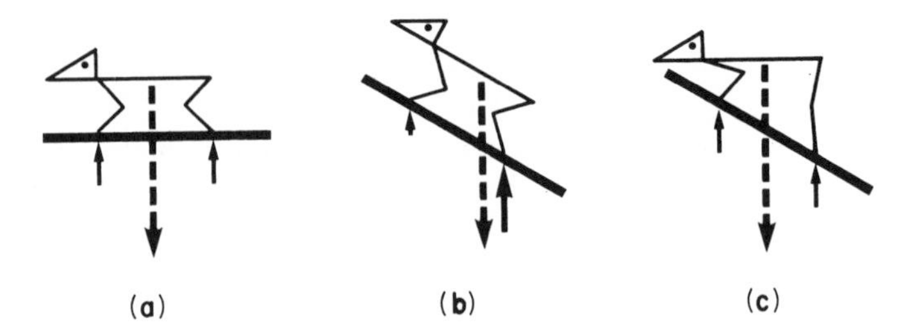

Fig. 11.4 How the positive supporting reaction results in good posture. (**a**) Animal standing on level ground, with projection of centre of gravity centrally placed between the supports (**b**) Facing up a slope: the weight is now unevenly distributed between front and back legs, resulting (**c**) in front-leg flexion and back-leg extension, and a better position of the projection of the centre of gravity.

the animal in effect leans against the experimenter (compare Fig. 11.3). It seems likely that analogous mechanisms may be used when the animal is lying down, involving information about differences of pressure on different parts of the body other than the feet. A blindfold animal, whose vestibular system has been destroyed (leaving cutaneous receptors as the only remaining source of postural information) will nevertheless right itself when laid on its side on the ground. But if a plank is laid on top of it that reduces the ratio of the pressures experienced by the two sides of the body, this body righting reaction is inhibited, a phenomenon sometimes used by veterinary surgeons to restrain an animal for operation.

All the mechanisms described so far assume that some sort of support is already present: there are other types of response that may be used to *find* postural support. If for some reason the projection of the centre of gravity has moved outside the critical area, then automatic *stepping reactions* are elicited that in effect move the feet in such a way as to track the centre of gravity: if one happens to be standing on one leg, the result is a *hopping reaction*. One can demonstrate these responses quite easily to oneself by first standing upright and then trying to fall over deliberately by leaning over: at some point, however hard one tries not to, reflex stepping or hopping comes into play and actual falling is prevented. Incidentally, if you try to fall over backwards in this way, you will also observe an involuntary upward flexion of the feet, as expected from the positive supporting reaction—though in these circumstances it is of no use whatever! It is also interesting to note that whereas in walking one is for most of the time in postural equilibrium—in the sense that one may 'freeze' at nearly any point of the walking cycle without falling over—this is not the case in running, when the centre of gravity is normally ahead of the critical area. One can think of running as being a series of regular and almost unconscious stepping reactions in response to the bent-forward posture that the runner maintains.

Other responses, called *placing reactions*, are used for acquiring postural support when none is present. If a blindfolded animal is suspended in the air and brought up to a table until the edge touches the backs of its paws, it will bring them smartly up to rest on the top of the table, in turn evoking a positive supporting reaction: a similar response may be triggered if the animal's whiskers are brought to touch the table. These responses are however rather more complex than those described above, and probably involve the cerebral cortex: unlike the stepping and supporting reactions, they cannot be shown in decerebrate animals.

Vestibular contribution to posture

The physiology of the vestibular system was described in Chapter 5. We saw that it provided two separate types of information: about angular velocity of the head, from the semicircular canals, and about its attitude relative to the effective direction of gravity, from the otolith organs. From the point of view of the control of posture it is of course the effective direction of gravity rather than its 'real' direction that matters: what determines, for instance, whether one falls over when one is standing in a bus that starts to accelerate is not the projection of the centre of gravity *vertically* relative to the critical area, but rather its projection in the direction of the vector formed by gravity and the horizontal linear acceleration acting together (Fig. 11.5), and this is what the utricle and saccule tell us. Each of these two divisions of the vestibular system gives rise to its own kinds of postural reactions, and so it is useful to distinguish between the static or *tonic* vestibular responses due to the otolith organs and the *dynamic* or phasic ones driven by the canals.

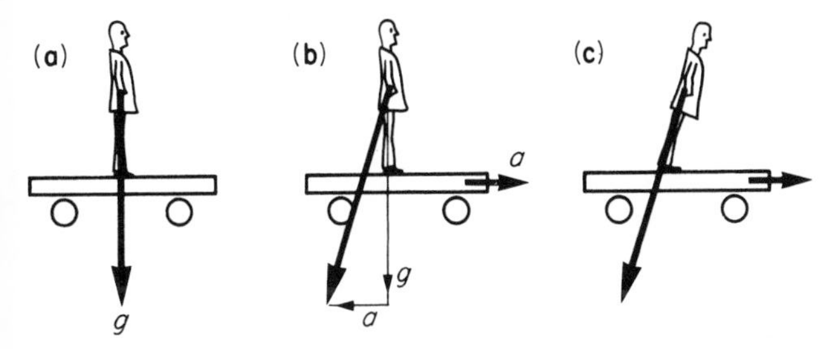

Fig. 11.5 Effective direction of gravity, and not true vertical, defines a good upright posture. (**a**) Man standing on a stationary platform, with vertical acceleration *g* due to gravity. (**b**) If the platform accelerates horizontally, the effective direction of gravity, as sensed by his otolith organs, is the vector sum of *g* and *a*, the acceleration of the platform. (**c**) It is the projection of the centre of gravity along *this* direction that must now be brought between his feet if he is not to fall over

The otolith organs produce on the whole rather less powerful postural responses than do the canals, particulary in higher animals. But in the long run they are the *only* source of information about the absolute position of the head in space, since the canals essentially signal only changes of position. One of their main functions is in fact to keep the head upright despite changes in the position of the body, through appropriate changes in the tone of the neck muscles; these are the *head righting reflexes*. If the head is forcibly tilted in different directions, so that the head righting reflexes cannot operate, one can also observe compensatory *static vestibulo-ocular reflexes* that similarly help to

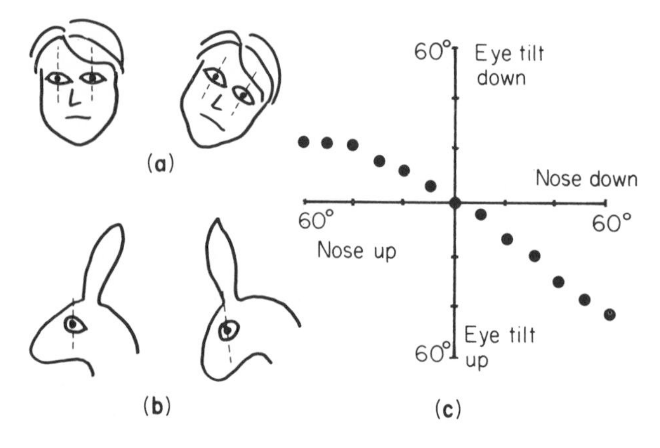

Fig. 11.6 Static vestibulo-ocular reflexes. Head tilt in Man (**a**) produces only a few degrees of ocular counter-rolling; in the rabbit (**b**), with its sideways-pointing eyes, vestibulo-ocular compensation is substantial over a wide range of angles (**c**). (Data from de Kleijn, 1921)

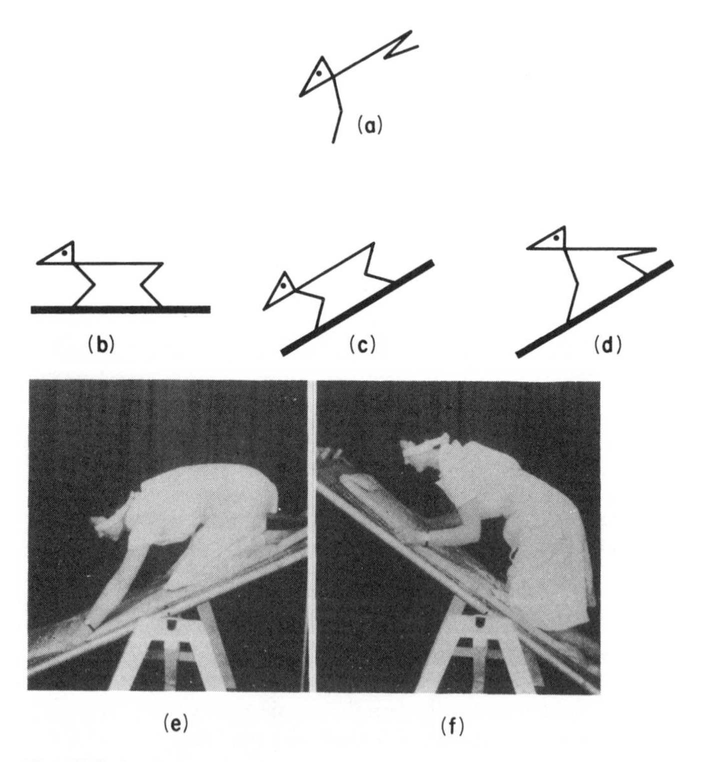

Fig. 11.7 Static vestibular righting reflexes. Tilting an animal's head and body nose-down (top) results in front-leg extension and rear-leg retraction; this response helps to produce a good posture when standing on a slope (*b,c,d*). Below, a human subject under similar conditions (Martin, 1967)

maintain the normal attitude of the eyes with respect to the outside world. In Man these eye reflexes cannot easily be demonstrated, and if the head is tilted to one side the resultant counter-rolling of the eyes is seldom of more than a few degrees and so cannot maintain the correct orientation of the retinal image. But in animals like the rabbit whose eyes essentially point sideways, tilting the head results in almost exact compensation over a wide range of angles of tilt (Fig. 11.6). While both these types of response obviously aid the sensory organs of the head by providing a stable 'platform' from which to operate, they are of course not strictly postural in the sense of contributing to the maintenance of equilibrium of the body as a whole. Reactions which are truly postural in this sense may be quite easily demonstrated in conscious animals, and are called *tonic postural vestibular reflexes*. If an animal is suspended in the air, and its head and body tilted nose downwards (Fig. 11.7), one observes extension of the front legs and retraction of the rear. Corresponding limb movements are found if the animal is tilted in other directions, for example to the side: in each case, there is extension of the limbs in the direction of downward tilt and retraction of the others. The function of this response is clear: if the animal is facing down a slope, this tonic vestibular response will assist the positive supporting reaction in shifting the centre of gravity backwards in relation to the feet, as may be seen in Fig. 11.7.

Dynamic vestibular reactions

Because the canals are velocity sensitive, they effectively give advance warning that one is *about* to fall over, possibly before the otolith organs have sensed that there is yet an actual error in head position. Perhaps for this reason, their responses are particularly fast, and generally bigger and more dramatic than the static vestibular reactions. The types of response they generate fall essentially into the same categories as the tonic ones: thus there is a dynamic component to the head righting reflex, that may be elicited by selective stimulation of the canals alone, and one may also demonstrate clear effects of canal stimulation on the eyes and limbs. If we seat someone on a rotating chair and record their eye movements (in the dark, so that there is no visual input to the oculomotor system), we find that the eyes move in the opposite direction to that of the head with a velocity that compensates almost exactly for the rotation, keeping the eye stationary with respect to the outside world. Clearly this *dynamic vestibulo-ocular reflex* cannot go on indefinitely, since sooner or later the eyes are going to reach the limit of their rotation in the orbit: what in fact is observed is that the smooth counter-rotation in one direction is interrupted at more or less regular intervals by a quick flick in the other direction, giving rise to a sawtooth-like eye movement called *vestibular nystagmus* (Fig. 11.8). The smooth, compensatory, movement is called the slow phase of the nystagmus, and the quick flick—which is essentially the same as an ordinary voluntary saccade—is called the quick phase; rather confusingly, it is the latter which is used in clinical practice to describe the direction of nystagmus, so that a subject turning to the right produces what would be called a nystagmus to the right, even though the more important, functional, component of the response

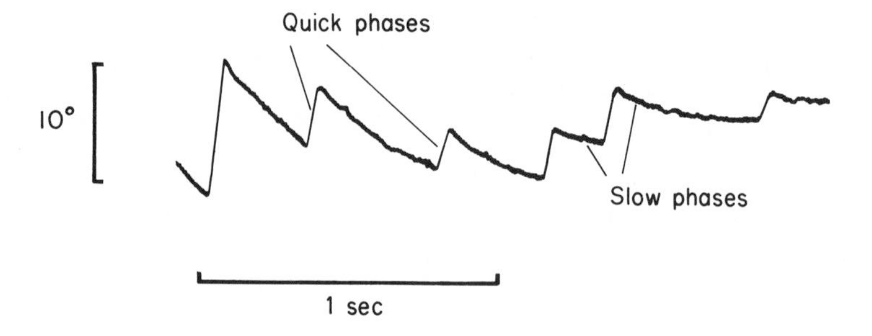

Fig. 11.8 A record of human vestibular nystagmus, showing slow and quick phases

is to the left. During rotation of this kind at constant angular velocity the response of the canals declines over a period of some 20 seconds on account of its natural adaptation (see Chapter 5), and so does the velocity of the slow phase. If the chair is suddenly stopped, so that the cupula is deflected in the opposite direction, a corresponding reversed nystagmus is seen which in turn declines with a time-course of some 20 seconds; these two types of nystagmus are respectively known as *per-rotatory* and *post-rotatory* nystagmus. Vestibular nystagmus provides a convenient means for clinical investigation of the functioning of the semicircular canals, since the eye movements produced by caloric stimulation (Chapter 5) of each of the two labyrinths may be examined separately to reveal imbalance of function on the two sides. Finally, *dynamic postural vestibular reflexes* exist, functionally equivalent to the static ones (head moving down gives front-leg extension, and so on), but much more powerful. In Man, unnatural stimulation of the canals may produce inappropriate postural responses that are vigorous enough to throw the subject to the floor—despite the action of all the other postural mechanisms in trying to keep him upright—as for example if one attempts to stand up after having been rotated in a revolving chair with one's head on one side for more than 20 seconds or so. The canals, because of their adaptation, then falsely signal that one is falling over: the consequent and extremely violent reflexes actually *make* one fall over, in the opposite direction.

Visual contributions to posture

The other receptors in the head that can provide information about head position and thus indirectly assist in the maintenance of posture are those of the retina. It turns out that there is a close parallel between the ways in which visual and vestibular information about head position are used in postural control, very probably because to a large extent they appear to share common pathways. One may again distinguish tonic effects from dynamic effects, and one also observes both effects on head and eye movements and also truly postural responses involving the limbs.

In so-called civilized surroundings, such as an urban street, our visual world is largely made up of horizontal and vertical elements, and our expectations about their orientation mean that we can in principle use our eyes to estimate head position. Many experiments have demonstrated that this *tonic* visual information is indeed used in making postural judgements and responses, and if a subject is seated on a tilting chair inside a dummy room which can itself be tilted to various angles (Fig. 11.9), it is found that his sense of the upright direction generally lies somewhere between the true upright and the apparent upright of the room. However, it is not clear that information of this sort is very readily available in more natural surroundings, such as jungles of the kind in which Man presumably evolved from his ancestors.

Fig. 11.9 Experiment with tilting room and tilting chair, to investigate relative contribution of visual and vestibular contribution to the sense of upright (Walsh, 1964, after Witkin, 1949)

We have already seen in Chapter 7 that at many levels of the visual system one may find large populations of neurones that are particularly sensitive to *movement* of the retinal image across the retina. If a sufficient number of these cells fire off simultaneously—in other words, if most of the visual field is in motion—the brain assumes, logically enough, that the world is in fact stationary and that it is the eye that is moving. This sense of self-movement is a very powerful one, as anyone who has had the experience of sitting in a train at a station while a neighbouring train moves off will agree. However, misleading circumstances like these are most infrequent in nature, and by and large when most of the visual field moves it is indeed because the head has moved relative to the visual surroundings, providing information that supplements what is provided by the semicircular canals, and is used in the same way. It is probably the main explanation for the postural instability associated with heights: if one is standing at ground level in a typical urban environment, most of one's visual field is filled with highly detailed visual texture, and the slightest head movement is likely to cause a brisk response from movement-detecting visual neurones. But on top of a mountain, things are very different. Most of the field is now occupied by clouds and sky, of low contrast and low spatial frequency, so that movements of the head are no longer so likely to be noticed by the visual system; the consequence is an increased instability, a tendency to sway about. Worse still, if one looks up, one's visual field is likely to contain nothing but the clouds moving past, which therefore give the visual system the illusion that one is falling over: the resulting postural compensation may well result in one *actually* falling over in the opposite direction.

Movement of the visual field also generates eye movements that are extremely similar to the dynamic vestibulo-ocular reflexes. If we seat someone in front of a rotating striped drum, close enough for it to fill a substantial part of his visual field, the result is a sawtooth-like movement of the eyes called *optokinetic nystagmus*, in which the eyes follow the moving stripes during the slow phase, and flick back again during the quick phase. One might think that this was merely the result of a conscious decision by the subject to track the stripes with his eyes: but this is not so, as in fact if the subject tries very hard to keep his eyes still and ignore the movement, the nystagmus is nevertheless still observed, and indeed is in some ways more pronounced and more regular.

How does the eye tell us about head movement?

All the visual system can tell us is that an image of an object in the outside world has moved relative to the retina: it cannot by itself tell us whether this was because the object itself moved, or because the eye moved relative to the head, or because head and eye moved together. We have already seen that movements of large areas of the visual field are generally interperted by the brain as being due to movement of oneself rather than of the world around us. But from the point of view of controlling posture, it is obviously very important to be able to disentangle the effects of movement of the head from those of movements of the eye, since it is the former that we really want to know about. How is this done? For a long time there were two rival theories. One, due to Sherrington, was that spindles in the eye muscles sent signals to the visual system telling it where the eye was in the orbit; this estimate of eye position relative to the head, which we may call E_H, would then in effect be subtracted from the estimate of eye position relative to visual space (E_S) provided by the visual system, to generate an estimate of head position in space, H_S (Fig. 11.10). Helmholtz, however, argued that it was unnecessary to have receptors in the eye muscles to tell one the position of one's eyes. Since the eyes are never subject to external forces in the way that our limbs are, we can always deduce their position from the motor commands that we send to them; thus in Helmholtz's scheme, a copy of the oculomotor commands—*efference copy*—is sent to the visual system to provide the required estimate of E_H. This theory, called the *outflow* theory in contradistinction to Sherrington's *inflow* theory, is supported by a number of pieces of experimental evidence. One of them is something you can do yourself: if you press on the side of your right eye with the left one shut, you will perceive an illustory movement of the outside world. This is exactly what would be expected from Helmholtz's model, but is not easy to explain with the inflow theory, since the muscle spindles ought still to be providing the necessary information to cancel the

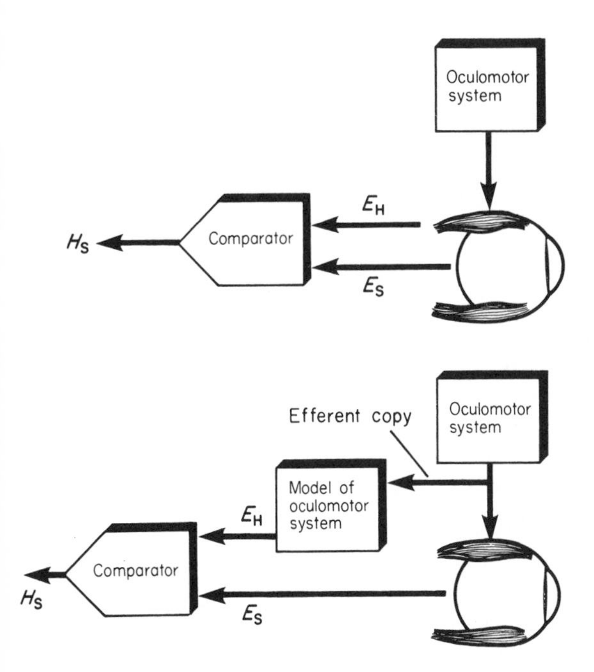

Fig. 11.10 The sense of eye position. Above, Sherrington's scheme: proprioceptors in eye muscles provide information about the position of the eye in the head (E_H), and this is then compared with visual information concerning the position of the eye relative to the visual world (E_S) to give H_S, the position of the head in space. Below, in Helmholtz's scheme, a copy of the oculomotor commands is used to calculate, from knowledge of the behaviour of the eye on past occasions, the expected value of E_H: this is then combined with E_S as before, to give H_S

signal produced by the movement of the retinal image. More convincingly, if the oculomotor commands are prevented from carrying out their usual effects on the eye muscles—for example, by some pathological condition affecting the oculomotor nerves or the eye muscles, or by artificial paralysis induced by local application of drugs like curare—then we would expect that every time the subject tried to make an eye movement, the E_H signal resulting from the efference copy would no longer be matched by the usual E_S signals from the retina, and the subject ought therefore to perceive an illusory movement of the visual world in the opposite direction; and this is precisely what is found. It seems therefore that efference copy is indeed the means by which the visual system can work out H_S, the position of the head in visual space, from the purely visual signal E_S.

Vestibular and visual interactions

We now have two quite different estimates of the position of the head in space, one provided by the vestibular apparatus, and one by the visual system. How are they combined? And what happens if they conflict with one another? We have already seen, from the tilting room experiment, that in fact the brain seems to take something like the weighted mean of these two estimates of head position in arriving at a final answer. This in turn raises the interesting question of how one type of information is *calibrated* in terms of the other. How does the brain know that one particular rate of firing of certain fibres in the vestibular nerve is equivalent to such-and-such a rate of firing of a movement-sensitive neurone in the visual system?

The answer seems to be that this correlation of the two inputs is one that is *learnt* by the brain as the result of experience, and that in fact the signals from the vestibular system are continually being checked against, and calibrated by, the signals from the eye. Thus if a man wears prisms in front of his eyes that effectively reverse his visual field, so that an object moving from left to right appears to him to be moving from right to left, one finds that after a surprisingly short space of time—a matter of days—his vestibulo-ocular reflexes follow suit by reversing as well, even when measured in the dark. A similar kind of long-term adaptation can be seen in the phenomenon of *Bechterew nystagmus*. Damage to the vestibular apparatus on one side of the head leads to an imbalance in the vestibular signals coming from each side: you may recall that vestibular fibres are tonically active, but that central neurones in the vestibular nuclei are excited by afferents from one side and inhibited by those from the other, so that at rest in the normal subject the tonic activity cancels itself out. After unilateral damage, the tonic discharge from one side is now unopposed by that from the other, and the result is a false sense of rotation (giddiness) combined with oculomotor and postural reactions, all of which may be extremely debilitating; it is the resultant unwanted nystagmus that is called Bechterew nystagmus. However, these effects gradually decline, and in a matter of weeks may have disappeared altogether. This seems to be because the visual system has informed the central vestibular pathways that the signals are false, resulting in some kind of tonic shift of activity that once again cancels out the continual signals from the unaffected side. That this is so may be demonstrated by subsequently cutting the vestibular nerve on the good side: the result is a Bechterew nystagmus in the opposite direction to the first—despite the fact that there is now *no* vestibular input to the system at all—presumably resulting from the central tonic activity that had been set up to correct the original imbalance; and this in turn gradually declines. There is some evidence that this plasticity may be controlled by the cerebellum, the vestibular division of which is supplied with fibres both from the vestibular apparatus and from the visual system; ablation of this region in the dog appears to abolish the vestibulo-ocular effects of prism reversal described earlier. We shall see in Chapter 12 that there are reasons for thinking that the cerebellum may play a part in other kinds of learnt motor responses as well.

Conflicts between visual and vestibular information about head position and movement also tend to give rise to *motion sickness*. It is a matter of common experience that motion sickness is most likely to occur when the visual field appears stationary, although it is actually moving, and sensed as such by the vestibular apparatus. Thus in a car one may feel perfectly well as long as one is looking out of the window, when the two estimates of head movement match, but feel sick when trying to read, when the visual field moves with the head.

One may also feel motion sickness under precisely the opposite conditions, for example sitting near the front of a cinema with a large screen, watching something like *Moby Dick* with its storm-tossed seas and heaving decks: our eyes tell us we are moving up and down, but our canals insist that we are stationary. Removal of the vestibulo-cerebellum in dogs is said to eliminate motion sickness entirely, presumably because it is in this region that the correlation of the two kinds of input is made.

Incidentally, it seems also that the sickness sometimes associated with alcohol intoxication is also, in effect, a kind of motion sickness. One of the effects of alcohol is to alter the relative density of the cupola and endolymph in the canals, so that they are no longer exactly equal. This means that the cupula is influenced by gravity—becomes, in effect, an otolith—and is deflected by static head position as well as by angular velocity. Under these conditions one may see what is called a *positional nystagmus*, a nystagmus that occurs spontaneously when the head is held in different positions. It is a matter of common experience that one of the effects of over-indulgence can be a sense of giddiness, particularly on changing head position, and if one goes to lie down there may often be a sensation of the bed turning head over heels.

Neck reflexes

Just as we need to know the position of the eye in the head before we can use vision to tell us about head position in space, so in exactly the same way we need to know the position of the head relative to the body (H_B) before we can use information about the position of the head in space (H_S) to tell us how our body is situated (B_S). And it is B_S that is the variable that we need to know about to control posture, since as has been repeatedly emphasized, postural control is all about keeping the centre of gravity of the body in the right relationship to its supports, and is *not* simply a matter of keeping the head upright.

This signal, H_B, comes from proprioceptors in the neck, mostly joint receptors around the vertebrae. Since H_B must be subtracted from H_S in order to produce B_S, we would expect the effects of bending the neck on its own to be as powerful as, and in the opposite sense to, the postural effects of changing the position of the head in space described earlier. This is found to be approximately true: the resultant responses are the *tonic neck reflexes* (Fig. 11.11). Thus if an animal's vestibular system is destroyed, and its head moved about while the body is suspended horizontally, we find that tilting the head up—resulting in dorsiflexion of the neck—gives front-leg extension and back-leg retraction. This, as you will recall, was the vestibular effect of tilting the head *down*. Thus corresponding to the vestibular mnemonic 'head down, front legs extend' we have the neck reflex mnemonic 'dorsiflexion, front legs extend'. As in the case of the vestibular reflexes, moving the head in other directions produces exactly analogous limb responses. The use of these reflexes in maintaining posture is clear. If an animal is standing facing down a slope, this time with its head in the normal horizontal position so that there is no vestibular stimulation, the result is exactly the same as if the head were

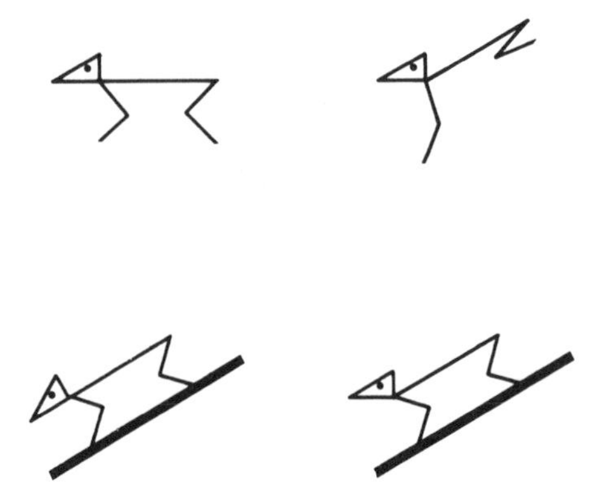

Fig. 11.11 Tonic neck reflex: if an animal is suspended with its head horizontal, dorsiflexion of the head (by tilting the back up) results in front-leg extension and rear-leg retraction (compare Fig. 11.7). Below, this means that the correct attitude is maintained on a slope, whatever the position of the head, because of the co-operation of neck and vestibular reflexes

pointing down. In fact, the animal will produce the same correct postural response *whatever* the position of the head. This consequence of the opposition of vestibular and neck reflexes is a very important one: if it were not the case, then every time a cow tried to bend down to eat some grass, its front legs would extend and prevent it reaching it! In other words, although the head is the single most important source of information about body position, nevertheless it can still move around freely without producing inappropriate postural responses. This is exactly analogous to the mechanism which, as we have seen, permits us to move our eyes around without at the same time perceiving apparent shifts of the visual world.

Figure 11.12 is intended to summarize this chapter by bringing together the various sources of postural information and the way they interact, in a hierarchical scheme that helps illustrate the parallels between the modes of operation of many of its parts. It is obvious that the scheme is highly redundant, in the sense that there are three separate channels from which B_S, body position in space, can be deduced. In fact it is clear that one can manage without any one of them, or even two, without losing one's ability to make postural adjustments. When swimming underwater, for example, only our vestibular system can provide us with information about body position; and in fact although patients with bilateral destruction of their vestibular apparatus can get about perfectly well in normal life by using the senses in their eyes and feet, it is precisely in circumstances like swimming underwater or running over a ploughed field that they may no longer be able to respond satisfactorily. Conversely, although one can do without any of the three inputs, a sufficiently strong stimulation of any one of them can cause disequilibrium despite the correct functioning of the other two. As already mentioned, a man may be thrown to the ground by his own vestibular reflexes after he has been spun in a

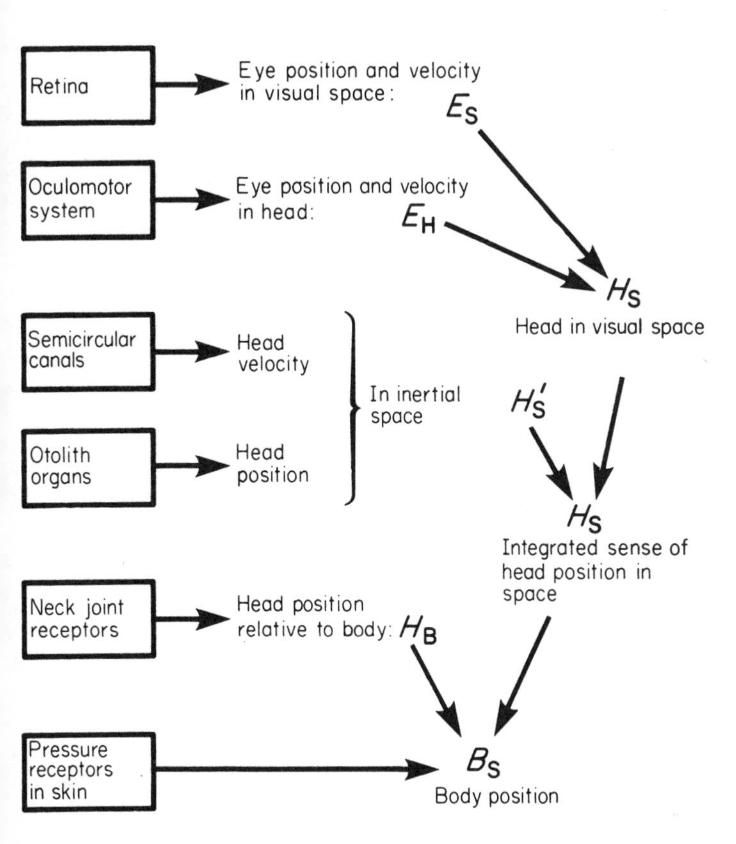

Fig. 11.12 Summary scheme of the sources of information used in the control of posture, and their inter-relations

revolving chair, and misleading visual information may similarly make him fall over if he stands looking up at the clouds. Finally, there is another source of information that is not sensory at all but the result of another kind of efference copy. Many of our voluntary actions are themselves threats to equilibrium, as for example if we pick up a heavy object and then throw it, or even for that matter when we walk along. In these cases the threat can of course be anticipated, and postural compensations initiated even before the disequilibrium has been sensed by the afferent pathways shown in Fig. 11.12.

12

Higher levels of motor control

We must now consider the levels of the motor system at which detailed information from the special senses first begins to be important: in ascending hierarchical order, the cerebral motor cortex, the cerebellum, and the basal ganglia.

Motor cortex

By the middle of the last century, there was an increasing interest in the interactions between electricity and living tissues, and particularly in the kinds of responses that could be obtained by electrical stimulation of the central nervous system. It turned out that there was a conveniently accessible region in the middle of the cerebral cortex, where stimulation evoked reproducible movements of parts of the opposite side of the body; since no other area of the cortex generated movements on stimulation, this region was called the *motor cortex*. At about the same time, the English neurologist Hughlings Jackson had been studying the relation between paralysis of different parts of the body due to stroke, and the post-mortem location of the associated cerebral lesions. He was also interested in a special variety of epilepsy called Jacksonian epilepsy in which there are characteristic motor signs such as spontaneous twitching, that start at one particular location—often an extremity such as a finger tip—and then move progressively and systematically, for example up the hand and arm, until they culminate in a more generalized convulsion. Putting these two types of observation together, he suggested that the progress of the epileptic signs over the body was the result of a 'march' of the region of pathological abnormality over the cortex itself. This is now known to be perfectly correct: an epileptic focus of overactivity in one cortical region tends to spread to the regions with which it communicates, a fact of great importance in trying to treat epilepsy by surgical means. The suggestion of an orderly representation of parts of the body in the cortex was wholly confirmed when accurate *motor maps* were made by investigators such as Sherrington in the monkey, and later by Penfield and others in human patients, in the course of operations to treat Jacksonian epilepsy. This area lies just in front of the central sulcus (Fig. 12.1), and corresponds to Brodmann's cytoarchitectonic areas 4 and 6. It thus lies alongside the primary somatosensory projection area (areas 3,2, and 1),

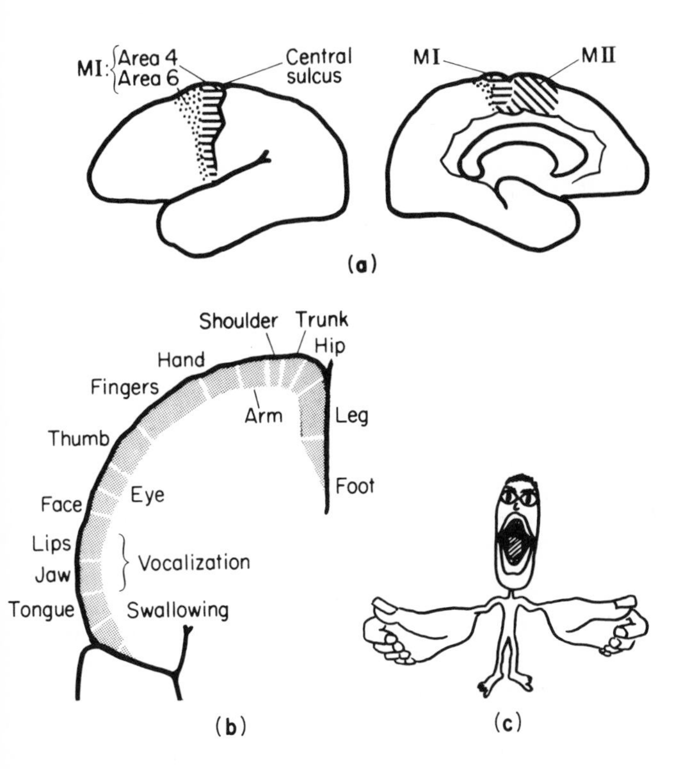

Fig. 12.1 (**a**) Lateral and medial views of human brain, showing the approximate positions of motor cortex, areas MI and MII. (**b**) Transverse section through MI, showing the distribution of areas devoted to different parts of the body. (**c**) Motor homunculus (see sensory homunculus, Fig. 4.6). (Partly after Penfield and Rasmussen, 1950)

and indeed shows a similar distribution of the representation of different parts of the body. Some regions of the body, such as the hands and mouth, have a disproportionately larger representation than others in this *motor homunculus* a possible reason for this will be considered later. Recent work has demonstrated a second motor area (called MII, the first being MI; Fig. 12.1) which, like the corresponding sensory area SII, is more bilaterally organized than the strictly contralateral MI.

At this point we need to break off in order to consider the neural structure of the cerebral cortex in general. In most mammals by far the largest part of the cortex is *neocortex*, a sheet made up of six layers of cells and fibres (Fig. 12.2) and elaborately wrinkled into the complex pattern of sulci and gyri that enables a large surface area to be stuffed into a relatively small volume. The main output from the cortex comes from the large *pyramidal cells* of layer V, forming what are known as the projection efferents, that go to subcortical destinations. The dendrites of these pyramidal cells extend vertically as well as branching sideways, so that the cells are capable of being influenced by all layers of the cortex, within a radius of half a millimetre or so. Corresponding to the projection efferents, on the input side there are the specific projection

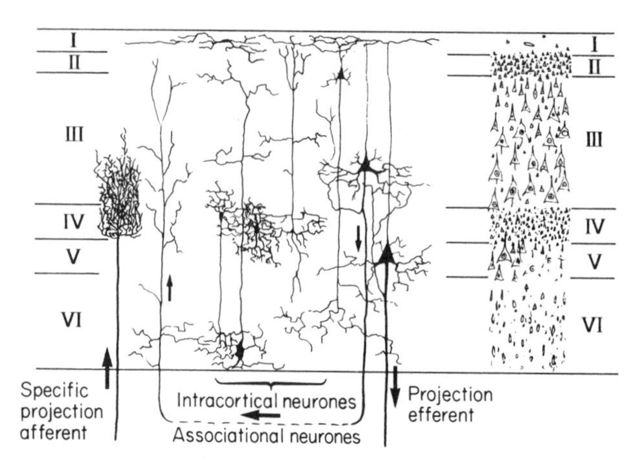

Fig. 12.2 Diagrammatic representation of the distribution of some types of neurone in the cerebral cortex, and their principal connections. On the left, afferent systems; and on the right, efferent. Far right, the general histological appearance of the six layers

afferents of subcortical origin (mostly in fact from the thalamus) that ramify into large terminal trees around layers III and IV. However, in a sense the whole point of the cortex is to bring many diverse types of input and output into functional association, and to a large extent this is brought about by associational neurones having cell bodies in layer III of one cortical area, that send off axons terminating in vertical columns in some other area. In addition to these primary types of neurone, there is also a host of different varieties of interneurone that communicate from layer to layer as well as horizontally, many of which are inhibitory in nature and probably carry out functions analogous to lateral inhibition.

It is important to appreciate that cortex and thalamus work together very much as a single unit, and that the latter is not merely a relay for fibres on their way to the cortex, but receives back from the cortex almost as many fibres as it sends there. Two thalamic nuclei are paired in this way with the cortical motor areas (Fig. 12.3): they are the *ventrolateral* nucleus which projects mostly to area 4, and receives fibres back from both area 4 and the somatosensory cortex; and the *ventroanterior* nucleus, which projects to area 6 and receives reciprocal connections from both area 4 and area 6. Unlike for example the ventroposterior nucleus, which sends spinal information to the somatosensory cortex, the ascending input to the motor nuclei of the thalamus is not sensory at all, but comes partly from the basal ganglia and cerebellum and partly from the reticular formation; these regions also project to the more diffuse centromedial thalamic nuclei, which communicate more widely with both somatosensory and motor cortex, and other areas as well. These relationships are summarized in Fig. 12.3. The *output* from the sensorimotor regions is diverse: some of course makes up the corticospinal tract, some can reach the cord indirectly via the red nucleus and the nucleus gigantocellularis of the reticular formation, but most projects to the basal ganglia and—via

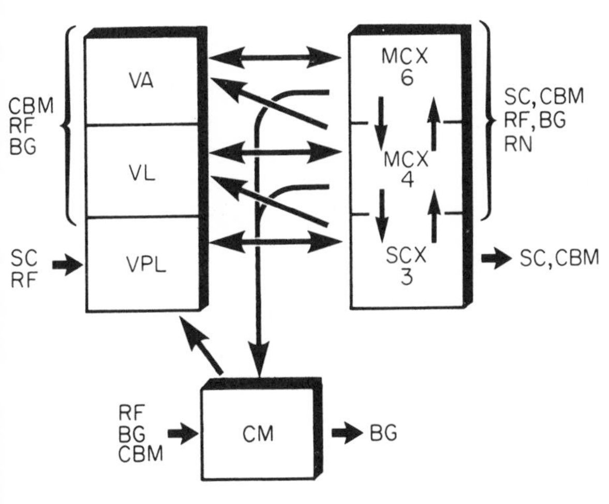

Fig. 12.3 Stylized representation of the main connections of motor cortex and thalamus. Cortical regions: MCX, motor; SCX, somatosensory. Thalamic nuclei: VA, ventroanterior; VL, ventrolateral; VPL, ventral posterolateral; CM, centromedian. Other regions: CBM, cerebellum; BG, basal ganglia; RF, reticular formation; SC, spinal cord; RN, red nucleus

relays in the pons and inferior olive—to the cerebellum. These indirect routes are sometimes lumped together under the heading of *extrapyramidal* pathways, those outputs that do not simply go straight down into the spinal cord.

Corticospinal concepts

Now area 4 is differentiated from all other areas of the cortex by having a number of particularly big cells in layer V, the giant *Betz* cells. Their size suggests that they might be the origin of the corticospinal tract, and for a long

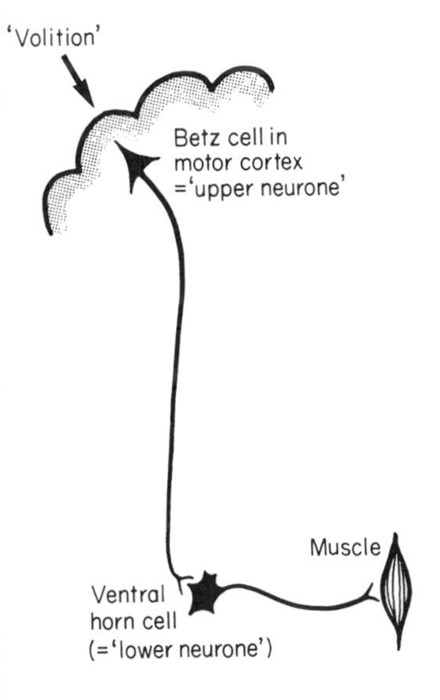

Fig. 12.4 Classical but misleading concept of 'upper' and 'lower' motor neurone

time the notion was current that the essence of the voluntary motor system was something like what is shown in Fig. 12.4, with 'volition' triggering off in some way the Betz cells of area 4, which in turn synapsed directly with spinal motor neurones: the former were called the 'upper motor neurones' and the latter the 'lower motor neurones', with the implication that the effects of stimulating the motor cortex were entirely due to stimulation of the cortical tract. This is an unhelpful and misleading picture in a number of ways. In the first place, it is clear that the Betz cells are not the sole origin of the corticospinal tract: for one thing, the latter contains about a million fibres, whereas there are only some 30 000 Betz cells. It is not even true that the tract comes only from the motor cortex: about 30 per cent comes from area 4, another 30 per cent from area 6, and the rest from all over the cortex, including 'sensory' areas like the visual and auditory cortex. Nor is it true that the effects of cortical stimulation are even mainly due to activation of corticospinal fibres. If in the monkey one traces the motor map by electrical stimulation both before and then after completely severing the pyramidal tract, the distribution of responses is found to be essentially unchanged (though they may tend to be a little slower and require a larger current to be evoked). And finally, as we have already seen, even in the primates only some of the corticospinal fibres end directly on motor neurones, while in other species none do: cortical control is probably rather of spinal circuits than of individual muscle units. Lesions of the corticospinal tract give similar, but not identical, effects to lesions of the cortex itself: generally hypotonia and weakness (paresis), loss of skilled movements, and a slowness and unwillingness to use the affected limb; lesions of area 4 are in general rather more severe, resulting at first in a wholly flaccid paralysis, and typically in greater functional loss than pyramidal section by itself. There are certain difficulties of interpretation, however, both by reason of the marked species variation that is seen—a cat can still apparently run about after complete resection of its pyramids—and also because gradual (though variable) recovery is a feature of lesions of both cortex and pyramidal tract. In time, the immediate flaccidity resulting from a lesion in area 4 usually develops into a spasticity, presumably through some kind of release phenomenon. Lesions that are large enough to encroach on area 6 as well as 4 also tend to produce spasticity, which might be interpreted as suggesting that area 6 has some kind of tonic inhibitory action, a notion reinforced by observations of the effects of stimulating area 6 during voluntary or evoked movements. At all events, the classical clinical description is that an 'upper motor lesion' gives a spastic paralysis, without muscle wasting, while a 'lower motor lesion' produces a flaccid paralysis.

Another, associated, early misconception was that the upper motor neurones coded for movements rather than for individual muscles, and that the translation from one to the other was a function of the manner in which the descending axones were distributed within the cord. It is certainly true that if extracellular electrodes are used to elicit movements by electrical stimulation of the motor cortex, then even at threshold one tends to observe not isolated twitches from separate muscles, but the simultaneous and apparently co-ordinated contraction of several muscles at once. But it is now clear that this is essentially because the current from extracellular electrodes of this kind spreads out over a wide area of cortex, and that one cannot avoid stimulating large numbers of pyramidal cells and interneurones simultaneously, even at the threshold for overt movement. More recent work with intracellular electrodes indicates clearly that the effect of stimulating one pyramidal tract neurone is the excitation of only one muscle at a time, generally the specific facilitation (or sometimes inhibition) of one particular stretch reflex. There is also a systematic representation of muscles in the cortex, analogous to that found in somatosensory or visual cortex, with a grouping of neurones corresponding to synergistic muscles in *columns*; it is presumably for this reason that extracellular stimulation gives the appearance of co-ordinated activation of synergists.

Co-ordination of somatosensory input with motor output

One of the most interesting findings comes from using the same micro-electrode for recording as well as stimulation. What is found is that pyramidal cells have very wide, multi-modal receptive fields, from joint receptors and tendon organs as well as skin receptors, and with a small representation of spindle afferents as well. If one looks at the relationship between what any particular cell does when it is stimulated and what it actually responds to, it is striking that frequently the cutaneous receptive fields represent areas of skin that are normally brought into contact with one another when that particular muscle contracts (Fig. 12.5). This suggests rather strongly that the cortex is the site of the kinds of sensorimotor correlation that are obviously necessary whenever grasping or touching or some other *manipulation* is being performed. That this is perhaps the most

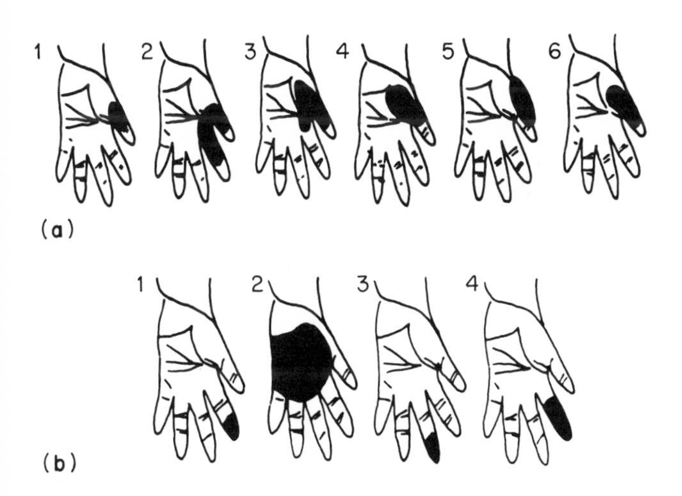

Fig. 12.5 Cutaneous sensory fields associated with pyramidal cells of monkey causing (**a**) thumb flexion, and (**b**) digital flexion on stimulation. The cutaneous fields correspond with areas that would be brought into contact by the resultant movement. (After Rosen and Asanuma, 1972)

important single function of the motor cortex is suggested also by the relative sizes of different parts of the body in the motor map. In Man, only two areas of the body are used to any large extent for tasks of this kind that require accurate feedback from the skin, namely the hands and the mouth; and both of these together form by far the greatest proportion of the motor map. In the monkey, which makes more use of its feet for handling objects, they have almost as large a representation as the hands, and relatively much larger than in Man. In most four-footed animals it is only the mouth and lips that are used for grasping and exploration, and their representation is correspondingly enlarged (Fig. 12.6); virtually the whole of the pig's motor cortex is said to be devoted to its snout! It is also perhaps significant that lesions in area 6—which we saw to be predominantly inhibitory in effect—often produce uninhibited grasping of the kind frequently observed in very young children: any object touched against the palm results in immediate forceful seizure, with an unwillingness to let go. Destruction of the motor cortex also abolishes the tactile placing reaction described in Chapter 11.

Fig. 12.6 Motor 'homunculi' of rabbit, monkey and Man, showing differences in the relative degree of representation of different regions (highly schematic: partly after Woolsey, 1958)

One further function of the cortex, of a related kind, is probably the control of the gain of stretch reflexes. We saw in Chapter 10 that there is evidence that this gain is altered according to the size of the load to be moved, and that one source of information about load that seems to be used to do this comes from pressure receptors in the skin; tendon organs, which as we have seen also project to pyramidal cells, are another likely source of information about load (see Fig. 10.14). Could it be the pyramidal tract fibres that are the route by which force commands are sent to the spinal cord? They certainly have the right kinds of information at their disposal to carry out such a function, and also appear to give the expected effects of facilitating stretch reflexes when stimulated. And if one implants an electrode in the motor cortex of a conscious freely moving animal, and observes the circumstances under which pyramidal fibres actually fire during voluntary movements, one finds that whereas the correlation between pyramidal activity and muscle length or velocity is poor, there *is* a clear correlation between frequency of firing and the *force* that is being produced at any moment to generate the movement, suggesting very strongly that this tract does indeed provide force commands. Finally, we have seen that the effect of lesions to motor cortex or pyramidal tract is not just a loss of the kinds of skilled movements that require close co-ordination between skin sensation and movement, but also a general weakness of the muscles, with a correspondingly increased sense of effort on the part of the patient. Sense of *effort* seems to be found in situations where the gain of stretch reflexes is insufficient, for example in the experiments of Fig. 10.13 when the pressure receptors in the subject's thumb were blocked by local anaesthesia, and may well be associated with voluntary excitation of motor cortex neurones. Partial paralysis tends to result in an apparent increase in the heaviness of a weight being lifted.

One might wonder, if it is true that the main function of the motor cortex is a relatively simple co-ordination between skin and other spinal afferents and the gain of stretch reflexes, why this function has to be carried out in the cortex and not in the cord itself, where the inputs and outputs actually are: why bother to go all the way up to the top of the head? There are probably two reasons why the spinal cord is essentially unsuited to the task. The first is that most spinal reflexes seem to be organized in a segmental manner. But the feedback from skin receptors resulting from contraction of a particular muscle will not always return to the cord via the dorsal roots of the same segment; in the cortex, however, information from different segments is brought into much closer approximation, and the large amount of convergence and divergence—in the monkey, each pyramidal cell receives some 60 000 synapses—means that connections from different inputs can presumably be attached more easily to their appropriate muscles, rather as subscribers' lines converge on a telephone exchange. The second point is that it is difficult to see how appropriate connections could ever be set up in the cord in the first place: how can an incoming sensory fibre from the skin *know* which of the hundreds of thousands of ventral horn cells are the ones to which they are supposed to make contact? It seems much more likely that such connections are established through *experience*, by frequent association of stimulation of a particular afferent with contraction of a particular muscle. This implies a richness and flexibility of connections, an ability to form functional contacts as the result of experience, that the cortex seems specifically adapted to perform, through the same types of memory-like mechanisms suggested in Chapter 7 in the case of the visual cortex. Finally, it should not be forgotten that if it is convenient for the control of movement to have cutaneous information readily available in areas 1, 2 and 3 right next door to the motor cortex, it is equally important in analysing cutaneous sensations to have information immediately on hand about what movements are being executed. It was emphasized in Chapter 4 that such common sensations as softness, resilience, roughness, stickiness, and the like rely on knowledge of what *forces* are being applied to the surface in question by the motor system, knowledge that is easily supplied by pyramidal tract neurones, through the rich interconnections between the two cortical areas. In human patients, electrical stimulation of the motor cortex produces sensory effects not very different from those resulting from stimulation of the somatosensory cortex: numbness, tingling, and a sense of movement which may or may not be accompanied by actual movement; if it is, the subject feels that he has been 'made' to carry out the action by the experimenter rather than 'wanting' to make it.

Cerebellum

The cerebellum and basal ganglia are, very broadly speaking, at a higher hierarchical level in the motor system than the cortex, in the sense that when they go wrong they tend to produce *disorders* of function, sometimes of a subtle kind, rather than discrete paralysis or weakness. We have already seen that anatomically they are considerably further from the final output, in that they send no fibres directly into the spinal cord. They are also older structures than neocortex, prominent in all vertebrates; in birds and reptiles the motor cortex or its equivalent does not appear to exist, so that one is forced to conclude that the cortex has developed more for refining actions than for generating them in the first place.

The cerebellum seems to have grown out of the brainstem as an adjunct to the vestibular system; but this oldest part of it, the *archicerebellum* or vestibular cerebellum, is now dwarfed by two newer areas: the *palaeocerebellum* associated with the spinal cord, and hence also sometimes called the spinocerebellum, and the large and central *neocerebellum* or corticocerebellum, whose development has been in step with that of the cerebral cortex from which most of its input is derived (Fig. 12.7). The cortex of the cerebellum has a very regular and beautiful neuronal structure quite unlike the chaos seen practically everywhere else in the central nervous system, which has led to it being called 'the neuronal machine', and has provoked more speculation than anywhere else—particularly from mathematicians and computer scientists—about how its circuits might work.

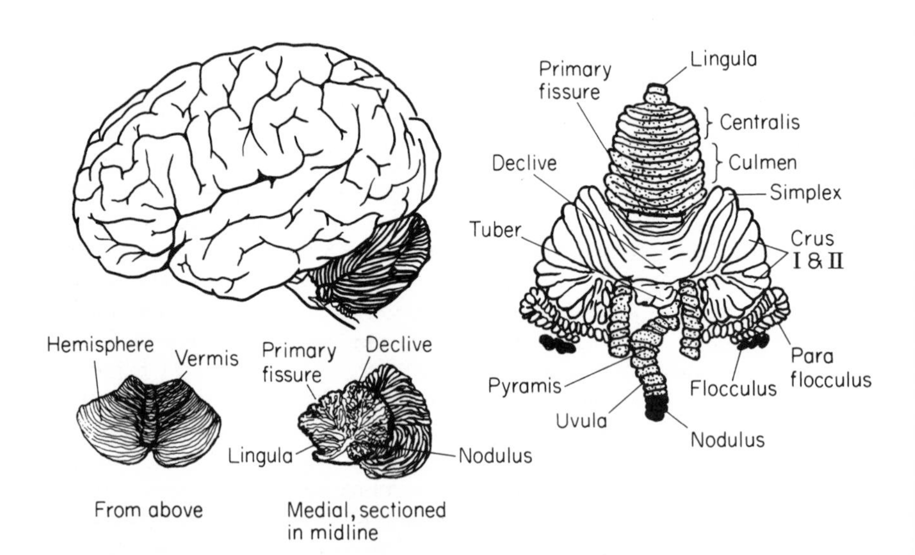

Fig. 12.7 Gross structure of the cerebellum. Left, viewed from various directions, and in sagittal section. Right, 'unrolled', showing the main anatomical divisions. Functionally, it may be divided into vestibulo-cerebellum (black), spinocerebellum (stippled) and neocerebellum (unshaded), though the divisions are not as clear-cut as this diagram implies

Cerebellar neurones and their connections

The most conspicuous type of cell in the cerebellar cortex is the *Purkinje cell*, which has a huge dendritic tree that is confined to a plane perpendicular to the surface and roughly anteroposterior (Fig. 12.8); the cells are lined up in soldierly rows over the whole of the cortex, with their dendritic trees thus stacked like a pack of playing cards. There are about 30 million of them, and each has a dendritic surface area equivalent to two average-sized front doors.

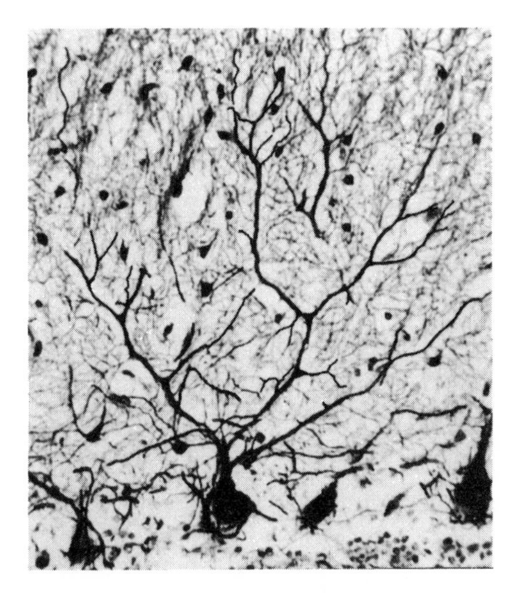

Fig. 12.8 Cerebellar cortex, stained by the Golgi silver method, to show a large part of the dendritic tree of a single Purkinje cell

They form the sole output of the cerebellar cortex, for their fibres run downwards to the *deep nuclei* that lie in the core of the whole structure. In primates there are four of these nuclei on each side (*fastigial, emboliform, globose* and *dentate*); in other species the second and third of these are fused into one, called the *interpositus*. The projection from the cerebellar cortex on to these nuclei is a systematic one: medial areas project to the fastigial region, lateral ones to the dentate, and the intervening region to the interpositus. A curious feature of the Purkinje cell is that although it is the only output from the cortex, it is entirely inhibitory on the deep nuclei. Finally, the deep nuclei themselves project to various other motor structures: the fastigial primarily to the vestibular nuclei, interpositus to the red nucleus, and dentate to the thalamus (mainly ventrolateral) and hence to the cerebral cortex; all probably also send fibres to the reticular formation of the brainstem. Thus there are several routes by which the cerebellum can influence the spinal cord.

Afferent information reaches the Purkinje cells by two quite distinct pathways (Fig. 12.9). In the first place, each Purkinje cell receives a unique

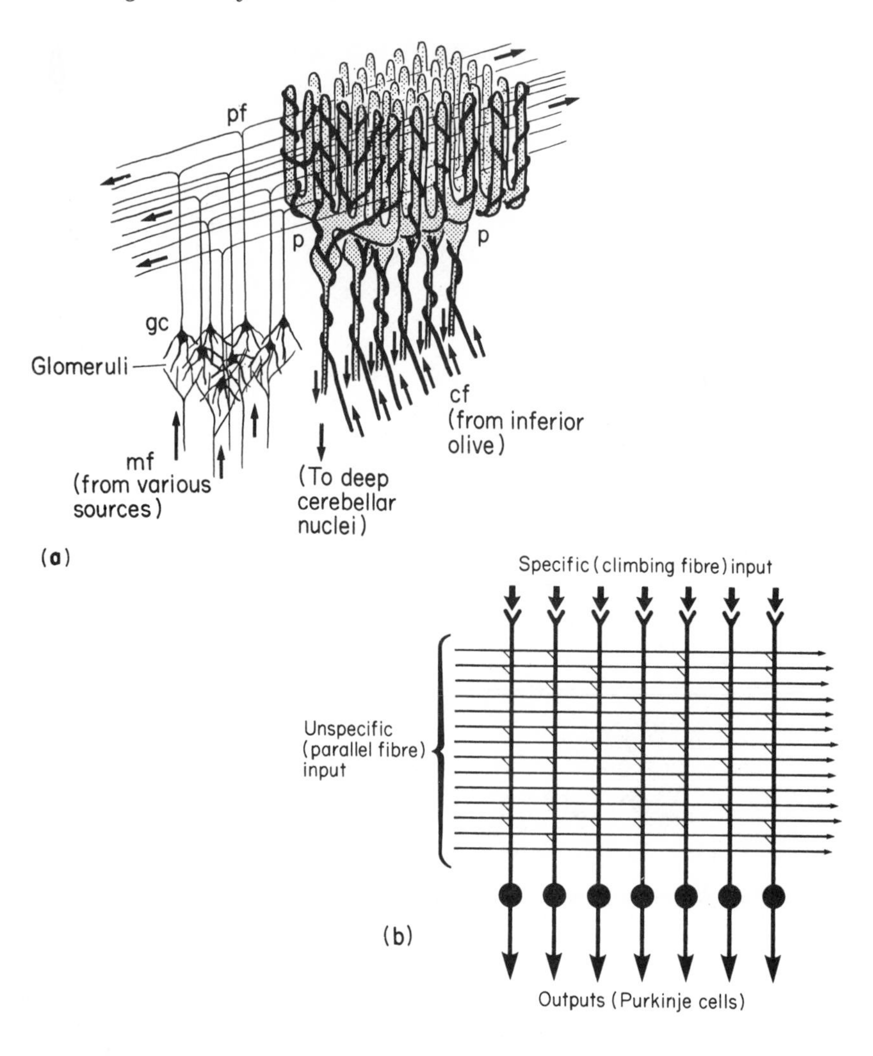

Fig. 12.9 Above, diagrammatic representation of the connections between mossy afferents mf (bottom left), granule cells gc and parallel fibres pf (left), Purkinje cells P and climbing fibres cf. Below, equivalent schematic arrangement

single fibre that climbs up its efferent axon and then branches to clamber all over its dendrites like ivy on a tree, forming synaptic contacts that are very strongly excitatory. These *climbing fibres* come from the *inferior olive* in the brainstem, which receives its input in turn mostly from the cerebral cortex, but also from the spinal cord and to some extent from the special senses (as for example from the visual system, in the vestibulocerebellum); as well as these contacts with Purkinje cells, they also excite the deep nuclei to which the Purkinje cells project. Such specificity of synaptic contact with a single cell is rather a rarity in the central nervous system, where wide divergence and convergence seems to be the general rule, and experimentally it is found that a single shock to a climbing fibre never fails to fire the associated Purkinje cell, sometimes repetitively. The other type of afferent system is utterly different. Here, the incoming fibres enter the lower layers of the cerebellar cortex and branch to form large terminal structures (giving them their name of *mossy fibres*) that synapse in *glomeruli* with a number of dendrites from *granule cells* in the cortex. These in turn send ascending axons to the surface, where they bifurcate and send two thin axons (called *parallel fibres*) in opposite directions, perpendicularly piercing the planes of the stacked Purkinje cells, to whose dendritic trees they form side-connections, like telephone wires on a telephone pole (Fig. 12.9). In this way, and in complete contrast to the highly specific climbing fibres, each parallel fibre can contact a large number of Purkinje cells; conversely, each Purkinje cell receives roughly a quarter of a million contacts from parallel fibres. Mossy fibres carry information from a very wide range of sources: spinal (notably from muscle spindles), visual, auditory, vestibular, reticular, and from the cortex via relays in the pons. The receptive fields of Purkinje cells are very large indeed—sometimes extending over a whole limb—and are remarkably multi-modal. This arrangement of parallel fibres and flat Purkinje cells is an efficient way of providing the largest possible number of output channels with access to the largest number of input sources, within the smallest possible space.

Thus the basic structure is essentially simple: it is slightly complicated by the existence—as everywhere else in the central nervous system—of various types of interneurone that mostly appear to provide lateral inhibition, sharpening up any spatial patterns of excitation that may be present. These include *basket cells*, excited by parallel fibres and inhibiting a parasagittal row of Purkinje cells, and *Golgi cells*, also excited by parallel fibres, but inhibiting granule cells instead and thus acting at the input rather than at the output (Fig. 12.10).

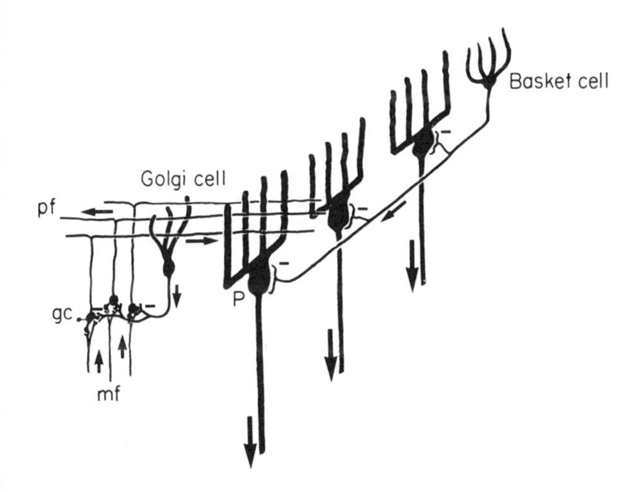

Fig. 12.10 Inhibitory interneurones in cerebellar cortex. Left, a Golgi cell inhibiting a cluster of granule cells (gc); right, a basket cell inhibiting a row of Purkinje cells (P). pf, parallel fibres; mf, mossy fibres

Disorders of the cerebellum

Although our knowledge of the neuronal structure of the cerebellum is quite precise, our knowledge of its *function* is rather more rudimentary, and almost entirely limited to the effects of cerebellar lesions and other kinds of damage, which are often quite specific and revealing. Damage to the vestibulocerebellum leads to difficulties of postural co-ordination that are similar to what is found with damage to the vestibular apparatus with which it is associated. There may be difficulty in standing upright, a tendency to dizziness, and sometimes a staggering gait when walking. It is very likely that this area acts as the centre of co-ordination for the various postural mechanisms described in the previous chapter. In some species, as was noted earlier, visual information enters the vestibulocerebellum through the climbing fibres, and vestibular fibres via the mossies; it seems probable that it is here that the comparison and integration of postural information from these two sources takes place. As mentioned earlier, cerebellar ablation in dogs leads not only to abolition of the effects of prism reversal on vestibular reflexes, but also to freedom from motion sickness.

Difficulties of gait—*ataxia*—are also found after damage to other cerebellar areas, but the effects are then found to be more generalized and not just postural: a lack of co-ordination of all kinds of movement, often *hypotonia*, and described as *asynergia*. It is worth considering some specific examples of these defects in more detail, since they reveal a good deal about the nature of cerebellar disability. Many can be explained in terms of the patient's motor system taking too long to respond to sensory information, of added delay round a feedback loop. Thus *dysmetria* or overshoot may be seen: when the patient reaches out to touch something, his hand goes too far, presumably because the command to stop the movement is sent out too late. A consequence of this is *intention tremor*, in which the overshoot is subsequently corrected by a movement in the opposite direction which then itself overshoots, resulting in a new correction, and so on—the result being an oscillation or tremor around the desired position. The tremor is not seen at rest, but only when the patient is aiming to achieve a particular limb position. This slowness to react to changed circumstances is seen also in *rebound*: if for example the patient is asked to flex his arm against a force, which is then suddenly removed, whereas in the normal patient the resultant inward movement of the hand is quickly checked, in the cerebellar patient it is not, and may strike his body with considerable violence. A related defect is *adiadochokinesis*: the patient is unable to make rapid alternating movements, as for example rapid oscillatory rotation of his wrist between pronation and supination; he cannot apparently issue the command to reverse a movement sufficiently soon after having sent the command to start it (Fig. 12.11). In the same way, he may show *scanning speech*: whereas a normal person does not have to think about the sequence of mouth and tongue actions that he makes while speaking, the cerebellar patient seems unable to generate the series of commands sufficiently rapidly, and appears to have to think about the formation of each separate phoneme. Altogether, in fact, the patient has to bring enormously more *conscious* control into his movements. A normal man can walk along, pick things up and so on without thinking much beyond merely willing the final outcome; but a cerebellar patient has to plan and think about the details not just of what to do, but *how* to do it. This is perhaps most clearly demonstrated in another dysfunction called *decomposition of movement*: complex movements that require the temporal co-ordination of several different muscles are simplified by being broken down into their components, being executed in effect by one muscle group at a time. Normal people can see for themselves what it is like to be in this condition by the simple expedient of getting drunk: alcohol seems to have a particularly noticeable effect on the cerebellum, and in such circumstances one does indeed sense the need to think consciously about putting one foot in front of the other in order to walk, and one's conversation may begin to approximate to scanning speech. The difficulty is essentially in using stored programs to carry out motor sequences that are usually automatic: one's actions are performed as if one was learning them for the first time. In fact the execution of any *new* task by a normal subject is strikingly similar to what is seen in cerebellar patients all the time. A simple experiment of this sort that you can do yourself is to try drawing whilst looking at what you are doing not directly, but in a mirror. If you attempt to move your pencil smartly towards a particular point on the page, you will see both dysmetria and intention tremor; more complex manoeuvres are only achieved by decomposition of movement; and all the time one is painfully aware of the need for continual thought about the details of the movements one is making. (Fig. 12.11). With enough practice, of course, one would in time learn to execute mirror-drawing without these defects, and without the need for continual conscious

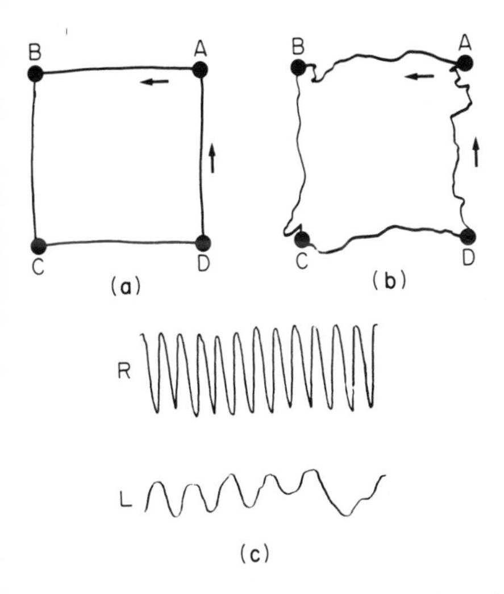

Fig. 12.11 Illustrations of some signs of cerebellar damage. Above, the (normal) subject was instructed to draw a pencil line linking the dots ABCD in the order shown (**a**) is with normal vision, (**b**) looking in a mirror. Dysmetria, intention tremor and decomposition of movement, of a kind similar to the signs of cerebellar damage, is obvious. Below, adiadochokinesis: records from a patient with damage to his left cerebellum, who can make rapid alternations of pronation and supination with his right arm (R) but not with his left (L). The duration of the trace is 5 seconds. (Holmes, 1922)

intervention; presumably some part of the brain is then carrying out automatically, perhaps by means of stored programs like those of Fig. 9.5, or some kind of internal model, as in Fig. 9.6, what previously had to be thought about. It seems increasingly probable that it is the cerebellum that is the part of the brain that carries out this kind of motor learning; that it acts, in a sense, as the body's autopilot.

Theories of cerebellar action

There are a number of theories as to how the computer-like neuronal circuitry of the cerebellar cortex might actually do this; none is wholly acceptable or confirmed by electrophysiology. One that is more specific than most in the way in which it relates structure to function, and that explains most readily the kinds of defects associated with cerebellar damage, is due to David Marr. His suggestion is that the cerebellum is capable of storing and executing specific sequences of actions, by means of a gradual process in which the sequences are first generated consciously while the subject is attempting to master the task, but as the result of repetition are gradually taken over by the cerebellum itself. In its final form the theory is a complex one, that requires advanced mathematics to be fully understood, but in essence is something like this. Consider how we might learn to play a scale on the piano. To simplify things, let us begin by supposing that there is one Purkinje cell that corresponds to each of the fingers playing 'C, D, E, F' (Fig. 12.12), and that when the cell fires, it causes that finger to play the corresponding note. (The fact that the Purkinje output is inhibitory need not be an embarrassment: if we inhibit an inhibitory cell, the result is excitation, and neuronal circuits within the CNS can work equally well whether we consider either an increase or a decrease in firing rate to represent a 'positive' signal: one need only think of the hyperpolarization of certain retinal receptors in response to light.) Now we have seen that the Purkinje cells are powerfully and specifically excited by their climbing fibres, and that one important source of these fibres, via the inferior olive, is the cerebral cortex. What is suggested is that during the initial learning phase, the Purkinje cells are driven by these climbing fibres—activated by some kind of volitional process—in the correct sequence: C,D,E,F; under these circumstances the cerebellum is doing nothing more elaborate than simply relaying this sequence of commands to some lower level. But consider what meanwhile is happening to the parallel fibres: it has been repeatedly emphasized that every time we carry out a motor act, it necessarily results in a kind of echo that comes back to us through our senses. When we play a note on the piano, we get feedback not only from proprioceptors, the muscle spindles, tendon organs and joint receptors, but also from endings in the skin, not to mention the visual stimulus of seeing the finger move and the note go down, and the auditory stimulus of hearing the result. Each note that is played consequently generates a particular pattern of sensory feedback that will be quite specific to that particular action and to no other. This pattern will be reflected in the pattern of activity of the parallel fibres, which we saw to convey information of the most diverse kinds to the dendrites of the Purkinje cells. Thus when we play the note D, having just

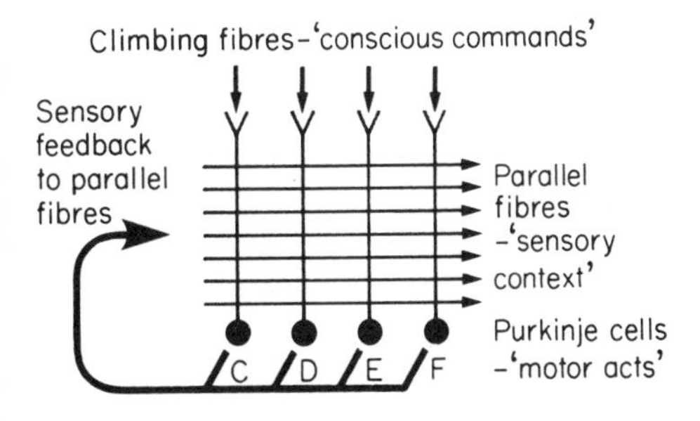

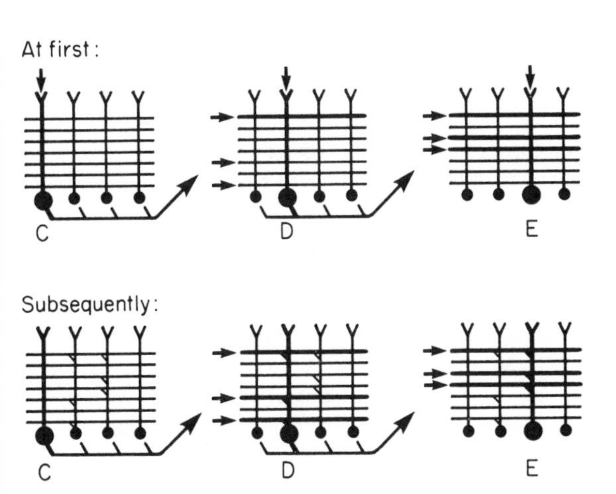

Fig. 12.12 A model of how the cerebellum may learn sequences of motor actions. Above, highly simplified diagram of cerebellar inputs and outputs (see Fig. 12.8) showing feedback from motor actions to the parallel fibres. Below, the hypothetical sequence of events before and after learning to play a piano scale C,D,E,... (for explanation, see text)

played C, the Purkinje cell corresponding to D is activated by its climbing fibre during what may be called a *sensory context* that is quite specific to the state of having just played C, and is quite literally present at its dendritic branches in the form of a particular and specific pattern of parallel fibre activity. We now have to make one more assumption—one that is common to virtually all theories of neuronal learning—by supposing that the synapses between parallel fibres and Purkinje cell are *plastic*: that is, that they may get more or less effective, depending on the relationship that has been experienced over a period between their own activity and that of the Purkinje cell. In particular, if we suppose that the condition for these synapses getting stronger is that the parallel fibre should often fire at the same time as the Purkinje cell, then we have a system that will learn to recognize the context associated with a particular action, and eventually respond to it automatically by generating the action itself. For if we play the sequence C-D over and over again, each time we do it the parallel fibres that are activated by the sensory feedback from C will fire at the same time as Purkinje cell D, so that their synaptic contacts with the latter will gradually get stronger and stronger. Eventually they will get so strong that they can fire D off even in the *absence* of a volitional command from the climbing fibre: D will then be produced spontaneously simply as the natural result of having played C, with no conscious intervention. In the same way, E will come to follow automatically from D, F from E, and G from F; and in the end all the subject needs to do is to initiate the sequence, and it will follow automatically—and more rapidly. Such a model explains the phenomenon of adiadochokinesis particularly simply. If we imagine just two Purkinje cells, one for supination and one for pronation, then what we are doing when we learn to make the rapid alternation of hand position is to connect the two cells up reciprocally so that the context produced by one eventually comes to fire the other, resulting in almost automatic oscillation; those familiar with electronics will recognize that we have in effect built an astable multivibrator out of our cerebellar components. That these alternating movements have to be learnt in the first place is clear if you try and execute them with some less familiar part of the body: most people—unless they've been practising!—suffer from adiadochokinesis of the toes, as you can easily verify for yourself. Finally, it is perhaps worth mentioning that Marr's model has already found a potentially useful application in the field of industrial robots: machines have been built that incorporate similar circuits and are equipped with sensors from the work area, and will learn to perform complex sequences of operations by first being driven 'consciously' (by a human operator, in fact) and then gradually recognizing the patterns of sensory input that are to act as triggers for particular items of motor output.

With slight modifications, the same model can also form the basis of the other two types of motor learning discussed in Chapter 9, namely the storage of ballistic programs, and the prediction of expected results from copies of motor commands, by means of a stored model of the behaviour of the body. In the first case, we need only assume that a part of the mossy fibre input comes from other motor areas at a lower hierarchical level, rather than from sensory receptors (as indeed is the case, particularly in the neocerebellum). Then instead of relying on actual feedback from the results of any particular item of a motor sequence, the *command* for one such item can trigger the next, producing ballistic sequences of motor acts that do not have to wait for actual feedback from results. Under these circumstances, the climbing fibre input (whose function in Marr's model is in effect to say to the Purkinje cells 'now learn this!') could be used to provide parametric feedback in order to improve ballistic performance through experience. For example, we have already seen that in the vestibulocerebellum there is evidence that visual information enters through climbing fibres and vestibular through the mossy fibres. In Marr's model, this would imply that visual information would not only drive postural responses such as eye movements directly, but also strengthen those *vestibular* connections to Purkinje cells that were appropriate, in the sense that they were in close correspondence with the visual signal. Such a mechanism would explain very nicely the way in which visual information appears to be capable of continually calibrating the vestibular input in such situations as the prism-induced reversal of vestibulo-ocular reflexes mentioned earlier. And lastly, if we imagine the parallel fibres to convey copies of motor commands, and the climbing fibres to be activated by actual sensory feedback from the results of those commands, then we have a system in which Purkinje cell output will as a result of experience come to provide an estimate of the result of motor commands, of the kind required in internal feedback systems like that of Fig. 9.6. Marr's model is thus a very versatile one, capable of embodying almost any kind of motor learning by defining its inputs and outputs in different ways. Many would say in fact that it explains almost too much, and is consequently difficult to test; and it has to be admitted that elegant and powerful though it is, there is as yet no hard neurophysiological evidence to support it. Its value at present is perhaps essentially explanatory, in that it helps to tie together in a coherent way both the observed effects of cerebellar dysfunction and what is known of cerebellar microanatomy and function.

The basal ganglia

The functions of the basal ganglia are as uncertain as their structure is complex; consequently they are not at present rewarding objects of study. A prominent component of the basal ganglia is the *corpus striatum* lying in the mesencephalon at the level of the thalamus; in higher animals it has suffered the fate that frequently falls to older structures in the brain—as we saw in the case of the archicerebellum—in that it has been elbowed out of the way by newer structures that have consequently distorted it into an even more tortuous shape than is altogether necessary (Fig. 12.3). Thus what was originally a relatively compact mass of cells has been disrupted by the arrival of the internal capsule—like a motorway through a village—with the result that it is now an elongated structure that twists its way round the newer ascending fibres. The stripes seen in cross-section, that give it its name, divide it into a number of different regions: of these, the most important distinction is between the older part, the *palaeostriatum* or *globus pallidus*, which lies on the inside, and the outer *putamen*, which is continuous with the long arc of the *caudate nucleus*, the two together forming the *neostriatum* or simply 'striatum'. There are other nuclei which conventionally are also considered part of the basal ganglia, though not everyone agrees as to what should or should not be included. They include the *subthalamus* lying below the thalamus, and the *substantia nigra* (so called because certain of its cells are darkly pigmented with melanin). Both of these structures are highly developed in Man, but less so in other animals; and some authors also include the red nucleus. The connections between these structures are very complex and of unknown functional significance. The important flow of information seems to be a projection from the cerebral cortex to the putamen and caudate nucleus, thence to the globus pallidus, then to VA and VL of the thalamus, and thus back to the cortex again, forming a large feedback loop (Fig. 12.13). The subthalamus has connections to and from the globus pallidus, and the substantia nigra receives an input from the putamen, while sending fibres back to both parts of the corpus striatum (Fig. 12.14). Other connections are difficult to establish, though it is likely that there are indirect connections from the older areas of the brain such as the limbic system (which is concerned with motivation and emotion), and at least in fish one may demonstrate a

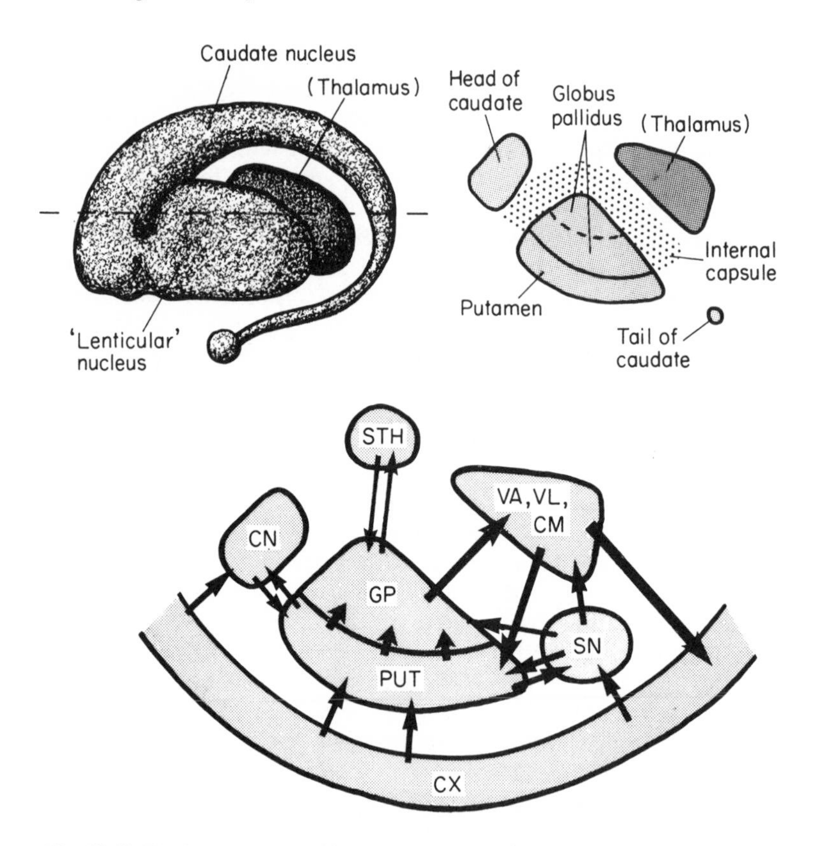

Fig. 12.13 The basal ganglia. Above, lateral view of the principal components of the basal ganglia, together with the thalamus; right, a horizontal section at the level of the dotted line, showing in addition the fibres of the internal capsule pushing their way through. Below, schematized representation of the principal connections. CN, = caudate nucleus; GP, globus pallidus; PUT, putamen; SN, substantia nigra; STH, subthalamus; CX, cerebral cortex; VA,VL,CM, ventroanterior, ventrolateral and centromedian thalamic nuclei. (Partly after Netter, 1962; copyright CIBA Pharmaceutical Company.)

functional projection from the olfactory system, which as we have seen is one of the oldest senses and one that is particularly associated with limbic functions.

As with the cerebellum, such meagre information as we have about the functions of the basal ganglia is almost entirely derived from clinical observations of the effects of damage in Man. The best known of these disorders is *Parkinsonism* or *paralysis agitans,* associated especially with damage to the pathways linking the substantia nigra and the putamen. The main feature of classical Parkinsonism is a general *poverty of movement (akinesia).* Expressive movements, such as the normal mobility of the face, may be absent, giving the patient a lifeless and apathetic appearance, and there may be a loss of *associated movements*, movements that normally occur in conjunction with a particular primary activity, but are not strictly necessary (such as swinging the arms when walking). The patient may blink less often than a normal subject; there is often a shuffling gait, and he may be very slow in walking about. None of these things are defects of the peripheral motor apparatus, for under the right circumstances, especially under strong emotional stimulation, quite normal movements may be made. Thus a Parkinson patient may be shuffling his way across the road when a car comes: he then runs briskly to the other side, only to continue his slow shuffle along the pavement. Other common features of Parkinsonism are a general *rigidity* and *slowness* (bradykinesia), vague *postural difficulties*, and a *tremor* that is the exact opposite of the intention tremor of cerebellar damage, being present only at rest, and disappearing as soon as some voluntary action is attempted.

If Parkinsonism is essentially a state of poverty of movement, other disorders of the basal ganglia, by contrast, result in the spontaneous production of unwanted movements. Lesions of the subthalamus in particular give rise to *ballismus* in which the patient may throw his limbs about in a violent manner. In other regions, damage may give rise to less energetic spontaneous movements called *chorea* ('dancing'), for example continual shaking or twitching, or to slow writhing movements known as *athetosis*. Some of these symptoms get worse as the patient tries to reach a particular goal. There may also be an exaggeration of associated movements. With the exception of the ballismus that can be produced in monkeys as well as Man by damage to the subthalamus, these effects are not clearly associated with lesions of specific areas of the basal ganglia, and indeed it has not proved possible to simulate them very closely in experimental animals. Consequently, treatment of these disorders is generally of a rather rough-and-ready kind: it is sometimes found, for example, that a Parkinson patient's rigidity and tremor may be alleviated by actually making further lesions in other parts of the basal ganglia. Another treatment for Parkinsonism that is often helpful and has the appearance of being more scientific is to treat the patient with DOPA, a precursor of dopamine, the transmitter in the projection from substantia nigra to putamen; it is possible that it is some defect in the production of transmitter here that gives rise to the condition in the first place.

At all events, it is certainly not yet possible to use the existence of these clinical disorders to deduce the detailed functioning of the basal ganglia. To suggest, as some of the older accounts do, that because damage to a certain area gives rise to tremor or to sudden violent movements, the *function* of that area is to reduce tremor or smooth movements out in some way, is only slightly less absurd than the analogy presented in Chapter 1, of removing a circuit board from a hi-fi and deducing that its function was to inhibit whistling. But one *can* deduce something about the hierarchical level at which the basal ganglia operate. In all these cases, whether there is loss of voluntary initiation of movements that can be evoked involuntarily, or the intrusion of unwanted movements that are, in their way, quite well executed (or even elegant, as in athetosis), it is clear that we are at a very high level in the motor system. Whereas lesions of the cerebellum give rise to defects of execution but not of initiation, and lesions of the cortex lead to even 'lower' defects like weakness or frank paralysis, damage here clearly interferes with the level at which movements are strategically planned and initiated. So if we had to represent, in general terms, what the relative hierarchical positions of these three higher levels were, we would put the basal ganglia at the top, responsible for the initiation and perhaps the large-scale or strategic planning of movements; the cerebellum underneath, automatically translating these commands into sequences of unitary actions, with reference to feedback from the periphery; and at the bottom the motor cortex, where commands concerning the desired positions of the limbs and so forth may be converted into yet more detailed instructions governing the forces required from moment to moment in order to achieve these results.

One might well ask what, if anything, lies above the basal ganglia in such a scheme. The problem here is partly one of terminology. In the general representation of the brain presented at the beginning of Chapter 9, it was emphasized that there is in effect a gradual series of neuronal levels that convert sensory information into motor movements. When we are considering areas near the centre of such a scheme, at the highest hierarchical levels, the distinction between 'sensory' and 'motor' becomes a somewhat arbitrary one. (Of course, one might cut through this problem at one stroke by introducing a 'ghost in the machine' which is both conscious of incoming sensory information and also capable of willing volitional movements: by definition, any structure upstream of such an entity is sensory, and downstream, motor. The question of whether such a notion is necessary in understanding human behaviour is postponed until Chapter 14.) Meanwhile, all one can usefully say is that it is simply a matter of convention that the inputs that provide the drive for motor acts are normally reckoned not to be part of the motor system. There are in fact two distinct *types* of input that must be considered; first of all the sensory information about the environment without which one clearly cannot make decisions about how to plan one's motor acts, even on a large scale; and secondly, the neural machanisms that decide *what* is to be done, that choose between all the various possible courses of action that are open to the brain at any particular moment. The first type of input, that of high-level integrated sensory information about the environment, forms the subject of the next chapter; the second type, which is what *motivates* the motor system, and requires information not just about the outside world but also about one's internal environment as well, one's state of *need*, forms the subject of Chapter 14.

13

Analysis and storage of information by the cerebral cortex

The processes covered by this chapter—unlike those discussed in any other chapter in this book—are the ones that we humans are best at, and to which we owe any temporary biological success that we may have achieved. Many animals are more agile and better co-ordinated than we are, can hear better or see better, and practically all of them have more sensitive noses. Our special virtue is that we are quite good at *storing* and *processing* such sensory information as gets through to us, so that we use it better in making effective responses to our environment. There is no sharp distinction between us and other creatures; these functions are not particularly mysterious or magical, and are after all performed even better by some computers. It is simply that the parts of the brain that carry out these functions in a rudimentary way in other species have in us been very greatly expanded and developed. If we compare the cerebral hemispheres of a series of animals from different evolutionary stages, what is striking is not just the expansion in the absolute mass of neural tissue by the time one reaches Man (Fig. 13.1), but also the changes in the relative *proportions* of the cortex devoted to different functions. Very little of a rat's cortex is not either primary motor or a projection area for one of the senses; in Man, by contrast, most of the cortex neither responds electrically to sensory stimulation, nor produces movements when electrically activated: these are the *silent areas*. Now these are precisely the properties we would expect from neural levels somewhere in the middle of the model of the brain that was first discussed in Chapter 9 and is shown again in Fig. 13.2: because each neurone at any level is only activated by a particular pattern of activity in the preceding layer, as we penetrate deeper into the sensory side we find that individual neurones become fussier and fussier about what they respond to, until eventually the chance of our finding out, in an experiment of finite duration, what they actually *do* respond to becomes extremely small; and on the output side, unless we happen to stimulate them in a pattern that makes some kind of neural sense, nothing will happen at all. Equally, because these areas integrate or associate information from diverse sources which cut across the conventional divisions of sensory modality (for which reason the corresponding cortical areas are also known as *association areas*), lesions in them are unlikely to have the same kind of circumscribed effects that are found, for example, in damage to the primary visual or motor cortex (Fig. 13.2). Nor, at

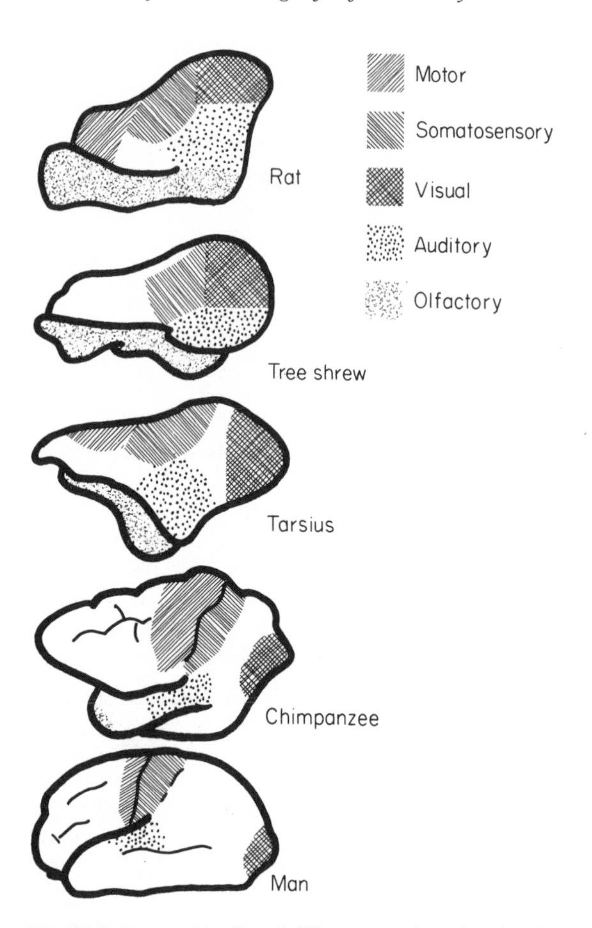

Fig. 13.1 Series of brains of different species, showing increase in extent of 'silent', associational, cortex (unshaded) in the course of evolution. (After Stanley Cobb, in Penfield, 1967)

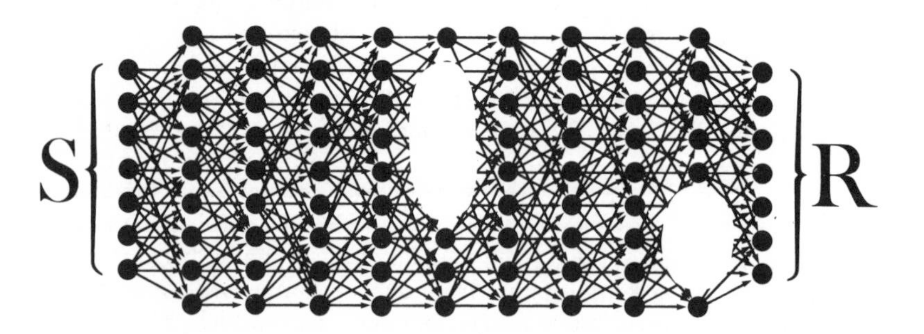

Fig. 13.2 Representation of the brain as a series of neuronal levels (as in Fig. 9.1) showing the differing effects of a lesion at a high hierarchical level (centre), and one nearer the periphery

this level, is there likely to remain much topological orderliness of the kind found at more peripheral levels. This situation is not unlike what happens in a telephone exchange: at the periphery—the region where the incoming cables arrive—there is a systematic relationship between a subscriber's number and the position of his particular connection, but the selector switches in the heart of the exchange that set up the circuits and form, in effect, associations between different subscribers, are shared by all of them and used to set up different circuits on different occasions. The capacity of an exchange—the number of associations it can make at any one time—is thus simply proportional to the quantity of this common switching equipment it contains. Might the neural elements of associational cortex also be in some sense shared in this way? Such a notion, of associational cortex being uncommitted to any particular task, but providing a reserve of computing power that can be applied to whatever job is on hand, was originally suggested by the experiments of Karl Lashley described on p.17. Lashley's 'Law of Mass Action'—that the effect of lesions in associational cortex depends more on how large they are than on their exact location—is now less in favour: it is clear from clinical observations in particular that discrete lesions in associational cortex can lead to fairly circumscribed functional defects, rather than something like a generalized loss of 'intelligence'. What *is* true, as we shall see, is that these functional defects may be of the wide-ranging and subtle kind that is characteristic of damage at a high hierarchical level—for instance the loss of the ability to speak French, while spoken English is unimpaired—and also that there is very little reproducibility from subject to subject, in the sense that a

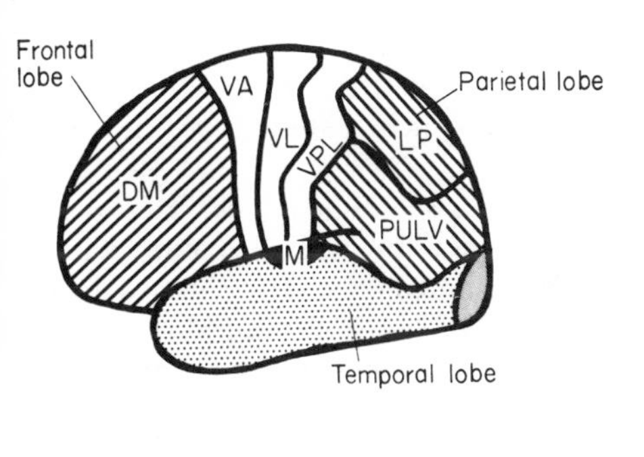

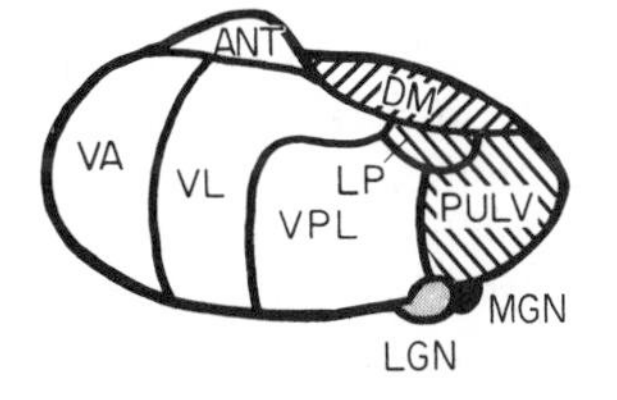

Fig. 13.3 Schematic lateral views of cerebral cortex and thalamus, showing corresponding regions of each. DM, dorsomedial; LP, lateroposterior; PULV, pulvinar; VA, VL, VPL, ventro-anterior,-lateral, and -posterolateral; MGN,LGN, medial and lateral geniculate nuclei; ANT, anterior. VA also projects diffusely to the frontal lobe. (After Carpenter, 1976)

lesion in a particular place in one person may have a completely different effect in another person. The idea of an uncommitted pool of 'brain-power' is almost certainly wrong; the neurones *are* specialized in their function, but at a high level in the hierarchy: it is this that gives lesions in associational cortex their subtle and unpredictable quality.

In primates there are essentially three distinct regions of associational cortex (Fig. 13.3): they are the *frontal lobe* (strictly, prefrontal), occupying the entire region anterior to the motor cortex; the *parietal lobe*, bounded by somatosensory, visual and auditory cortex; and the *temporal lobe*, bounded above by visual and auditory cortex, and occupying the rest of the inferior portion of the cerebral cortex—the thumb of the cerebral boxing glove.

Frontal cortex

We have already seen that it is not just sensory projection areas that have afferent fibres relayed through the thalamus: the frontal lobes form the largest single division of the cortex in Man, and receive fibres mostly from the equally enlarged *dorsomedial nucleus* of the thalamus (Fig. 13.3). This nucleus in turn receives fibres not only back from the frontal lobe, but also from the hypothalamus and other parts of the *limbic system*, an area predominantly associated with such functions as emotion and motivation, to be discussed in Chapter 14. These are very old parts of the brain indeed, found practically unchanged throughout the animal kingdom, and it is striking that their cortical projection is to the very newest part of that new area, the cerebral cortex. Man is in fact distinguished most from primates by the absolute and relative size of his frontal lobes; until a century or so ago it was assumed that they must therefore be the seat of the very highest functions—intelligence, morality, religion, etc.—those that were thought to differentiate Man most clearly from the apes.

The case of Phineas Gage in 1848 came as a rude shock. He was an American mining engineer, who was one day tamping down dynamite with an iron bar (one cannot help wondering about his intelligence *before* his accident!). The consequence can be imagined (Fig. 13.4). Astonishingly, when one considers contemporary standards of medical care, he survived, and made a precarious living for many years by exhibiting himself in public, together with the bar. Post-mortem examination showed that most of the frontal cortex had been destroyed; what was extraordinary was how little effect this seemed to have on him: far from turning into an ape, or losing his powers of reason, he seemed to suffer little more than a slight change in personality. His doctor described him as 'fitful, irreverent, indulging at times in the grossest profanity (which was not previously his custom), manifesting but little deference for his fellows, impatient of restraint or advice when it conflicts with his desires, at times pertinaciously obstinate, yet capricious and vacillating'—what, in fact, we would now call perfectly normal! Indeed his friends said that in some ways he was actually happer—more carefree and less inhibited—after the accident than before.

As a result of this dramatic demonstration, controlled experiments were

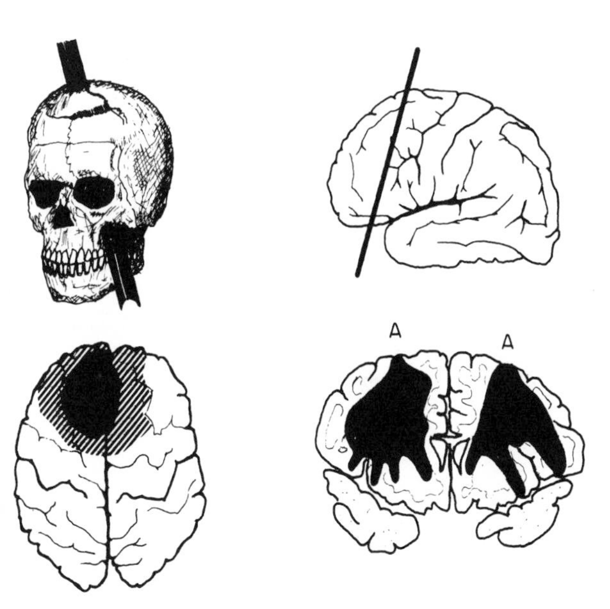

Fig. 13.4 Prefrontal damage. Left, Phineas Gage's skull, with crowbar; horizontal section below indicates the subsequent area of destruction (black) and probable neural damage (shaded) (after Cobb, 1946). Right, deliberate frontal leucotomy, showing (above) the approximate plane of the cut, and below, cross-section of an actual cut of this kind. The operation is performed by means of a spatula inserted through skull openings at the points marked A on each side (after Freeman and Watts, 1948)

performed on animals, with the same conclusion: that lesions in the frontal lobes seem generally to reduce *anxiety*—monkeys worry less when they make mistakes in learning tasks—and inevitably the idea developed that such a procedure might even be of benefit to depressive patients or anxious schizophrenics. This operation, *frontal leucotomy* or *lobotomy* (Fig. 13.4) began to be practised around 1935 and remained popular until the introduction of pharmacological agents doing much the same thing in a more reversible manner, in the early 1960s. There is no doubt that the operation gave a great deal of relief: not only alleviation of tension and anxiety but better adjustment to work, and increased weight and energy. A difficulty was that the changes in personality might go too far, developing into euphoria, tactlessness, an over-lackadaisical approach to life, and a lack of social inhibitions such as a tendency to urinate in the fireplace. One circumstance where these side-effects seemed worth putting up with was in the treatment of intractable pain, not easily dealt with by other measures. The result in this case was not so much loss of the objective knowledge of the pain—not in other words analgesia—but rather a loss of the *affect* of the pain, its unpleasant or emotional quality. Thus when asked what the pain was like, a patient might reply 'Oh doctor, it's absolutely appalling, unbearable', yet would be smiling as he said it and not—apparently—really *feeling* it despite being able to *sense* it.

What is perhaps most notable about the effects of frontal lesions is how little defect of ordinary intelligence occurs, with one exception: there are almost always difficulties in carrying out two programs of activity simultaneously. Thus a patient may be asked to recite the letters of the alphabet 'A,B,C,D', and then be interrupted and asked to add together 13 and 15: '28'; if then told to carry on, the original task is forgotten and the response may be '29,30,31,...'. Similarly, there may be an inability to organize actions in proper temporal sequence; this may become apparent when trying to prepare a meal, where one has of course to plan well ahead, bearing in mind the different times that different things take to cook, so that everything is ready simultaneously. These effects can be demonstrated in monkeys by means of the *delayed reaction test*. Here a monkey sits behind a glass partition in a cage, in front of which are two boxes, one of them containing a reward such as a banana. The doors of the boxes are first opened to show what is in them, then closed again; after an interval of perhaps 10 minutes, the partition is raised, allowing the monkey to go and open the correct box and receive his reward. Normal animals can do this very well: frontal animals cannot, unless they spend the waiting period doing nothing except sitting and concentrating single-mindedly on the correct door.

These seemingly miscellaneous observations on the effects of frontal lobe damage can be unified quite satisfactorily once it is appreciated that each involves a defect in the ability to store a program of action, for further rather than immediate use. Anxiety is of course a side-effect of the sense that something has to be done in the future, and lack of anxiety may sometimes merely indicate a lack of forethought: worry, if rational, is a thoroughly good thing. Thus it was anxiety that presumably made our ancestors save some of their seed harvest to plant for next year, despite their immediate needs, and it is perhaps not going too far to suggest that the development of useful anxiety of this kind is indeed what separates us most from the other primates. Finally, it may also be that the unpleasantness of pain, particularly when it results from terminal illnesses (it is this type of pain that frontal leucotomy seems best at alleviating) is at least in part due to the anxiety it causes us by reminding us of our impending death: the painfulness of an injury depends very much on the significance that we attach to it (see Chapter 4). By stripping pain of its meaning for the future, we also relieve its emotional threat.

The parietal lobe

The parietal lobes occupy a central position in the cortex (Fig. 13.3), and one might therefore expect them to be concerned with the co-ordination of information from the visual, auditory, somatosensory and motor areas which surround them; and by and large this seems to be true. There are massive fibre bundles connecting these neighbouring cortical regions with the parietal region, and it also receives a projection from the *pulvinar* and *lateral posterior* nuclei of the thalamus. The pulvinar in turn receives sensory information from visual areas 18 and 19, and from the lateral and medial geniculate bodies; in addition it receives the usual reciprocal fibres from the parietal cortex itself. The lateral posterior nucleus obtains its input partly from the pulvinar and partly from the (somatosensory) ventral posterolateral thalamic nucleus (Fig. 13.3).

Practically all that we know of the parietal cortex comes from the effects of damage to it; there are very many kinds of defects that can arise from such damage, and not all are well localized. They fall essentially into three groups:

agnosia—disorders of high-level sensory analysis;
apraxia—disorders of high-level motor co-ordination and appropriateness;
aphasia—disorders in communicating and using symbols.

Agnosia

One kind of agnosia has already been mentioned in Chapter 4: lesions of parietal cortex near the somatosensory region may give *tactile agnosia*. Here there is no appreciable peripheral disorder—the subject has normal sensitivity to touch or temperature, and his acuity as measured by the two-point discrimination test may be unimpaired—but what is lacking is the ability to *use* this sensory data properly in order to recognize and respond to objects that are sensed by the skin. Such a patient may not recognize a matchbox when he is given one to hold, but can do so if he is allowed to see it; such difficulties in feeling the shape of an object in the hand are sometimes called *astereognosia*. (The -gnosia root, incidentally, means 'knowledge': 'astereognosia' means 'no-shape-knowledge'. A little familiarity with some Greek roots helps one to make some sense out of the forbidding jargon characteristic of clinical descriptions of parietal lobe defects.) A comparable defect of vision is called *visual agnosia*. Again, simple tests of visual performance reveal no abnormality—acuity, colour vision and sensitivity may all be normal—but the subject cannot always *appreciate* what he sees, and recognition of objects and places may be difficult. As in all agnosias, it is generally the most difficult tasks that are most affected, and in this case a difficulty in recognizing people's faces may be the first sign that something is wrong. A related but distinct defect is *spatial agnosia*: the subject has difficulty in appreciating the spatial relationships between objects, tends to get disorientated more easily, or may have difficulty in trying to draw a map or sketch a complicated object like a bicycle (Fig. 13.5). Very commonly the defect is unilateral, as a result of one-sided brain damage, and the disability is then confined to half of the visual field, which may often show lack of use or *neglect*. A curious feature of many of these high-level defects is that the subject may often be strikingly unaware that anything much is wrong, and resent suggestions to the contrary; the defective field is simply ignored—much as we ignore our own blind spot—and it may require specially designed neurological tests to reveal the disorder. One particularly bizarre example of this is when the agnosia takes the specific form of defects in the perception of one's own *body image*. Such a subject may emphatically deny that a particular part of his body such as a leg actually exists, and disown it when it is forcibly brought to his attention, as in the following Pinter-like dialogue:

Doctor: Is this your hand?
Patient: Not mine, doctor.
Doctor: Yes it is. Look at that ring: whose is it?
Patient: That's my ring. You've got my ring, doctor!

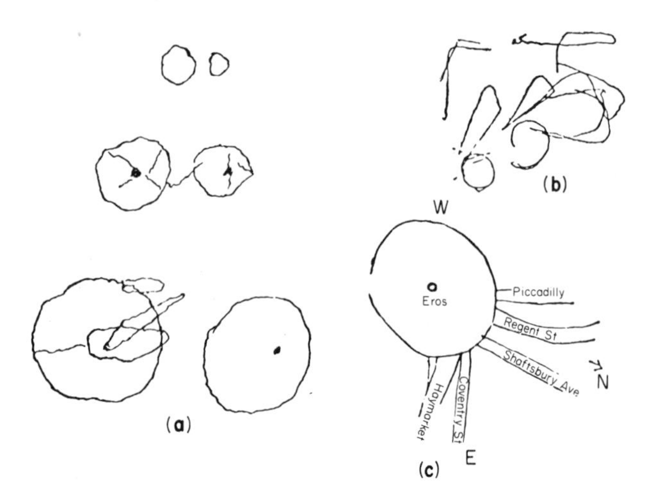

Fig. 13.5 (**a**) Attempts at drawing a bicycle: left parietal lesion. (**b**) Another attempt: biparietal vascular lesion. (**c**) An attempt at a map of Piccadilly Circus, showing neglect of top and left: right parietal lesions. (Critchley, 1971)

This is not merely a conscious fabrication on the part of the patient: he may be completely consistent in his attitude to the limb, not drying it after a bath, not bothering to dress it, tending to bump it against door-frames, and the like. Nor can one talk about any lack of intelligence in the normal sense: rather the lack of a certain kind of synthesis between somatosensory and visual inputs.

Apraxia

Apraxia implies clumsiness, but of a kind that is much more specific for *particular* tasks than the more general impairment associated with lesions of the cerebellum or motor cortex. It may be especially noticeable when the subject had previously been extremely skilled at using a particular tool or carrying out a highly trained sequence of actions, as in the case of a fish-filleter whose biparietal lesion led her to forget how to do it: although she 'knew in her mind' how to set about filleting a fish, she was unable actually to execute the manoeuvres that she wanted, and was sent home by the foreman for 'mutilating fish'. Sometimes a patient cannot produce specific actions on command—for example, gestures such as beckoning or saluting—but can do so spontaneously in appropriate circumstances. A more specific variety of apraxia is *constructional apraxia*, a sort of motor version of spatial agnosia: though the patient may seem to perceive spatial relationships quite readily, he may for example find it difficult to put building blocks together to make a particular shape, or construct a simple jigsaw puzzle.

Aphasia

There are many different ways in which aphasia may be manifested: a useful classification is into *sensory aphasia, motor aphasia* (these being in effect agnosia and apraxia in the particular field of language and communication), and *central aphasia*.

In sensory aphasia, the patient's sense of hearing may for example be perfectly normal, and the sounds of speech are heard, but they make no sense: he may complain that everything sounds like a foreign language. This is what is meant by *word-deafness*, and may indirectly lead to speech defects as well, since the patient can no longer monitor effectively the words he is producing. A similar disability specifically affecting reading is word-blindness or *alexia* (dyslexia in milder forms). These sensory aphasias are generally associated with a relatively localized region that borders on both visual and auditory cortex, called *Wernicke's area* (Fig. 13.6).

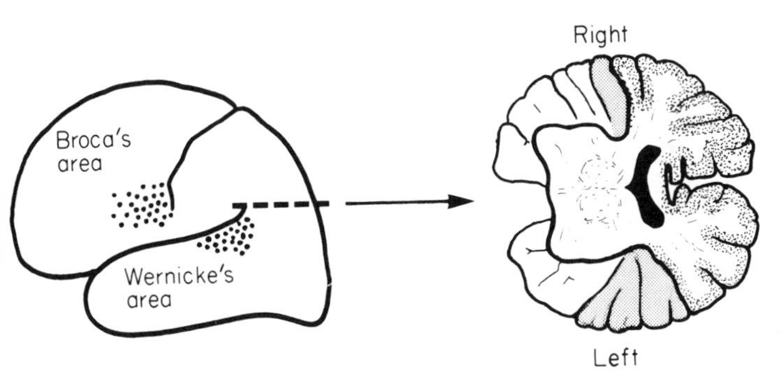

Fig. 13.6 Left: lateral view of left hemisphere, showing approximate location of Broca's and Wernicke's areas. Right: section along the dotted line above showing relative enlargement of the left planum temporale (approximating to Wernicke's area, shaded) in comparison with the right. (After Geschwind and Levitsky, 1968; copyright AAAS.)

In motor aphasia the subject can show by his actions that he understands what is said to him or what he reads, but has difficulty in initiating such communications himself. Thus one may find *agraphia*, an inability to write, and apart from the actual absence of speech (expressive or Broca's aphasia) there may be less severe disabilities such as stuttering and other more generalized defects of articulation *(dysarthria)*. That this is not simply a peripheral motor defect is shown by the fact that emotional expression is sometimes unaffected—swearing may continue unabated—and stutterers often find that under sufficient duress, when they are not thinking consciously about what they are saying, the disability may suddenly vanish. Lesions specifically affecting speech and articulation are associated with another cortical area, *Broca's area* (Fig. 13.6), which is close to the tongue and mouth regions of the motor cortex, and is not actually in the parietal lobe at all, but in the posterior frontal lobe. Damage here can frequently result from stroke due to a vascular accident; a classical example is that of Dr. Samuel Johnson, who has left a vivid account of what it is like to experience such a stroke. Johnson was a stutterer before this episode, and was notoriously clumsy: it is possible that he may in fact have suffered some slight brain damage in early life:

> 'I went to bed, and in a short time waked and sat up. I felt a confusion and indistinctness in my head that lasted, I suppose, about half a minute. I was alarmed, and prayed God, that however he might afflict my body, he would spare my understanding. This prayer, that I might try the integrity of my faculties, I made in Latin verse. The lines were not very good, but I knew them not to be very good: I made them easily, and concluded myself to be unimpaired in my faculties. . . . Soon after I perceived that I had suffered a paralytick stroke, and that my speech was taken from me. Though God stopped my speech, he left me my hand. My first note was necessarily to my servant, who came in talking, and could not immediately understand why he should read what I put into his hands. In penning this note I had some difficulty; my hand, I knew not how nor why, made wrong letters. My physicians are very friendly, and give me great hopes; I have so far recovered my vocal powers as to repeat the Lord's Prayer with no very imperfect articulation'.

A number of features of this account are interesting: the aphasia was clearly predominantly expressive, with a little disturbance of writing, but no disorder either of the ability to comprehend speech, or to formulate it in the mind—even in Latin verse!—and certainly no evidence of any general impairment of intelligent thought.

Central aphasia

This term covers a number of miscellaneous conditions in which the defect is not primarily either sensory or motor, but involves the mental mechanisms for forming concepts, for understanding symbols, and making sentences. A patient may be shown a common object such as a knife and be unable to name it: yet he can use it, and by employing a paraphrase—'what you use to cut with'—he may show that he can designate it in speech. Nor is the defect simply motor, since he can repeat the word 'knife' when told to do so; what seem to be at fault are the normal central *connections* that ought to link the sight of the object to the utterance of its name. Sometimes such a patient may use the wrong word for something without realizing it: given a pair of scissors he promptly describes them as a nail-file, and on being corrected may say 'no, of course it's not a nail-file, it's a nail-file'. He may produce speech sounds that are correctly executed and sound grammatical but actually make no sense; for instance, shown a bunch of keys, may come out with: 'Indication of measurement or intimating the cost of apparatus in various forms'. Such a response, often with much repetition of meaningless phrases, is described technically as *jargon*. All these defects may be quite specific for only one category of symbolization: thus in bilingual patients, only one of the languages may be affected. An interesting case of specificity of this kind occurred in the composer Maurice Ravel, who was affected by aphasia in later life, yet though unable to speak or write, could still sing and play and compose music. Other specialized aphasias of the central kind that have been described include *acalculia*, an inability to perform arithmetical operations, and *amusia*, an inability to appreciate music. Conversely, individuals are not infrequently found with extraordinary development of these same faculties—the *idiots savants* or calculating prodigies, infant musicians, and those remarkable people who seem to find it no trouble at all to learn 20 or 30 different languages: but these are not normally reckoned to be disorders!

It is in fact important to appreciate that *normal* people suffer from all types of aphasia on occasion. Not everyone can guarantee to complete the *Times* crossword puzzle; we all sometimes stutter or stumble over words; we are often at a loss for the name that goes with a face we know well, or for something that a moment ago was 'on the tip of our tongue'; and all of us are guilty from time to time of generating jargon, especially in social situations where we are compelled to speak, but have nothing to say. In people of limited education, one may observe a tendency for remarks to be repeated endlessly with only slight variations, or for a small number of concise adjectives to be applied indiscriminately; at a more exalted level we find 'ongoing situations', 'meaningful scenarios' and so forth. Equally, we all suffer at times from more or less severe attacks of agnosia: few normal people are really very proficient at, say, drawing a map of the town where they live; and untrained drawings of the human face reveal obvious distortions of the proportions between the various parts that amount to a kind of disorder of body-image perception (Fig. 13.7). All this suggests that the perietal lobes are a fruitful area for human improvement, that might well become better developed in the course of future evolution. We'd all like to be able to speak several languages, to be 'good with our hands', to have a good ear for music, to be able to recollect everyone we meet, and be quick at doing mental arithmetic: but it seems that our cerebral cortex is just not up to doing all these things well at once. What we mean by 'intelligence' is perhaps no more than a relative freedom from the more obvious kinds of aphasia!

Fig. 13.7 Left, young child's drawing of a man; middle, untrained adult's, showing incorrect proportions, especially in the position of the eyes, compared with reality (right): the eyes are half-way down the face

Left–right asymmetry in the brain

One important point of interest in connection with the aphasias is that they show a functional asymmetry between the left and right halves of the brain. Although the agnosias can on the whole be found with lesions of either hemisphere, aphasia is nearly always associated with lesions of the left hemisphere (that governs the right side of the body), at least in right-handed people. This asymmetry is reflected in the relative anatomical size of certain parts of the cerebral cortex on the two sides, notably in Wernicke's area (Fig 13.6). In living subjects this cerebral *dominance* (the dominant hemisphere being the one associated with aphasia) may be demonstrated by injecting a substance such as sodium amytal into the carotid artery on one side or the other, while the subject is carrying out some such task as reciting the letters of the alphabet. If the injection is on the dominant side, the recitation is interrupted for a short time and then continues; on the non-dominant side, very little is observed or felt by the subject. An exciting technique recently developed enables both dominance and other aspects of cerebral localization to be shown in an extremely dramatic manner (Fig. 13.8). One consequence of

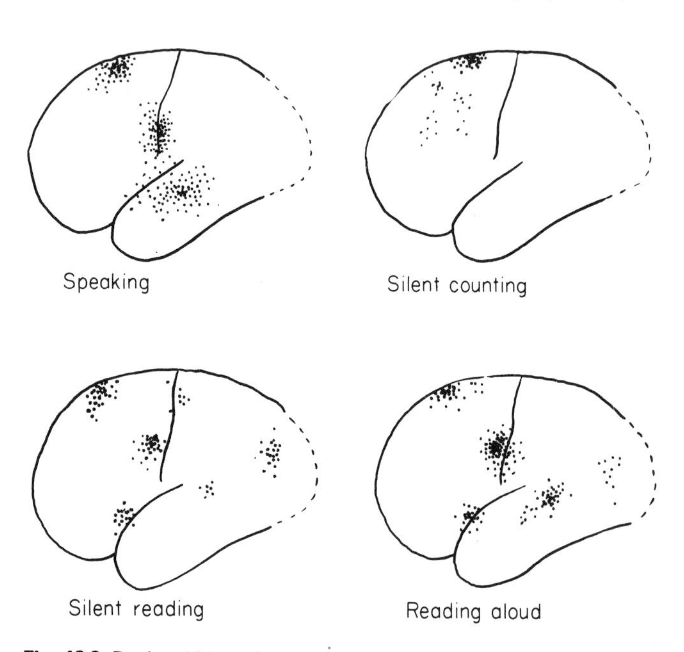

Fig. 13.8 Regional blood flow in the cerebral cortex of a conscious human subject, revealed by a radioactive marker, under the various conditions shown. (After Lassen *et al.*, 1978)

activity in any particular region of the cortex is that blood flow is locally increased; by using a radioactive marker in the blood, and a large number of detectors in an array over the skull, an image may be displayed on a television monitor that indicates which regions of the cortex are most active. In this way one may establish not only which half of the brain is the dominant one, but also see directly the changing patterns of activity associated with different types of mental process.

In most people with left-hemisphere dominance, one finds that not only is the right hand used preferentially for writing and other skilled tasks, but often the subject is right-legged and right-eyed as well: one may discover this by observing which foot is used to kick a ball, or which eye looks through a peephole. Other, more unconscious, actions may be revealing: thus the right leg may be crossed over the left when sitting, or if asked to fold his arms, the right arm may be placed on the left. But in the 7 per cent or so of the population who have right-hemisphere dominance, most (but not all) are found to be left-handed. The statistics relating to this correlation between dominance and handedness are shown in Fig. 13.9. It is clear that although there is a strong correlation between the two, it is not an absolute one; one factor that tends to distort such figures is that there are considerable social pressures from school and family for 'natural' left-handers to learn to use their right hands in preference, producing an artificial shift of the distribution towards right-handedness, shown by the horizontal arrows. It is likely in fact that in the

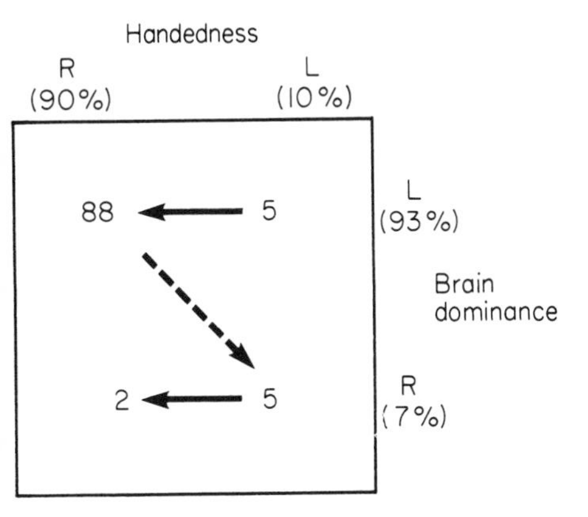

Fig. 13.9 Incidence of right and left handedness and brain dominance, expressed as percentages of the whole population. The horizontal arrows indicate the social pressures tending to turn natural left-handers into apparent right-handers; the dashed arrow indicates the likely effect of early damage to the left hemisphere. The data is derived from observations of the incidence of aphasia after unilateral brain lesions in right- and left-handers (Zangwill, 1967) and is therefore necessarily somewhat approximate

absence of such pressures the number of left-handers in the population would be rather more than the 10 per cent or so usually reported. It also appears from these statistics that there are essentially two distinct ways in which left-handedness can come about. The first is what might be called 'normal' left-handedness, and is probably genetically determined and essentially independent of dominance. The second type may be the result of slight brain damage to the left hemisphere early in development, which causes both speech and handedness to shift to the other hemisphere, as indicated by the downward diagonal arrow in the figure; of these, some are again converted to apparent right-handedness by social pressures. Left-handers in this category may often show vague disabilities of speech such as stuttering, or mild forms of apraxia or agnosia, a fact that has given left-handers as a whole—including the 50 per cent of left-handers who are in every way perfectly normal—a bad name: literally so, when one considers the etymology of words like 'dextrous' and 'sinister', not to mention 'right'!

One interesting consequence of the lateralization of speech occurs in patients who have undergone surgical section of the fibres of the *corpus callosum* (see Fig. 1.5), an operation sometimes performed when there is an epileptic focus on one side of the cortex, in an attempt to prevent its spread to the mirror-image position on the other side by the mechanisms described earlier. What is most surprising in such cases is the apparent lack of ill-effects, despite the severance of a massive fibre bundle containing more than 200 million fibres: a fact that prompted the facetious suggestion that the function of the corpus callosum is to allow the spread of epilepsy across the brain! However, although in the course of everyday life the patient may not perceive that much is amiss, more careful testing in the laboratory reveals an extremely interesting state of affairs: each half of the brain now appears to act independently, receiving information from the opposite half of the visual field, and controlling the opposite half of the body; but only the left hemisphere is able to speak and thus tell you what it is thinking. Each half can perform matching tasks, so long as input and output are both on the same side—an object in the left visual field can be chosen by the left hand to match a picture shown on the left, but not by the right hand—but tests involving a *comparison* of left and right cannot be done. The non-dominant side is not entirely aphasic, for it can read: the left hand will pick up a cup to correspond with the word 'CUP' presented in the left visual field, but the patient will be unable to

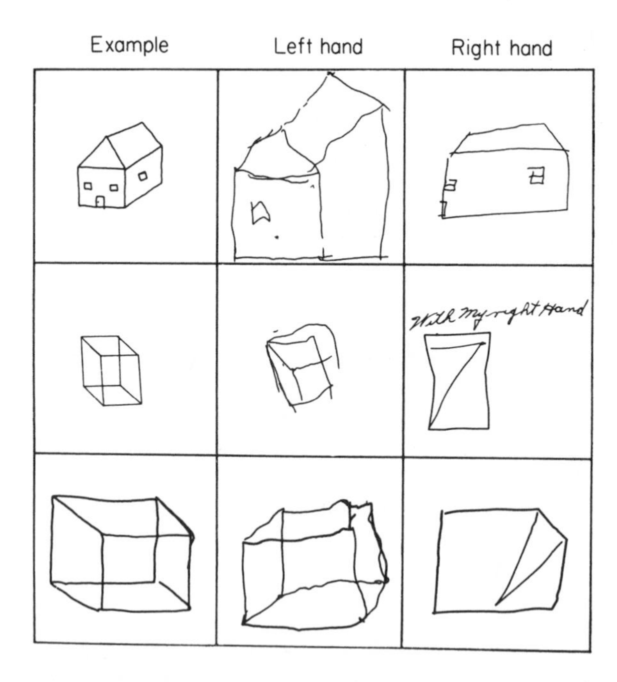

Fig. 13.10 Right- and left-handed attempts by a split-brain patient to copy the series of drawings on the left; the superiority of the left hand in dealing with the implied three-dimensional relationships is evident, although the right hand is slightly better at carrying out the drawing *movements* (Gazzaniga, 1967; copyright Scientific American, Inc.)

name the object he has just selected, because the specific function of speech is wholly localized in the other hemisphere. Nor can the non-dominant hemisphere do things like selecting from a list of alternatives the word needed to complete the sentence 'THE CAT SAT ON THE . . .', indicating that it suffers from central as well as motor aphasia. Some communication appears to be possible between the hemispheres, but of a subconscious, emotional kind rather than of 'facts'. Thus a patient whose non-dominant hemisphere is allowed to see a pornographic picture may blush or giggle, and asked what was there may indicate the awareness of the emotion without being able to describe exactly what was seen. Some functions—for example spatiovisual tasks like drawing, and probably musical appreciation as well—appear to be performed better by the non-dominant hemisphere (Fig. 13.10), which has led to a certain amount of semi-mystical speculation about the possibility of a fundamental split in Man's psyche between the rational and factual left hemisphere and the intuitive, artistic, right, and the importance of not allowing one hemisphere to develop at the expense of the other. Finally, one may sometimes observe in such split-brain patients the effects of evident struggles between the two sides about what should be done, one hand perhaps trying to tie up the patient's shoelaces while the other unties them. Such observations raise rather difficult problems concerning the nature of consciousness and its relation to the brain: do we here have two minds in one body? The one thing we cannot do, of course, is to ask the patient what *he* thinks is going on: only the dominant hemisphere will reply!

The temporal lobe: neo- and archicortex

There is no very clear distinction between temporal and parietal cortex, and it will already have been noticed that some of the areas mentioned in the preceding section—Wernicke's area, for one—lie partly in the temporal lobe. There is however a specific type of disability associated with damage to the temporal cerebral hemispheres, *amnesia*, which is quite different in kind from anything seen with damage to frontal or parietal cortex. Unfortunately, it is now recognized that the classical 'temporal lobe' disabilities are probably more related to structures forming part of the *limbic system*, and lying within the temporal cortex itself. To make clear the distinction between these two quite different areas that are often lumped together as 'temporal lobe' it is necessary to begin by outlining the structure and development of those deeper and older structures that form the limbic system, and which on account of their close association with the olfactory sense, discussed in Chapter 8, are sometimes classed together as the *rhinencephalon* or 'nose-brain'.

As has been noted several times before, older structures of the brain tend to get elbowed out of the way by newer ones, and thus often become twisted into complex and at first sight incomprehensible shapes; this is markedly true of the limbic system. It consists of nuclei (notably the *amygdala, septal nuclei, mamillary body* and *hypothalamus*) and areas of cortex (in particular the *hippocampal gyrus, cingulate gyrus,* and *entorhinal, periamygdaloid* and *prepyriform cortex*, the latter two having particularly important olfactory connections), all

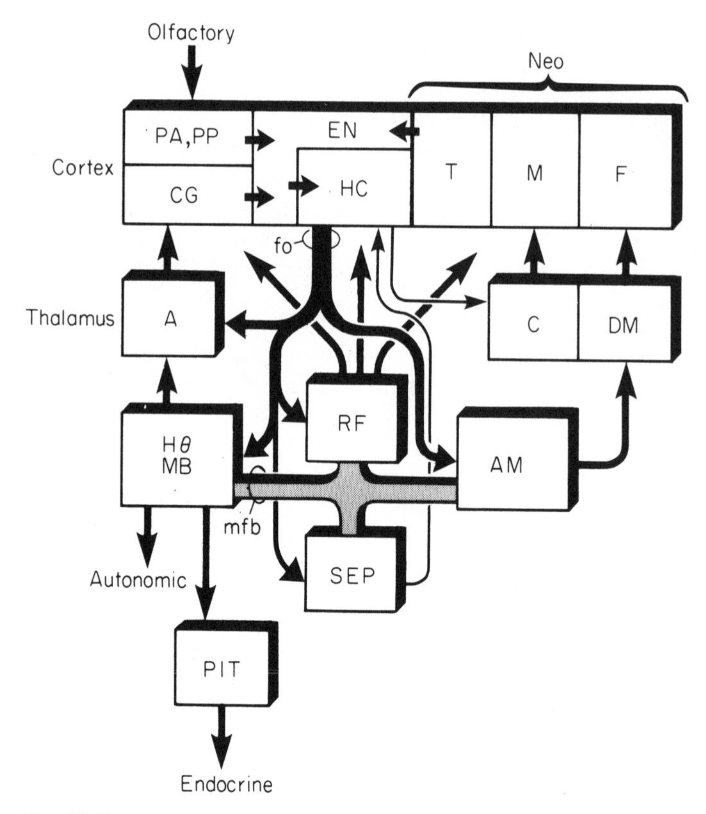

Fig. 13.11 Highly schematic and simplified representation of the principal areas of the limbic system and their connections to other structures. Cortical areas: PA,PP, periamygdaloid, prepyriform; CG, cingulate, EN, entorhinal; HC, hippocampal and subicular; T,M,F, temporal, motor and frontal neocortex. Thalamic nuclei: A, anterior; DM, dorsomedial; C, centromedian, intralaminar etc. Other regions: Hθ, hypothalamus; MB, mamillary bodies; AM, amygdala; SEP, septum; PIT, pituitary; mfb, medial forebrain bundle; fo, fornix

joined together by fibre tracts (for instance the *fornix, medial forebrain bundle*, and projections from the mamillary body to anterior thalamus, and from there in turn to the cingulate gyrus); these are all shown schematically in Fig. 13.11. Originally, it seems that the two main areas of limbic cortex, hippocampal and cingulate, formed almost the entire cortical surface of the brain, lying side by side immediately over the relatively compact group of their associated nuclei. But in the course of time, this *archicortex* was infiltrated by the newer *neocortex*, which by expanding almost explosively in the region separating the two areas, swelled into the modern balloon-like cerebral hemispheres, leaving the now dwarfed limbic cortex out on a limb (hence the name) round the edge (Fig. 13.12), and tucked away out of sight. Meanwhile, the massive fibre tracts required to link all this bulk of new cortex to the thalamus and other subcortical structures have forced their way between the older nuclei, cutting them off from one another and making their communicating nerve fibres wind

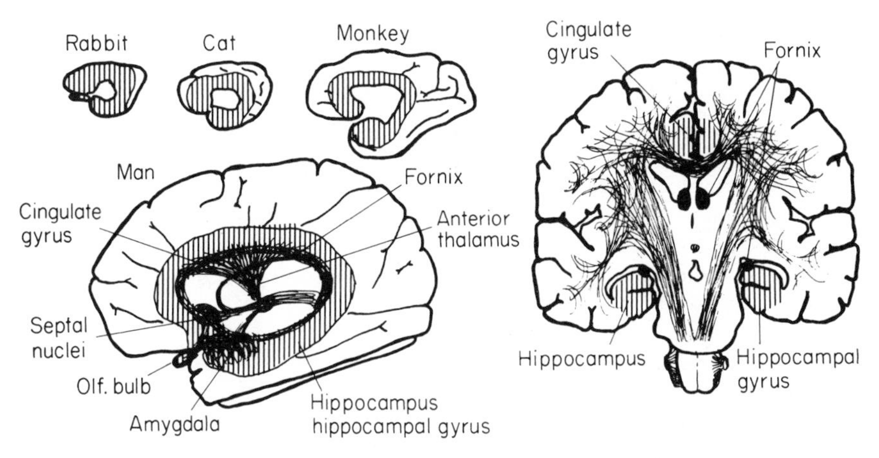

Fig. 13.12 The eclipse of the limbic cortex. Left: approximate area of limbic cortex (shaded) in rabbit, cat, monkey and Man, showing the relative growth of neocortex and consequent relegation of limbic structures to medial and central regions. (Partly after Ochs, 1965). Right: transverse section of human brain, showing limbic cortex (shaded) and fornix, in relation to massive fibre bundles serving the neocortex

their way right round the outside in circuitous fashion. At the same time, the amygdala, originally a structure on the wall of the hemisphere, became—like the corpus striatum—submerged beneath the incoming tide of neocortex, and ended up as an additional subcortical nucleus.

All this makes the neuroanatomy of the limbic system look a great deal more complex than it really is, and a schematic representation of the functional connections, as in Figure 13.11, is in many ways more helpful than trying to reproduce in one's mind all the three-dimensional muddle of its actual form. Most of the limbic system appears to be concerned with such functions as emotion and motivation, with the neural control of the body's internal environment, and to some extent with olfaction; these aspects will be dealt with in the next chapter. The cortical regions, and especially the hippocampus, seem on the other hand to be more concerned with learning and memory, and will be discussed here.

As we have seen, the hippocampus lies along the bottom edge of the temporal neocortex, and it is perhaps not too fanciful to think of it as a kind of cortical gutter: sensory information being increasingly analysed and refined as it trickles from neuronal level to neuronal level down from sensory projection areas, through the complex associational networks of parietal and temporal cortex, and finally draining into the hippocampus itself. Certainly in the more posterior region of the temporal neocortex, which borders on visual areas, the logical progression already noted in the visual cortex by which cells are found first with simple concentric fields, and then with more and more specificity in terms of such parameters as line orientation, movement, or width, appears here to be continued; thus cells have been found in the temporal cortex of monkeys which are claimed to respond preferentially to a stimulus as complex and specific as the appearance of the monkey's own hand. In Man, electrical stimulation in this region has been undertaken quite frequently as a preliminary to the surgical treatment of epileptic foci. What has sometimes been reported is that stimulation at particular sites gives rise not to the discrete flashes and spots of light characterisic of electrical stimulation of the human visual cortex, but rather to complex and repeatable hallucinations of an unusually realistic kind, sometimes apparently not static but moving in 'real time' (for example, of a tune played by an orchestra, to which the patient could beat time), and often producing an experience which is a synthesis of many sensory modalities at once. In one case a patient described a sense of it being Sunday morning, a bright summer day, the car being washed, children shouting, and so on. Such experiments have naturally been rare, and there must be some doubt as to how they should be interpreted: necessarily, they have not been carried out on normal people, and one can never be quite sure that the effect of electrical stimulation is not merely to evoke an epileptic discharge of some kind. Nevertheless, it does seem probable from electrical recording in animals that some kind of progressively more detailed analysis of sensory information, and its integration across sensory modalities, does occur on its way down to the archicortex that runs round the bottom edge of the temporal cortex.

A cross-section through this gutter—its seahorse-like shape giving rise to the name 'hippocampus'—reveals a surprisingly regular neuronal structure, almost as machine-like as the cerebellum (Fig. 13.13). Archicortex differs from neocortex in having only three layers instead of six; there is only one layer of pyramidal cells, with fibres running predominantly transversely above and below them, and making afferent synaptic contact with the pyramidal cell dendrites. There appears to be a regular sequential arrangement, with each pyramidal cell in the entorhinal cortex projecting to a long row of pyramidal cells in the dentate gyrus, each of these in turn projecting to a row of cells in the CA3 region of the hippocampal gyrus, and finally these cells in turn projecting to the pyramidal cells of the CA1 region. Branches of the CA3 cells form the large fibre bundle called the *fornix*, which projects to the mamillary bodies and septal nuclei and in Man contains more fibres than either the pyramidal tracts or optic nerves. The CA1 cells project to the neighbouring *subiculum* and thence, amongst other areas, to the anterior and central regions of the thalamus, by which route they may ultimately influence the basal ganglia and neocortical areas. Apart from receiving fibres from temporal neocortex, the entorhinal region has projections also from the neighbouring olfactory areas, prepyriform and periamygdaloid cortex, and septum. Thus the entire structure can be represented in the highly schematic form shown in Fig. 13.13**c**, a strongly hierarchical arrangement well adapted to integrating together information from neocortex and from the olfactory system, recognizing specific patterns of activity, and producing both motivational responses through the motor system and emotional responses via the limbic nuclei. The question of what its output actually does must wait until the next chapter; what we are concerned with here is how the output is derived from its input. That the hippocampus does indeed form in a sense the final output from the sensory analysers of the neocortex is clear from electrical recordings from its pyramidal cells. Ninety-five percent of the pyramidal cells of area CA3 have been described as totally multi-modal, responding to almost any

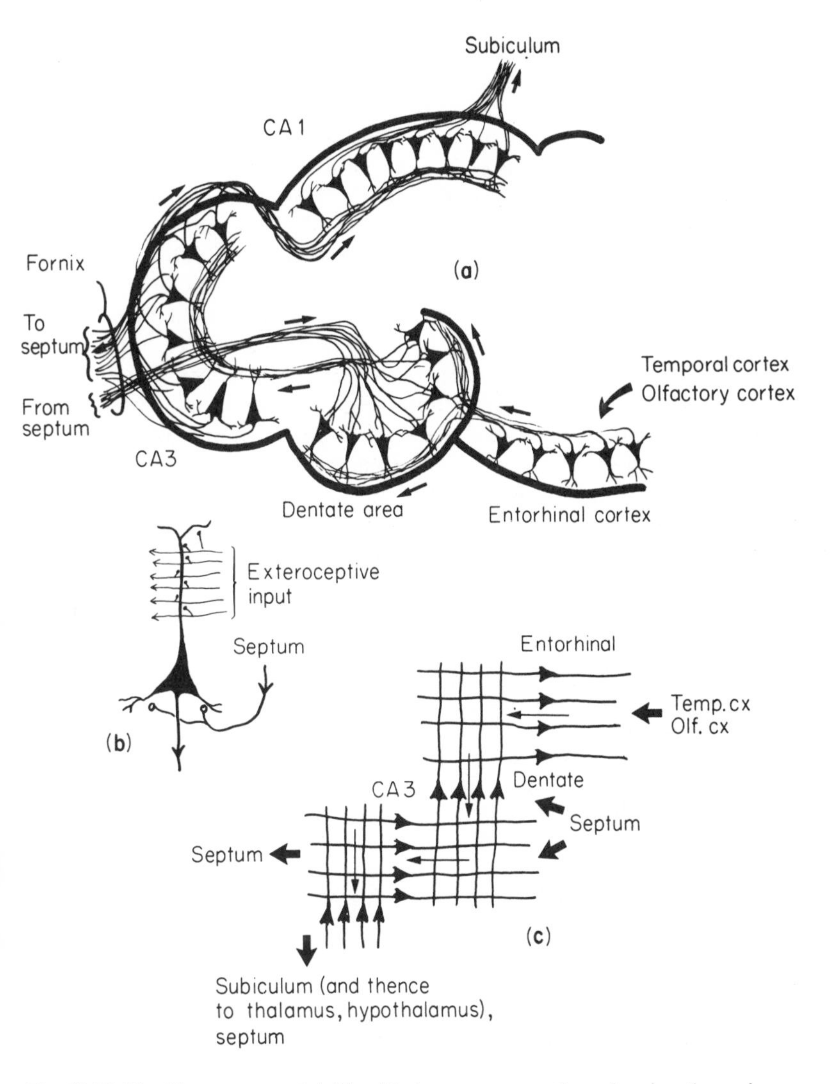

Fig. 13.13 The hippocampus. (**a**) Simplified transverse section, showing the main neuronal connections. (**b**) Schematic representation of single hippocampal pyramidal cell from CA3, showing different regions of termination of septal and 'exteroceptive' afferents (Maclean, 1975). (**c**) Stylized representation of neural circuitry of the hippocampus

combination of sensory modalities, and they are also described as being 'novelty-conscious'; that is, that they tend to show *habituation* to a stimulus if it is repeatedly presented, and respond more readily to things that are new; this in itself represents a kind of memory process. The main reason for believing that the limbic cortex is concerned with memory functions comes from the effects of lesions, and in particular with a somewhat rash operation that has sometimes been performed with a view to alleviating certain types of epilepsy, in which parts of one or both temporal lobes have been excised. The first operation of this kind, carried out by the American surgeon W. B. Scoville, consisted of the removal of the tips of the temporal lobes on both sides. This had the disastrous result that although the patient's memory for events that had occurred *before* the operation was good, he was in effect unable to remember for more than 10 minutes or so anything that had happened *after* it, a condition known as *anterograde amnesia* (Fig. 13.14). Subsequent work in animals has confirmed that it is damage to the hippocampal region that is responsible for this defect, and that it only happens when the lesion is bilateral, in which case the deficit is quite unspecific as to the nature of the material to be learnt: it can be recalled for some 5 – 10 minutes, but after that time, unless the subject can in some way rehearse it in his mind—as for example when trying not to forget a telephone number in the interval between looking it up in the book and dialling it—it is lost for good. Significantly, purely motor skills are unaffected: previously learnt ones are not lost, and new ones (like learning to type or to ride a bicycle) may be acquired: motor skills are learnt elsewhere, presumably either in cerebellum or neocortex. Unilateral lesions do not have the same dramatic effect, although some difficulty has been reported in learning verbal material if the lesion is on the dominant side.

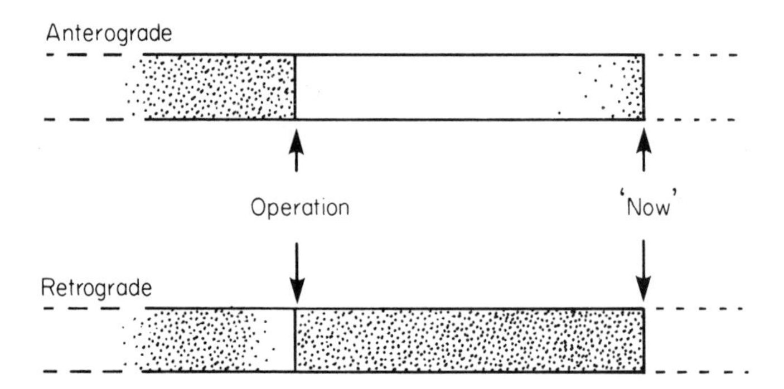

Fig. 13.14 Two kinds of amnesia. Above: anterograde amnesia: shaded areas represent the stretches of past experience that can be recalled (in this case only very recent events, or those before the operation). Below: in retrograde amnesia there is a loss of recall of occurrences just before the operation or other precipitating event

The commonest type of anterograde amnesia is seen as a result of chronic alcoholism, and is called the *Korsakov* syndrome. It is thought to be the result not so much of the effects of the alcohol itself as of the malnutrition that goes with it, in particular of thiamine deficiency. After death, one may see degenerative changes to various areas on the limbic system, notably the mamillary bodies and anterior thalamus, both of which lie on the output from the hippocampus; however one cannot of course be sure whether or not other regions, such as the hippocampus itself, may be functionally deranged even though their gross visual appearance may be normal. Again, long-term storage of memories is impaired and things cannot be remembered for more than a few minutes without conscious rehearsal. The victim consequently often seems to be stuck in a past era: if asked who the Prime Minister is, he might reply 'Harold Macmillan', and told to describe the latest fashions for men, would talk about winkle-pickers and drainpipe trousers. As in the agnosias associated with parietal cortex, the patient is often strikingly unaware that anything is wrong, and if confronted with facts that don't fit in to his private time-warp, may start to *confabulate*, making up elaborate fantasies to explain the discrepancies; he may also become paranoic and aggrieved, believing that there is some kind of global conspiracy directed against him.

The neural mechanism of memory

It is one thing to find a particular region of the brain that appears to be concerned with a particular function, but quite another to deduce exactly how the neurones of that region actually perform that function. In the case of memory, there is an additional complication in that there is not just one type of memory, but at least three, apparently associated with different regions of the brain. In the first place, we have the kind of sensory memory which is implied by the ability of higher-order sensory cells—for example in the visual cortex—to develop for themselves a selectivity to those particular patterns of input that actually occur in the environment (Chapter 7); in the second place we have the ability of the motor system to learn, through practice, to produce ballistic sequences of actions, using sensory feedback to modify the responses if they do not lead to success; and finally we have the kind of memory that seems to be lost in the Korsakov syndrome or after bilateral hippocampal damage: the ability to put together analysed information from different sources, attach some kind of significance to it, and then store it so that it can be recalled at will. Can we hope to find any common mechanism for each of these types of memory, performed by different areas of the brain? Certainly the first and third of these types are not conceptually very different: one can imagine a continuum of types of learning between, say, learning to associate patches of light on the retina into lines and edges, learning to associate these geometrical fragments into letters of the alphabet, and learning to recite a poem composed of the same letters. In each case, the key operation is one of forming *associations* between those elements of the stimulus that tend to recur together: if retinal units tend to fire in rows, we learn to recognize lines; if lines quite often lie in a certain relation to one another, we learn to recognize an 'E'; and after we have seen a particular configuration of such letters a few times, we have learnt 'Mary had a little lamb'. And as was emphasized in Chapter 1, sensory analysis and coding *necessarily* imply a degree of plasticity in the connections from one level to the next—imply in fact a kind of memory—if the brain is not to be of astronomical size. Furthermore, in the case of Marr's plausible model of motor learning by the cerebellum, examined in the previous chapter, it is the *associations* formed between sensory feedback patterns and ensuing fragments of action that ultimately result in the learning of motor sequences. Thus all learning by the brain amounts, in the end, to the formation of physical connections between neurones in such a way as to mirror the associations that exist in the real world between the stimuli that those same neurones code for.

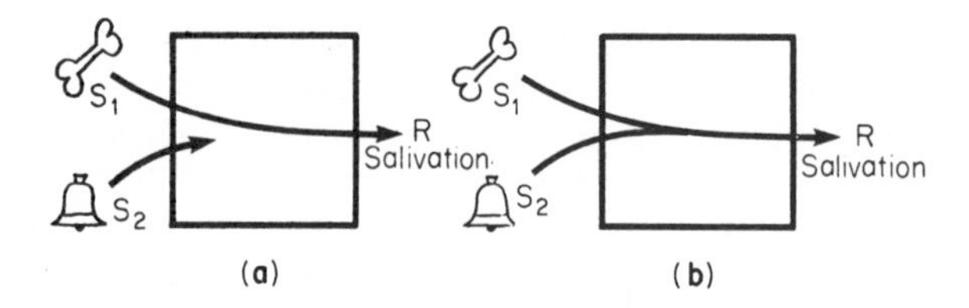

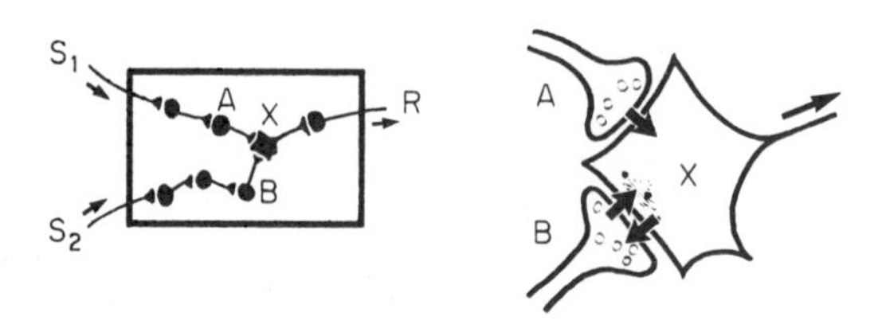

Fig. 13.15 Memory and neuronal connections. Above: schematic representation of functional pathways before (**a**) and after (**b**) Pavlovian conditioning of salivation (R) to the sound of a bell (S_2) by frequent pairing of bell and food (S_1). Below, simplified representation of the functional chains of neurones that must exist after conditioning: X (shown in more detail on right) is the first neurone common to both paths. It has an afferent A that is driven by S_1 and an afferent B, driven by S_2, whose effectiveness has increased because of frequent joint activity of X and B

Consider for example the classical example of the Pavlov dog, trained by frequent association of sound and food to salivate when a bell is rung (Fig. 13.15). What can we deduce about what must be going on in his brain? Here there are, in simplest terms, two stimuli or inputs (S_1, sight of food; S_2, sound of bell), and one output or response (R, salivation). In the end, since either input will produce the output, there must be at least one chain of neurones forming a functional pathway from S_1 to R, and another from S_2 to R. Before the period of training, the second pathway either does not exist, or perhaps exists in the structural sense but is functionally incapable of initiating salivation. It follows that learning the association between S_2 and R is brought about either by growth of new neuronal connections, or by the activation of pre-existing ones. The only questions that remain are, firstly 'What are the conditions under which such growth or activation occurs?' and secondly, 'What is the biophysical mechanism of these processes?'

Now there is one further point that may be deduced about the Pavlov dog's brain when it has finally learnt to make its conditioned response. There must be at least one cell—the neurone that actually innervates the salivary gland, if none other—that is common to *both* pathways; this is the cell X shown schematically in Fig. 13.15, and in the simplest case of all might have exactly one synapse (A) driven ultimately by S_1, and one (B) driven by S_2. Let us for the moment consider only the second of the two possibilities mentioned earlier, namely that both synapses are structurally in existence before the training period, but that the synapse B is in some kind of inactive, dormant state; we assume that synapse A on the other hand is always capable of firing

X and hence producing salivation. What we observe is that after sufficient pairings of food with bell, the bell alone eventually produces salivation; translating this into what is happening in the region of X, this means that the more often A (and hence X) fires *at the same time* as B, the stronger becomes the connection from B to X, until in the end B is able to fire X all by itself: the bell produces salivation. Note that it is the *associated* firing of B and X that is necessary to strengthen the synaptic connection: mere overactivity of B alone (if for example the unfortunate dog were to be subjected to continual bell-ringing *except* at meal-times) is not a sufficient condition. Cerebellar learning, in Marr's model, can be described in precisely similar terms: here it is the paired association of Purkinje cell firing with parallel fibre activity that results in strengthening of the connection from one to the other. In terms of Fig. 13.15, A is the climbing fibre, B is the parallel fibre, and X the Purkinje cell itself.

It may of course be objected that the notion that the connection from B to X already exists structurally before the period of training is an implausible one, even granted the amount of convergence and divergence of pathways that occurs in the brain, and the bringing together of diverse sources of information in such regions as the hippocampus. However, the model will still work without that assumption, if we imagine that paired firing of B and X results in some way in growth of B towards X and eventual functional contact (or alternatively, in growth of dendrites of X towards B). It is not difficult to think of plausible biophysical mechanisms that would behave in the required manner. The condition that B and X must be simultaneously active amounts to the same thing as saying that both must have channels in their membranes that are simultaneously open: this in turn suggests the notion that some substance Q might then and only then be able to pass from B into X or vice versa—Q might indeed be B's normal transmitter substance. If Q had the property that on penetrating *within* X (by entering channels that are open when X is active) it increases the sensitivity of X to the transmitter (perhaps by turning on the genes producing receptor sites, in a manner analogous to the interaction between afferent nerve fibres and acetylcholine receptors in muscle fibres described in Chapter 3), the required increase in synaptic effectiveness would occur. Alternatively, it might be that Q is released from X and in some way promotes the growth of B's terminals (but only when it succeeds in penetrating them). At all events, neither scheme proposes any particularly novel biophysical mechanisms, and meets quite satisfactorily the primary requirement that synaptic strengthening should only occur when both pre- and post-synaptic cells fire in association. The timing need not be very exact: if B fires a little before X, Q may still hang around for a while before finally diffusing away.

Circumstantial support for the existence of some mechanism of this kind comes from the fact that the output cells in precisely those three regions of the brain that are associated with learning of one sort of another—the pyramidal cells of neocortex and of hippocampus, and the Purkinje cells of the cerebellum—have a suggestive peculiarity of their dendrites that is not commonly seen elsewhere. Certain classes of afferent in each case terminate not directly on the soma or dendrite surface but rather on a sort of bud sticking out from it (the *dendritic spine*, Fig. 3.10), which contains a prominent Golgi apparatus, implying a specifically localized production of protein. It is tempting to speculate that the function of this apparatus is to initiate some structural or functional change in the post-synaptic membrane, doing it in a punctate, discrete way so that each spine can control its own receptiveness independently of others, presumably in response to some external and equally local stimulus. Some support for such an idea is given by the fact that the number of spines on visual cortical cells is greatly reduced by visual deprivation; but highly suggestive though the presence of these spines is, and plausible though the model of synaptic learning as a whole may be, the fact remains that there is as yet no hard neurophysiological evidence to back it up.

Almost all theories of memory imply some notion either of synaptic growth or the strengthening of local synaptic action by transmitter synthesis or by the generation or regulation of new receptor sites; whichever of these actually occurs, protein synthesis is likely to be needed, implying in turn some involvement of RNA. Many experiments have shown that there is a connection between RNA and memory: certain kinds of learning are associated with an increase in RNA synthesis in the brain, and in some cases with characteristic changes in the ratios of the RNA base pairs. Drugs such as puromycin that inhibit protein synthesis through their action on RNA can also in certain circumstances reduce an animal's learning ability. Some early investigators, enthused by findings of this kind, went so far as to suggest that memories could actually be *stored* in RNA molecules, by a process analogous to the storage of genetic information: particular experiences could be coded into corresponding sequences of RNA base pairs, and then decoded again when the experience was to be recalled. While it is true that some experiments were performed—the most famous being that of feeding flatworms with the flesh of other flatworms that had previously been taught some task, and seeing whether the memory was thus transferred—that seemed to support such an idea, they have often proved difficult to repeat, and few neurophysiologists now believe in this 'specific RNA' theory of memory. There are in any case very great conceptual difficulties in trying to imagine how a particular pattern of impulses in a neurone could possibly be converted into a sequence of base pairs, and even more in seeing how the code could ever be converted back into neuronal firing patterns. It seems more probable that the involvement of RNA in memory is a general rather than a specific one, in which a particular set of circumstances (for example, coincidence of pre- and post-synaptic firing) in some way triggers off the synthesis of new proteins that indirectly enhance the effectiveness of particular synapses.

Finally, it is clear that whereas a stimulus that one remembers for a life-time may only be present perhaps for less than a second, growth or strengthening of synapses must take some time to implement. There must therefore be a period of *consolidation* during which the event to be remembered is actually converted into some kind of semi-permanent structural change. There is in fact good evidence that there are really two distinct memory stores in the brain: a *long-term memory* (LTM) which takes the form of the kind of synaptic changes that we have been considering, and a *short-term memory* (STM) which retains information temporarily to cover the period—probably of the order of 20 minutes or so—during which consolidation takes place. It appears that STM is much more vulnerable than LTM, suggesting that the short-term store is a dynamic one, perhaps consisting of impulses continually circulating round looped chains of neurones. Sudden shocks of any kind—a blow on the head, or the passage of a large electrical current across the skull, as in electroconvulsive therapy—are sufficient to disrupt STM, and cause a characteristic type of amnesia called *retrograde amnesia*, in which the ability to recall events that occurred either after the shock, or long before it, are unimpaired, but a period of some 20 minutes or so before the shock remains more or less blank (Fig. 13.14). It seems as though memories need to be stored for a certain time in STM in order to make as it were a sufficient impression on the permanent memory trace, as suggested by the two-tank analogy of Fig. 13.16, and that violent disruption of the brain's activity through electroconvulsive therapy or some other shock simply empties the STM of its contents. Anterograde amnesia, characteristic of hippocampal damage and the Korsakov syndrome,

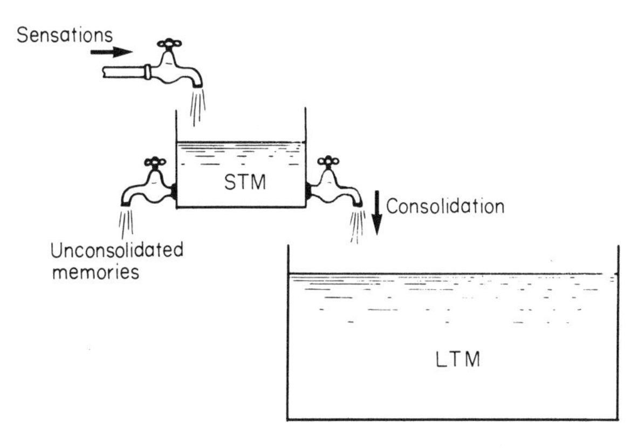

Fig. 13.16 The two-tank analogy of short-term and long-term memory (STM and LTM).

is in a sense exactly the opposite: it is as if the flow from the upper to the lower tank, from STM to LTM, had been permanently disconnected, leaving the patient with a functional STM but fossilized LTM. Finally, the leak in the STM tank in Fig. 13.16 is a reminder that not everything in STM—perhaps fortunately!—finds its way into permanent memory, and there is little conscious control, if any, over what is or is not permanently stored. Some unconscious control certainly does occur, since experiences with a strong emotional significance are almost always transferred to LTM. (A striking instance of this, for those of my own generation at least, is that nearly everyone remembers with unusual vividness exactly what they were doing when they heard the news, in 1963, that President Kennedy had been assassinated.) One complicating factor is that things may have been stored perfectly well in LTM, but cannot be recalled because the mechanism for *retrieval* is not working properly: this is particularly obvious in the case of experiences that are unpleasant, and psychiatric help may be required in order to bring such repressed memories to consciousness. In other cases, forgetting may be the result of learning new material. Since retrieval is essentially by association, memories that are linked together by too many associations may become irretrievably entangled. Unique and strange events are easy to recollect; boring things like telephone numbers are much more difficult, because of the vast number of pre-existing associations in our minds between each of the digits, the result of having remembered many other numbers in the past. Methods for improving one's memory that are commonly advertised generally work by translating each digit into a unique and vivid mental image: thus if 7 is 'elephant', 3 'cigar' and 4 'bicycle', the number 734 could be recalled by picturing an elephant smoking a cigar and riding a bicycle. The snag is of course obvious: after a while, there will be such a tangled knot of connections between elephants, bicycles and cigars and so forth as a result of learning one's friends' telephone numbers that new numbers will be just as difficult to remember as ever!

14

The highest levels of control

The concept of the hierarchical structure of the brain and of its organization in an ascending series of levels was introduced in Chapter 9. This chapter is concerned with the very highest levels of all, those that determine *what* is to be done, rather than how to do it. Actually, there is no very logical distinction between 'what' and 'how' in this sense: the task of deciding what to do amounts in the end to deciding *how* to stay alive, or at worst, how to immortalize one's genetic instructions. Thus the sensory inputs of this highest level come—somewhat paradoxically—not just from the special senses that tell us about the outside world, but also from those interoceptive senses that are usually considered so 'low' as to be beneath our conscious notice: information, that is, about the physiological well-being of the body, and state of the *milieu interieur*, and our distance from that final condition that awaits us all.

Motivational maps

Why, in fact, do we ever bother to do anything at all? Teleologically, the answer is obvious: even if an animal is at rest, it is using up energy which it must replenish, or perish. Unless it is completely sessile, relying on food that happens to pass by, this means that it must deliberately expend *some* of its energy on a sort of gamble in the hope of getting more back as a result, like a business investing part of its profits in the expectation of greater future returns. Whether or not to undertake the risk of such an investment is both the most important and the most difficult decision an animal has to make, and the whole of the recent expansion of the brain can be thought of as an attempt to reduce the risks involved, by making *predictions* about the likely outcomes of alternative courses of action on the basis of the closest possible analysis of more and more information about the outside world, and past experience of the relationship between actions and results, stored not just in our brains but in our books. But the decision process need not be as complicated as this. In a simple creature like an amoeba this fundamental mechanism of *motivation* is easy to appreciate: its decisions take the form of tropisms in response to gradients of chemical stimulation in its environment. On the one hand, attractive stimuli like food set up a positive gradient down which the animal moves; while on the other hand, poisons or other threatening conditions create

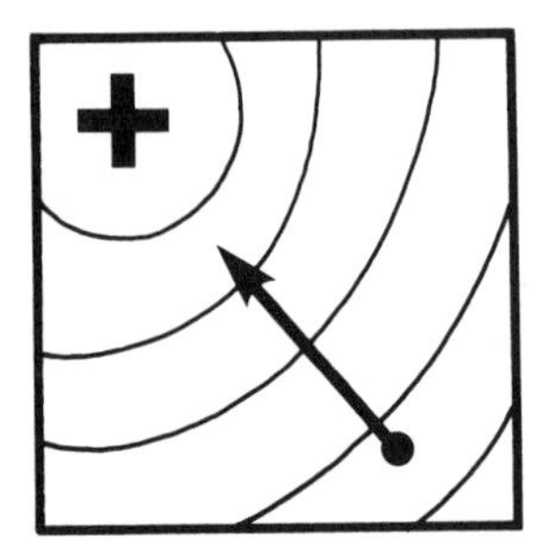

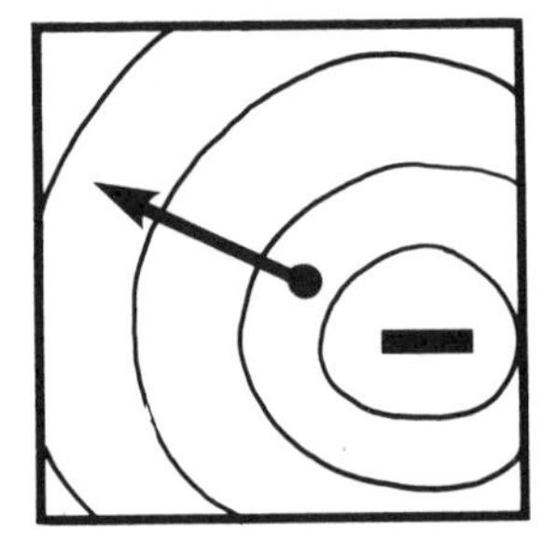

Fig. 14.1 Positive and negative motivational gradients

a negative gradient, and the animal moves away (Fig. 14.1). One can think of the amoeba's environment as a sort of motivational potential field, the amoeba itself acting like a little charged particle that moves about in response to local gradients, the path it traces out being a direct function of its environment.

Higher animals produce more complex behaviour, but the fundamental mechanism is essentially the same. The added complexity comes about for two reasons. First of all, because there are many more types of desirable and undesirable stimuli to which they may react, and many of them—perhaps most—are learnt: these are the secondary motivators (like money) that through experience become associated with other more self-evidently desirable goals. Consequently each individual has its own classification of stimuli into desirable and undesirable categories, unique because it is the result of that individual's own personal experience. Secondly, because the relative desirability of different attractants and repellants is constantly changing, in response to the organism's *needs*; thus changing patterns of need give rise to changing patterns of action even though the environment itself is the same, in a way that may seem to an outsider to be unpredictable. Thus Cambridge for me consists of a large number of separate gradient or contour maps, each corresponding to a different need: one for food, with high points at all the food shops and restaurants, one for money, centred on my bank, one for newspapers, for tobacco, and so on. Which one is operative at any particular moment depends on my need at that moment, rather like those electrical maps sometimes seen at tourist resorts, with bulbs that light up when you press one of a set of buttons marked 'parking', 'pubs', 'post offices', and so on. In other words, the central mechanism of the motivational system of the brain can be thought of as a sort of 'yellow pages' connecting particular gratifications to particular parts of one's internal representation of the outside world: like the tourist map, it translates information about need into the kind of tropistic data that can in turn be turned by the higher levels of the motor system into actual patterns of activity (Fig. 14.2). It seems very likely that these 'yellow pages' are embodied in the hippocampus, and that it is the hypothalamus, as the centre to which autonomic afferents project, and which itself monitors such physiological states of the blood as glucose concentration, temperature and osmolarity, as well as levels of circulating hormones, that defines one's state of

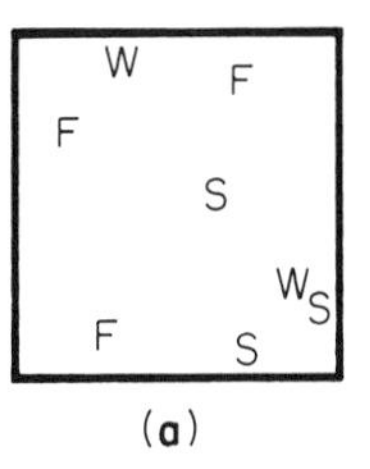

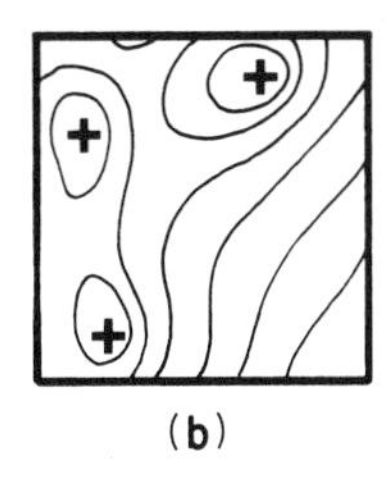

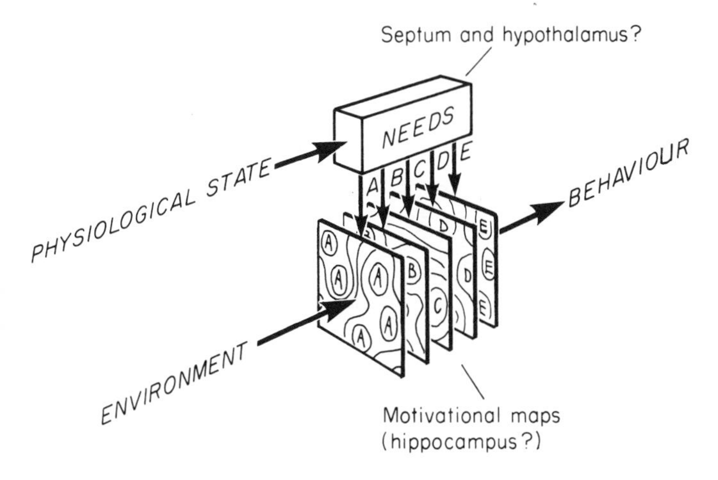

Fig. 14.2 Motivational maps. Above, (**a**), a neutral environmental map showing the location of food (F), water (W) and shade (S); and the corresponding motivational maps when the animal is hungry (**b**), or thirsty (**c**). Below, hypothetical model of mechanism for computing directed behaviour from needs sensed by monitoring the body's physiological state. A,B,C etc. are separate needs, as for example hunger, thirst, etc., and each has its own stored motivational map that is activated in appropriate circumstances

need. It is also in the hypothalamus that primary consummatory responses such as eating and drinking may be triggered off by electrical stimulation.

Perhaps it is difficult for us to accept the notion that our own richly complex lives, with the apparent wealth of choices open to us, and our sense of liberty to choose among them, could possibly be determined by so simple a mechanism. But as Herbert Simon has said, human behaviour is really rather simple, but because most people live in very complex physical, man-made and social environments, their actual behaviour *appears* extremely complicated; thus the path traced out by an ant moving over rough ground may be very complex in appearance, even though its behaviour is simply directed at getting back to its nest. In fact by averaging over large numbers of individuals, it is not difficult to measure quite directly the same kinds of tropistic gradients for us humans, that work so well in describing what an amoeba does. If you take a group of people and ask them the very simple question 'Where in Britain would you like to be?', it is possible to obtain contour maps of average

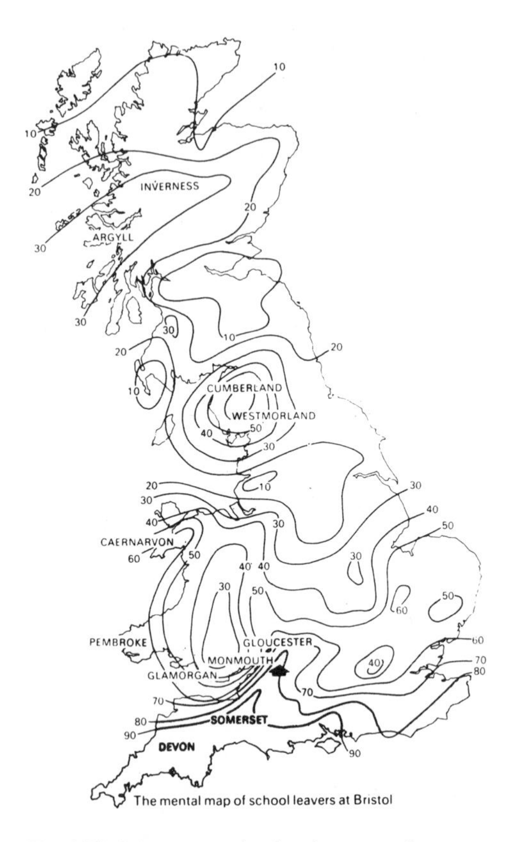

Fig. 14.3 A human motivational map: preference contours defining relative desirabilities of different parts of Britain (the question that subjects were asked was simply 'where would you like to be?'). (Gould and White, 1974)

preferences (Fig. 14.3) which presumably, if the individuals had the means to do it, would be translated into actual migratory behaviour not very different in essence from our amoeba moving blindly down its tropistic gradient.

Emotion

Apart from motivational tropism, another way in which behaviour is controlled is by switching on patterns of activity that are *not* directed at particular goals in the way that the tropisms are, but are rather in some sense *preparatory* or generally useful as an adjunct to the actual directed behaviour. This is what is meant in the widest sense by emotional behaviour; the emotions that we may *feel* at the same time are the sensory side-effects of this undirected behaviour (Fig. 14.4). There are as many types of emotional behaviour as there are types of motivational goal. Salivation, for example, is in this sense an emotional response accompanying the directed behaviour of

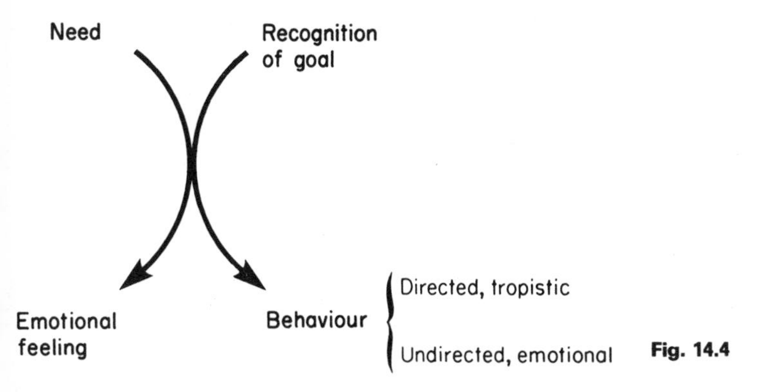

Fig. 14.4

getting food and eating it; and penile erection is an obvious preparatory response to another kind of goal. One may also include such internal responses as the release of hormones in this general category, as for example the surge of LH that triggers ovulation in response to copulation in some species, or the release of adrenaline associated with the need for sudden exertion. As Man has a richer set of possible needs and goals, including abstract or even spiritual ones, so his types of emotional behaviour and emotional sensations seem more varied and complex. But there are two basic emotional patterns found throughout the animal kingdom, and perfectly evident in Man as well, associated with tropisms of *any* kind: these are arousal and conservation.

Two basic emotional states

Arousal signifies the emotional state associated with a steep tropistic gradient, which may be either towards a desirable goal or away from a source of threat (Fig. 14.1): the state often described by physiologists as 'fight, fright or flight', that results in an increase in the general activity of the sympathetic system, and the release of adrenaline. The consequent bodily responses are all of more or less obvious use in preparing the body for the expenditure of the energy used to achieve the goal: blood flow through the muscles is increased, the heart rate is raised, glucose is released into the blood, the bronchioles and pupils dilate, the electrical activity of the brain increases, reaction times get quicker, and there is an associated feeling of general excitement. All of this of course involves a certain expenditure of energy, and would be a drain on the body's resources if kept up for a long time: but much is now at stake, and the gamble is one worth taking.

Conservation or withdrawal is in a sense the opposite of arousal. In a situation like that shown in Fig. 14.5, when every possible action is unpleasant—like standing in the middle of a minefield!—the sensible response is to conserve one's resources, and do nothing at all, in the hope that the difficulties will go away of their own accord. The result is inactivity and stupor, a loss of muscle tone, sleep or even hibernation; if the situation is a sudden one, there may be abrupt immobility or freezing—the animal thus incidentally making itself inconspicuous and feigning death (a common response to oncoming motor

Fig. 14.5 The kind of motivational map for which the only appropriate response is conservation

cars, but not a particularly helpful one). By all these means the rate of energy expenditure is greatly reduced, enabling the animal to ride out what may be only a temporary state of siege. The associated feelings are of apathy, tiredness and weakness: because of the reduction in muscle tone, one may actually feel heavier, pressed to the ground—the origin of the word 'depression'. Loss of muscle tone in the face produces a characteristic sagging of the lower jaw and of the corners of the mouth, and bowed head. In mild forms, the conservation state occurs only too commonly when a person feels that nothing is worth doing and circumstances are against him, giving rise to reactive depression. A curious feature of such chronic depression, though one that is readily understandable in terms of motivational maps, is that in times of severe and particular stress, as in war, the incidence of this kind of emotional state actually decreases. The more acute form of conservation is fortunately only rarely seen in civilized societies, except in response to cataclysmic disasters.

Fig. 14.6 (Engel and Schmale, 1972)

The woman in Fig. 14.6 has just emerged from shelter after an earthquake that has destroyed most of the town in which she lived. The objective signs of conservation are obvious: the stooped posture, the hand lifted to the face to support the dropped jaw, the immobile staring eyes. In such circumstances one may find a general state of apathy and inactivity that continues for a long time and is not conducive to survival.

Neither of these two kinds of emotional state is often seen in its pure form. Real objects tend to be both attractive and repellant: a hunting animal's prey may both be desirable as food and also dangerous, and in many species even the sexual act is a risky undertaking for the male. It can in fact be illuminating to think of emotions in terms of a continuum of types distributed around the two axes of arousal and conservation (Fig. 14.7), emphasizing the ambivalent

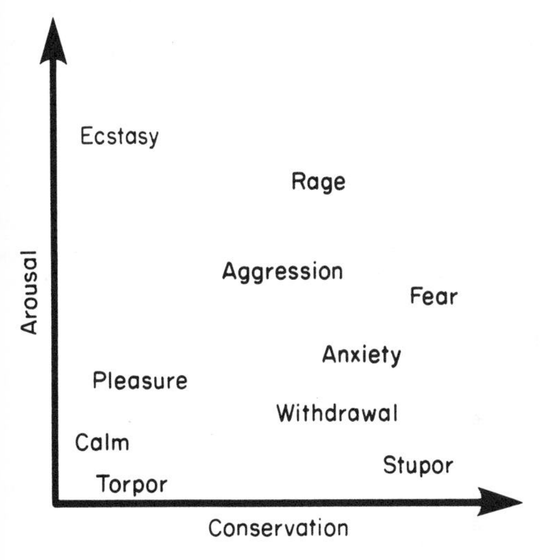

Fig. 14.7 Mixed emotional state as a result of stimuli that are partly attractive and partly repellant

nature of such states as rage and fear, the knife-edge between attack and retreat. Food does not usually have quite this effect on humans—dining is rarely a frightening experience in modern society—but rage can easily be elicited in situations of frustration, when the positive and negative aspects of a possible goal are nicely balanced. The generality of these two states can also be seen in human sexual behaviour, where sexual aggression shades off imperceptibly into sadism, and where affection may be expressed by licking and biting and other responses more appropriate to a food drive. Of course any such scheme is over-simplistic; for one thing, it ignores the important part that memory, especially the kind of anticipatory memory discussed in the previous chapter in connection with the frontal lobes, may play in introducing an extra *temporal* dimension into our emotions: such emotional states as hope, worry, confidence and regret clearly involve an element of this kind. But it may help us to remember that there is nothing particularly recherché or highfalutin' about human emotional responses, and that there is no reason to suppose that they are produced by fundamentally different mechanisms from those generating the remarkably similar patterns of behaviour seen in animals.

These neural mechanisms are not in fact well understood. The existence in the hypothalamus of neurones that detect need is of course well established. The postulated role of the hippocampus in forming associations between particular locations or objects and more basic motivational stimuli capable of satisfying these needs is suggested both by its general involvement in learning, discussed in the previous chapter, but also by more specific observations in animals. Hippocampal neurones have been found in rats that respond specifically when the animal is at a particular point in its environment, for example within a maze the rat has learnt, supporting the idea of hippocampal maps of the outside world; fish, who cannot really be said to have an environment of a mappable kind, since little in their world is fixed, are said to have no hippocampus at all. And its general relationship with the other areas of the brain—its output both to higher motor regions (for motivation) and to the hypothalamus (for emotional responses), and inputs from hypothalamus (need), olfactory areas, and other special senses—is exactly what is required

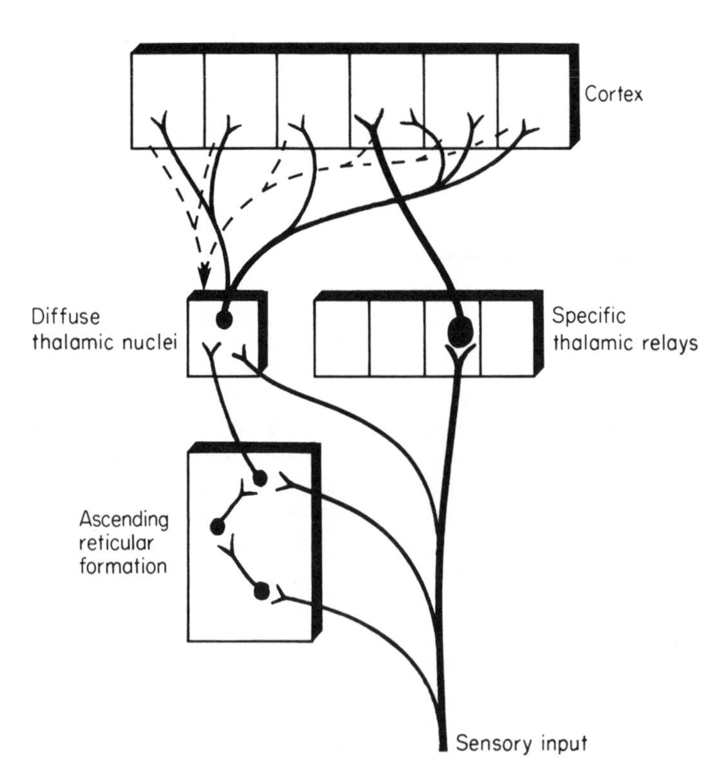

Fig. 14.8 Specific (thick lines) and non-specific (thin) routes by which sensory information can reach the cortex

for it to carry out such functions: one cannot say much more than that without indulging in the purest speculation. As far as the mechanisms of arousal and conservation are concerned, it has been known for a long time that in parallel with the specific ascending pathways that project from the sensory receptors via thalamus to the cerebral cortex, there is a second, more diffuse, ascending system, consisting mostly of the ascending reticular formation and the diffuse nuclei of the thalamus (Fig. 14.8) that seems to be closely associated with the control of generalized states of activity, particularly of the cortex. This *reticular activating system* has extremely widespread inputs from all over the brain and from collaterals of ascending fibres, integrated over neurones with astonishingly large fields of projection (Fig. 14.9); through the thalamic relays

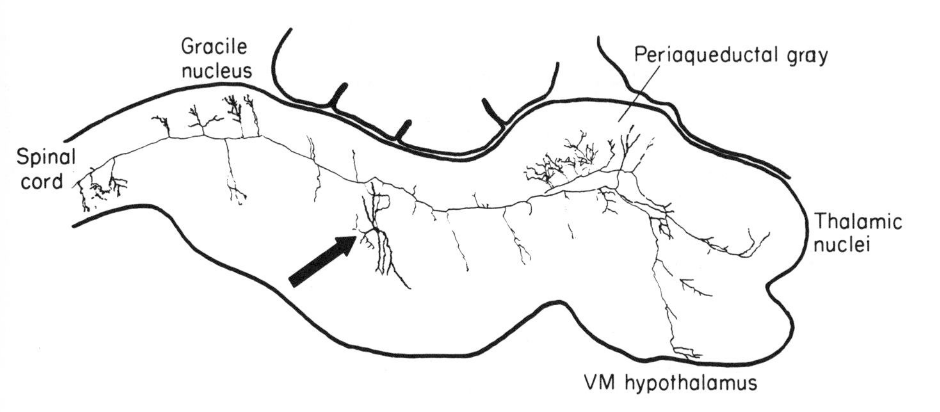

Fig. 14.9 Single cell (arrowed) from the nucleus magnocellularis of rat reticular formation, showing the distribution of branches of its axon to widespread areas of the brain. (Scheibel and Scheibel, 1957)

it seems to project in an unspecific manner on the neural circuits of the cerebral cortex, altering their level of activity by some process of general facilitation. This activation may on some occasions be relatively local, acting rather like a spotlight to focus attention on a particular cortical region—as when a sudden sound at night alerts our auditory system—or more widespread, in the generalized states of arousal described earlier. It may well also have a role in regulating the average level of activity in the cortex: as we have seen, the presence of such a large degree of convergence and divergence amongst the neurones of the cerebral cortex (the average number of synapses on a neurone in the monkey's motor cortex is around 60 000) means that they are liable to fire each other off in a kind of explosive chain reaction: one can think of the reticular formation as acting rather like the damping rods in a nuclear reactor, altering the thresholds of the neurones in step with the level of incoming sensory activity, in such a way as to maintain a sufficient degree of sensitivity without triggering off the kind of neural explosion that is seen in epilepsy.

Sleep

The general level of activity of the cortex may be measured by means of large electrodes attached to the scalp, which pick up the average electrical activity of very large numbers of cortical cells at once; a record of these potentials is called the *electroencephalogram*, or EEG. Paradoxically, the largest potentials are not those recorded when the brain is active, but when it is at rest. The reason seems to be that because cortical cells are so richly interconnected with one another, with multiple opportunities for feedback circuits that loop back on themselves, if left to their own devices they tend to lock into rhythmic oscillation, giving rise to waves of potential that run across the cortical surface like ripples on a pond. In a state of conscious quiet relaxation, these waves have a frequency of around 10 Hz, and are known as *alpha waves* (Fig. 14.10),

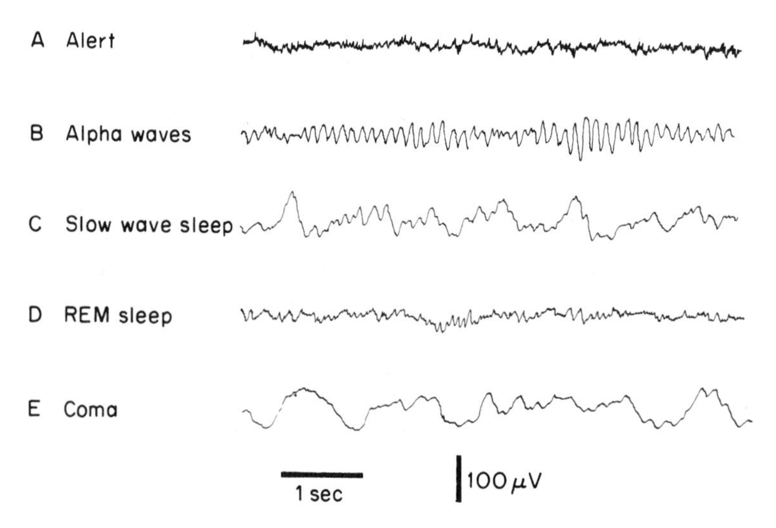

Fig. 14.10 Human electroencephalograms in different arousal states. (Penfield and Jasper, 1954; Roche, 1968)

most prominent near the occipital region. If the brain is aroused as a result of outside stimulation, this idling pattern is broken up into essentially random fluctuations of no particular frequency and small amplitude, just as the even flow of waves across a pond is disrupted by a shower of rain; the resultant EEG is then described as *desynchronized*. In sleep, which in many ways can be thought of as an emotional state akin to conservation, the EEG shows even more synchrony than that of the alpha rhythm in conscious rest: high-voltage waves of very low frequency, sometimes combined with bursts of high-frequency activity called spindles, are now seen (Fig. 14.10), a state called *slow-wave sleep* (SW sleep). The muscular tone of the body is much reduced, and the lack of cortical control is evident from the fact that primitive spinal responses like the Babinski sign may sometimes be evoked. However, if one examines the EEG continuously throughout the course of a whole night's sleep, one finds that the slow-wave pattern of activity is not present all the time: there is in fact a slow cycle of changes in the type of EEG, in which the slow-wave activity is interrupted every 2 hours or so by long episodes in which the EEG resembles closely what is normally recorded in the waking state. Yet the patient is actually more profoundly asleep—in the sense that he is harder to wake—during these episodes than he is in SW sleep, and for this reason these phases are called *paradoxical sleep*. About 20 per cent of adult human sleep is of this kind, and this proportion is much increased in children and babies. Associated with the desynchronization of paradoxical sleep is a marked increase in bodily movement; the eyes in particular make large and rapid excursions, giving paradoxical sleep the alternative name of rapid-eye-movement, or *REM sleep*. If the subject is woken during REM sleep, he usually reports that he has been dreaming: the movements of the eyes and body may well reflect the contents of the dreams. In sleeping animals, stimulation of the reticular activating system produces waking, and if continued, a marked state of general arousal. Lesions, on the other hand, put the animal in a permanent state of coma, with large-amplitude cortical waves; many general anaesthetics and sedatives act primarily on the reticular formation, as do such stimulants as amphetamine.

The generation of the two types of sleep is not thought to be reticular, however, but under the control of two brainstem nuclei, the *locus ceruleus*, and the nuclei of the *raphe* (Fig. 14.11). The cells of the locus ceruleus secrete noradrenaline; those of the raphe, serotonin. If the former are destroyed, the result is the abolition of REM sleep, but not of SW sleep; lesions of the latter, on the other hand, reduce both types of sleep and result in insomnia. What the *functions* of sleep are is far from clear; there are several well attested instances of people who have succeeded in ridding themselves of the habit of regular periods of sleep, though it is generally believed that in such cases, instead of having all his sleep in one daily dose, the subject tends to drop off continually for periods of perhaps a few seconds without noticing it. When sleep deprivation is properly enforced, there is a steady decrease in the brain's efficiency, together with marked irritability, and eventually a state resembling psychosis results. Teleologically, it makes sense for an animal which cannot function efficiently at night to harbour its resources by entering a conservation state (just as hibernating mammals do during the winter); what is not clear is what the particular function of REM sleep, and its associated dreaming, might be: it is evident that less conservation of energy is taking place during it. It may be that dreaming is in some way *needed* by the brain, for if a subject is specifically deprived of REM sleep—by waking him up as soon as his EEG shows desynchronization—it is found that for several subsequent nights the proportion of time spent in REM sleep is increased to make up for it.

Other regions of the limbic system have been shown to be concerned with arousal, conservation, and emotions more specifically associated with specific motivational drives. One set of structures, in particular, seems to correspond quite well with the two fundamental types of emotion, namely the dorsomedial and lateral nuclei of the *amygdala*, and their connections through the medial forebrain bundle with the lateral and ventromedial nuclei of the hypothalamus (Fig. 14.12). The dorsomedial amygdala projects directly on to the lateral

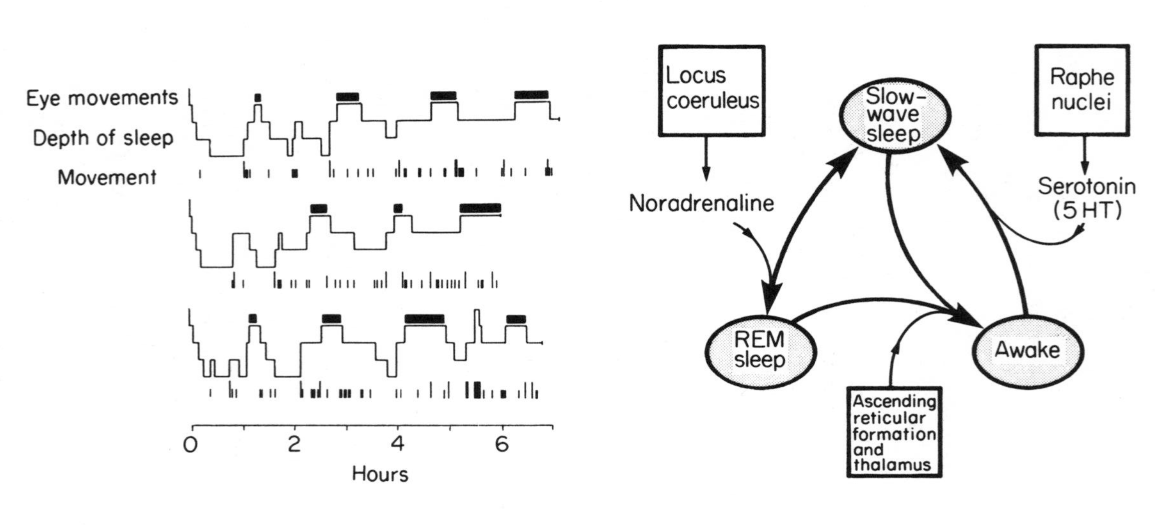

Fig. 14.11 Left, cyclic changes in a human subject during three different nights' sleep, showing correlation of periods of rapid eye movement with those of lightest sleep as determined by the EEG and increasing shallowness of the cycles as the night progresses (after Dement and Kleitman, 1957). Right, somewhat hypothetical relationship between the states of sleep and the activity of certain central areas

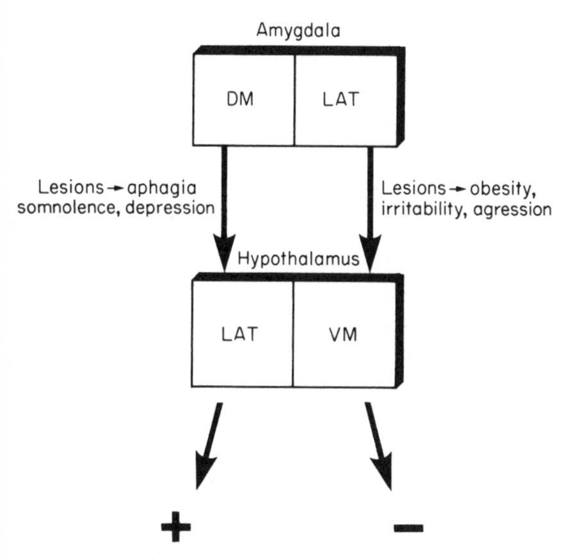

Fig. 14.12 Corresponding regions of amygdala and hypothalamus, broadly associated with arousal (+) and conservation (−). DM, dorsomedial; VM, ventromedial; LAT, lateral. (From data of Fonberg, 1972)

hypothalamus, and both regions seem to correspond in their actions with the vertical axis of Fig. 14.7. The lateral hypothalamus is well known as a 'feeding centre', in the sense that lesions here produce aphagia; but they also produce apathy and a general depression: dogs with lateral hypothalamic lesions are described as having a sad appearance, and are listless and somnolent. Lesions of the dorsomedial amygdala have similar effects, with perhaps more of the general, affective, component; stimulation in this region may produce hissing and growling and other signs of positive arousal. The properties of the other half of this system, the lateral amygdala and its projection to the ventromedial hypothalamus, are completely different, and correspond more to the horizontal axis of Fig. 14.7. The ventromedial hypothalamus has long been known as a 'satiety centre': lesions here produce over-eating and obesity, again with a more general affective change as well, an increase in irritability and aggressiveness. Stimulation of the lateral hypothalamus produces passivity, and even sleep or stupor. Large bilateral lesions of the amygdala and surrounding areas create an animal in which tropistic behaviour is greatly exaggerated (the *Kluver–Bucy* syndrome): everything in the environment seems indiscriminately attractive, and the monkey will compulsively examine and try to eat such things as the bars of its cage and its own faeces, and even things like snakes that would terrify a normal animal. The same kind of hypertropism is seen in its sexual activity: the animal is markedly hypersexual, and may try to copulate with members of its own sex, as well as inanimate objects.

Elsewhere in the limbic system, in a variety of regions, one may obtain fragments of emotional behaviour through electrical stimulation. In the hypothalamus in particular, because of its role as the relay through which autonomic responses are produced by the brain, one may obtain more or less co-ordinated bits of emotional repertoire such as piloerection, sweating, freezing, blushing, and generalized sympathetic activity as a result of electrical stimulation. The cingulate gyrus is a region where some specifically sexual responses may be elicited, as well as more general items of emotional expression such as pupil dilatation, changes in facial expression, and so on. Other regions of the limbic system are described as *pleasure centres*, in the sense that if an electrode is implanted in, for example, the septal nuclei, and connected up so that when the animal presses a lever in its cage it receives a pulse of electrical stimulation through the electrode, then as soon as the animal discovers what pressing the lever does, it will go on pressing it repeatedly, often in preference to 'really' pleasant stimuli like food or sex. Of course one cannot tell whether it is *feeling* pleasure as a result: but it is clear that the electrode must in a sense be bypassing the normal motivational processes of hypothalamus and hippocampus, and in some way activating the tropistic input to the motor system directly. Other sites have been found to produce direct motivation in this way, including parts of the amygdala; but other regions of the amygdala, like the dorsomedial thalamic nucleus, have exactly the opposite property: once the lever is pressed, it is never pressed again, presumably because the stimulus is activating avoidance rather than positive tropism; but one has to be sure in such cases that the animal is not merely feeling pain.

'Mind' and consciousness

'Nothing puzzles me more than time and space; and yet nothing troubles me less, as I never think about them' (Charles Lamb)—a reaction not very different from that of most neurophysiologists to problems of mind, brain, and consciousness. This is of course a field that has been thoroughly dug over since the days of Descartes and Hume and indeed long before: and philosophers have every right to question whether mere empirical physiologists can add much to such a hoary debate, in which the various arguments have been rehearsed so exhaustively. But recent developments both in neurophysiology and in computer science—the fact that I can now go into a chain-store with £20 in my pocket and come away with an electronic device hardly bigger than a box of cigarettes, that is the intellectual superior of half the animal kingdom—have so enlarged our notions of what classes of operation a physical system may in principle be capable of, that a great deal of earlier thought on the subject is now merely irrelevant. In a nutshell, 'brain versus mind' is no longer a matter for much argument. Functions such as speech and memory, which not so long ago were generally held to be inexplicable in physical terms, have now been irrefutably demonstrated as being carried out by particular parts of the brain, and to a large extent imitable by suitably programmed computers. So far has brain encroached on mind that it is now simply superfluous to invoke anything other than neural circuits to explain every aspect of Man's overt behaviour. Where Descartes' dualism proposed some

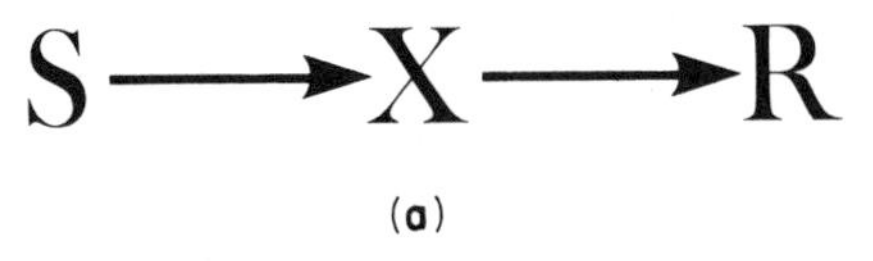

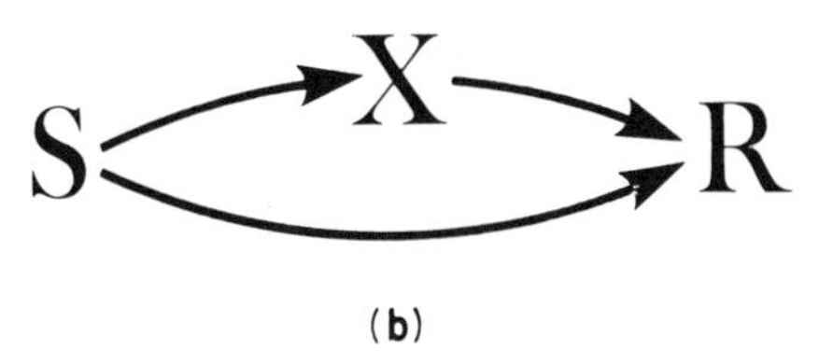

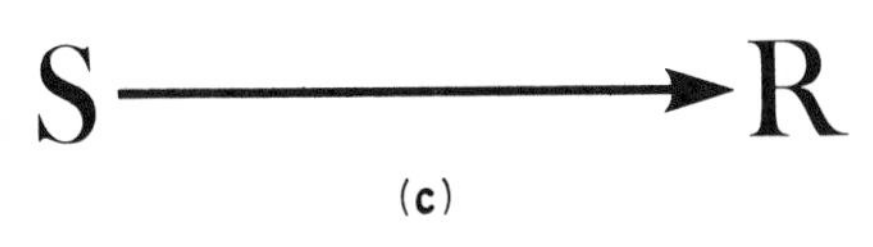

Fig. 14.13

non-material entity—the 'ghost in the machine'—that was provided with sense data directly by the sensory nerves, analysed them within itself, and then responded with appropriate actions by acting directly on motor nerves (the mind thus having the same relation to the body as a driver to his car: Fig. 14.13**a**) and later to be modified to include the existence of certain automatic reflexes that clearly do not pass through the mind (Fig. 14.13**b**), modern neurophysiology admits of *no* other path between stimulus S and response R than unbroken chains of neural connections (Fig. 14.13**c**): X, the ghost in the machine, has finally been laid to rest.

So is there still a problem, or have the philosophers been wasting their time? Indeed there is: that problem is *consciousness*. However sure I may be that (**c**) is a fair representation of *your* brain, there remains the obstinate and unshakable conviction that *my* brain is like (**a**). Though I might perhaps concede that (**b**) is a better representation, and particularly after reading Freud may reluctantly agree that a great deal of what I do is not consciously willed, nevertheless that there is *no* X is simply inconceivable. Now philosophers can have a great deal of fun with beliefs such as these, since the existence of my own consciousness is not something I can prove to other people in the way I can, for example, prove that I have hands. Clearly its outward manifestations could easily be imitated by a machine (as Hebb's example of the calculator programmed to say 'I am multiplying' every time it multiplies). But this kind of scepticism is so self-consistent as to be utterly tautological: if, like Wittgenstein, we decide that the only criterion for consciousness must be overt, public, behaviour, then of *course* we have nothing to say about it, because we have defined it out of existence. And for lazy neurophysiologists it provides a veneer of philosophical respectability for their unwillingness to think about the subject at all: Pavlov used to fine his students when he caught them using the words 'voluntary' or 'conscious'.

Yet to evade the problem by such specious materialism is perhaps no worse than to take the opposite extreme and accept consciousness as something much too mysterious and wonderful for a scientist even to be able to begin to think about. Both attitudes contribute to the evident intellectual muddle that surrounds the whole subject. So how would a brash and simple-minded physiologist proceed? Once he had accepted the reality of the phenomenon, he might go on to relate it to the fabric of the brain in much the same way as he would in the case, say, of the sense of sight. It is clear, for example, that loss of a limb does not lead to blindness, whereas loss of the eyes does; and by the use of inductive reasoning hardly more sophisticated than that, one may proceed into the brain itself and map out, almost neurone by neurone, the mechanism of the visual pathways. This kind of work has not of course been carried out systematically in the case of consciousness, if only because experiments of this sort on animals are useless to us. All the same, it is clear that we do in fact already know quite a lot about the functional anatomy of consciousness, even if we have little idea what consciousness actually *is*. We know, for instance, that while massive lesions of the cerebral cortex and its underlying fibres may blunt our perceptions, paralyse our limbs, impair our intelligence or even—as in the case of Phineas Gage—our morality, they have little effect on consciousness itself. Conversely, relatively slight injuries—perhaps a blow on the chin—that affect an area in the core of our brainstem (the same region of the reticular formation that is associated with arousal and sleep) can produce complete unconsciousness even though the whole of the rest of the brain is unimpaired. At a different level of description, it is clear that we are conscious of some kinds of brain activity but not others, and that the boundaries of this zone of awareness are not fixed, but vary on occasion. By and large it is what goes on at moderately high hierarchical levels that we are conscious of, although by introspection we can often learn to increase our awareness of lower levels. Curiously enough, we also tend to be relatively unconscious of the highest levels of all, those that control our motivations; and further, even the most complex mental processes can sometimes be carried out without being conscious of the fact at all. While reading through a difficult score at the piano, I have suddenly had the realization that for several bars I have been thinking of something entirely different, yet my brain has been getting on with the complex task of translating printed notes into finger movements perfectly well without me. Often we may find the solution to a problem that has baffled the most energetic conscious cerebration quite suddenly and unexpectedly when we were not thinking of it at all—perhaps, like Archimedes, in the bath; or, like Coleridge with *Kubla Khan* asleep! L. S. Kubie has gone as far as to say that there is nothing we can do consciously that we cannot also do unconsciously. The natural conclusion must surely be that since consciousness is more associated with 'higher' functions than 'lower', yet not particularly affected by damage to precisely those regions of the brain that we know to carry out those higher functions, nor is it necessary to be conscious for those functions to be carried out perfectly adequately, that the ghost in the machine is not an executive ghost, as it is in (**a**) and (**b**), Fig. 14.13, but rather a *spectator*, watching from its seat in the brainstem the play of activity on the cortex above it, able in some way to direct its attention from one area of

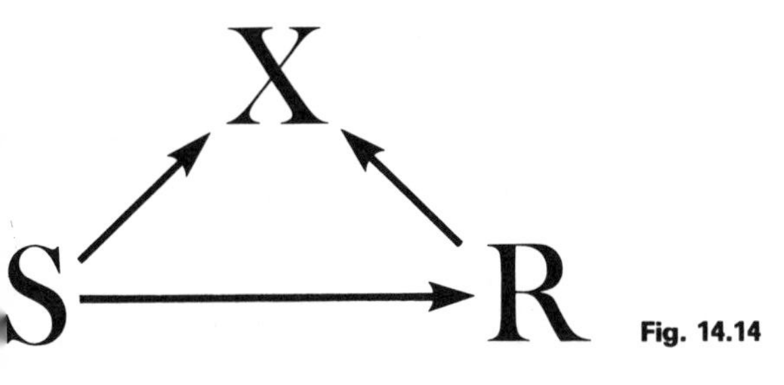

Fig. 14.14

interest to another, but not able to influence what is going on (Fig. 14.14).

But what about free will? The ghost in such a scheme would observe the body's actions being planned, and see the commands being sent off to the muscles before the actions themselves began, and so one can well imagine how it might develop the illusion that because it knew what was going to happen, that it was itself the cause. For X, the distinction between 'I lift my arms' and 'My arms go up', in which Wittgenstein epitomized the notion of voluntary action, would amount simply to the distinction between those actions which it observed being planned, and those—such as reflex withdrawal from a hot object—which it did not. There is no implied necessity here for us to be *deterministic* in our actions—to an outsider we may appear to have free will—since the physical processes linking S and R can be as random and essentially unpredictable as we please. Such a scheme seems more intellectually satisfying than (**a**) or (**b**) (Fig. 14.13) without conflicting with our own feelings about ourselves; and unlike (**c**), does not merely evade the issue. The most serious objection to it is perhaps that it is difficult to see what on earth X is *for*, since it can't actually do anything. Perhaps it does occasionally intervene. But in any case, what is the audience at a concert for? Or the spectators at a football match? The idea that I am being carried round by my body as a kind of perpetual tourist, a spectator of the world's stage, is not—on reflection—so very unattractive. The moral is clear: enjoy your trip!

References and Further Reading

Chapter 1 Studying the brain

References

Buchsbaum, Ralph (1971). *Animals without Backbones.* Penguin, Harmondsworth.

Fowler, O. S. *Human Science of Phrenology.*

Reading

Brodal, A. (1981). *Neurological Anatomy in Relation to Clinical Medicine.* Oxford University Press, Oxford.

Carpenter, M. B. (1976). *Human Neuroanatomy.* Williams & Wilkins, Baltimore. (Two general accounts of the anatomy of the brain).

Chandler Elliott, H. (1963). *Textbook of Neuroanatomy.* Pitman, London. (Valuable for its clear three-dimensional illustrations of brain topology).

Gaze, R. M. (1970). *The Formation of Nerve Connections.* Academic Press, London.

Hofstadter, D. R. (1980). *Gödel, Escher, Bach: an Eternal Golden Braid.* Penguin, Harmondsworth. (For the mathematically-inclined, an imaginative and thoughtful discussion of fundamental problems in understanding the brain).

Lashley, K. S. (1929). *Brain Mechanisms and Intelligence.* University of Chicago, Chicago. (A reference to the idea of 'mass action').

Lund, R. D. (1978). *Development and Plasticity of the Brain.* Oxford University Press, Oxford. (Describes neuroanatomical techniques as well.)

Sherrington, Sir Charles (1941). *Man on his Nature.* Cambridge University Press, Cambridge. (A reflective, even poetic, appreciation of the brain by a distinguished neurophysiological pioneer).

Stevens, P. S. (1976). *Patterns in Nature.* Peregrine Books, Harmondsworth (Beautiful examples of the way in which simple rules may give rise to complex patterns).

Walsh, E. G. (1964). *Physiology of the Nervous System.* Longman, Harlow. (A highly intelligent and readable account, with ideas and examples not found in more recent general textbooks).

Chapter 2 Nerves

References

Erlanger, J. and Gasser, H. S. (1938). *Electrical Signs of Nervous Activity.* University of Pennsylvania Press, Philadelphia.

Fabre, P. (1927). L'excitation neuro-musculaire par les courants progressifs chez l'home. *Complex Rendu de l'Academie des Sciences, Paris* **184**, 699–701.

Hodgkin, A. L. (1938). The subthreshold potentials in a crustacean nerve fibre. *Proceedings of the Royal Society B* **126**, 87–121.

Hodgkin, A. L. (1939). The relation between conduction velocity and the electrical resistance outside a nerve. *Journal of Physiology* **94**, 560–570.

Hodgkin, A. L. (1958). Ionic movements and electrical activity in giant nerve fibres. *Proceedings of the Royal Society B* **148**, 1–37.

Hodgkin, A. L. and Horowicz, P. (1959). The influence of potassium and chloride ions on the membrane potential of single muscle fibres. *Journal of Physiology* **8**, 127–160.

Hodgkin, A. L. and Huxley, A. F. (1952*a*). The components of membrane conductance in the giant axon of *Loligo*. *Journal of Physiology* **116**, 473–496.

Hodgkin, A. L. and Huxley, A. F. (1952*b*). A quantitative description of membrane current and its application to conduction and excitation in nerve. *Journal of Physiology* **117**, 500–544.

Hodgkin, A. L. and Katz, B. (1949). The effect of sodium ions on the electrical activity of the giant axon of the squid. *Journal of Physiology* **108**, 37–77.

Katz, B. (1966). *Nerve, Muscle and Synapse.* McGraw-Hill, New York.

Rushton, W. A. H. (1951). A theory of the effects of fibre size in medullated nerve. *Journal of Physiology* **115**, 101–122.

Reading

Aidley, D. J. (1978). *The Physiology of Excitable Cells.* Cambridge University Press, Cambridge. (A well-written account of nerve conduction, synaptic action, and other topics).

Hall, J. L. and Baker, D. A. (1977). *Cell Membranes and Ion Transport.* Longman, Harlow.

Hodgkin, A. L. (1964). *The Conduction of the Nervous Impulse.* Liverpool University Press, Liverpool.

Hodgkin, A. L. (1977). Chance and design in electrophysiology: an informal account of certain experiments on nerve carried out between 1934 and 1952. In *The Pursuit of Nature.* Cambridge University Press, Cambridge. (Conveys much of the atmosphere of the classical experimental work on nerve conduction).

Kuffler, S. W. and Nicholls, J. G. (1977). *From Neuron to Brain.* Sinauer Associates Inc, Mass. (A particularly lucid account of the electrical behaviour of circuits and cell membranes).

Chapter 3 Receptors and synapses

References

Araki, T., Eccles, J. C., and Ito, M. (1960). Correlation of the inhibitory postsynaptic potential of motoneurones with the latency and time course of inhibition of monosynaptic reflexes. *Journal of Physiology* **154**, 354–377.

Burke, R. E. (1967). Composite nature of the monosynaptic excitatory postsynaptic potential. *Journal of Neurophysiology* **30**, 1114–1137.

Coombs, J. S., Eccles, J. C. and Fatt, P. (1955). Excitatory synaptic action in motoneurones. *Journal of Physiology* **130**, 374–395.

Curtis, D. R. and Eccles, J. C. (1959). The time courses of excitatory and inhibitory synaptic actions. *Journal of Physiology* **145**, 529–546.

Eccles, J. C. (1963). Presynaptic and postsynaptic inhibitory actions in the spinal cord. In: *Brain Mechanisms*, ed. G. Moruzzi. Elsevier, Amsterdam.

Eccles J. C., Eccles, R. M. and Magni, F. (1961). Central inhibitory action attributable to presynaptic depolarization produced by muscle afferent volleys. *Journal of Physiology* **159**, 147–166.

Eccles, J. C., Eccles, R. M., Shealy, C. N. and Willis, W. D. (1962*a*). Experiments utilizing monosynaptic excitatory action on motoneurones for testing hypotheses relating to specificity of neuronal connection. *Journal of Neurophysiology* **25**, 559–579.

Eccles, J. C., Schmidt, R. F. and Willis, W. D. (1962*b*). Presynaptic inhibition of the spinal monosynaptic reflex pathway. *Journal of Physiology* **161**, 282–297.

Granit, R., Kernell, D. and Shortess, G. K. (1963). Quantitative aspects of firing of mammalian motoneurones, caused by injected currents. *Journal of Physiology* **168**, 911–931.

Hubbard, S. J. (1963). Repetitive stimulation of the mammalian neuromuscular junction, and the mobilisation of transmitter. *Journal of Physiology* **169**, 641–662.

Hunt, C. C. and McIntyre, A. K. (1960). An analysis of fibre diameter and receptor characteristics of myelinated cutaneous afferent fibres in cat. *Journal of Physiology* **153**, 99–112.

Ito, M., Kostyuk, P. G. and Oshima, T. (1962). Further study on anion permeability in cat spinal motoneurones. *Journal of Physiology* **164**, 150–156.

Loewenstein, W. R. and Mendelsohn, M. (1965). Components of receptor adaptation in a Pacinian corpuscle. *Journal of Physiology* **177**, 377–397.

Lundberg, A. and Quilisch, H. (1953). On the effect of calcium on presynaptic potentiation and depression at the neuromuscular junction. *Acta Physiologica Scandinavica Suppl.* **111**, 121–129.

Mountcastle, V. B. (1966). The neural replication of sensory events. In *Brain and Conscious Experience*, Ed. J. C. Eccles. Springer, Berlin.

Oshima, K. (1969). Studies of pyramidal tract cells. In *Basic Mechanisms of the Epilepsies*, Eds. H. H. Jasper, A. A. Ward and A. Pope. Little, Brown, Boston.

Reading

Eccles, J. C. (1964). *The Physiology of Synapses.* Springer, Berlin. (A classical account of synaptic transmission).

Katz, B. (1969). *The Release of Neural Transmitter Substances.* Liverpool University Press, Liverpool.
Pierce, J. R. (1962). *Symbols, Signals and Noise.* Hutchinson, London. (A very readable general introduction to information theory).
Shepherd, G. M. (1979). *The Synaptic Organisation of the Brain.* Oxford University Press, Oxford.

Chapter 4 Skin sense

References

Keegan, J. J. and Garrett, F. D. (1948). The segmental distribution of the cutaneous nerves in the limbs of man. *Anatomical Records* **102**, 409–437.
Kenshalo, D. R. (1976). Correlations of temperature sensitivity in Man and monkey, a first approximation. In *Sensory Functions of the Skin in Primates,* Ed. Y. Zotterman, pp. 305–330. Oxford University Press, Oxford.
Penfield, W. and Rasmussen, T. (1950). *The Cerebral Cortex of Man.* Macmillan, London.

Reading

Iggo, A. and Andres, K. H. (1982). Morphology of cutaneous receptors *Annual Review of Neuroscience* **5**, 1–31.
Melzack, R. and Wall, P. D. (1962). On the nature of cutaneous sensory mechanisms. *Brain* **85**, 331–356. (An intelligent discussion of the concepts of modality and specificity in cutaneous sensation.)
Melzack, R. and Wall, P. D. (1982). *The Challenge of Pain.* Penguin, Harmondsworth.
Sinclair, D. (1981). *Mechanisms of Cutaneous Sensation.* Oxford University Press, Oxford.
Willis, W. D. and Coggeshall, R. E. (1978). *Sensory Mechanisms of the Spinal Cord.* Wiley, New York.
Zotterman, Y. (1976). *Sensory Functions of the Skin in Primates.* Oxford University Press, Oxford.

Chapter 5 Proprioception

References

Barker, D. (1948). The innervation of the muscle spindle. *Quarterly Journal of Microscopical Science* **89**, 143–186.
Boyd, E. A. and Roberts, T. D. M. (1953). Proprioceptive discharges from stretch-receptors in the knee-joint of the cat. *Journal of Physiology* **122**, 38–58.
Matthews, P. B. C. (1964). Muscle spindles and their control. *Physiological Review* **44**, 219–288.
Whitteridge, D. (1959). The effect of stimulation of intrafusal muscle fibres on sensitivity to stretch of extraocular muscle spindles. *Quarterly Journal of Experimental Physiology* **44**, 385–393.

Reading

Matthews, P. B. C. (1972). *Mammalian Muscle Receptors and Their Central Actions.* Edward Arnold, London.

Wilson, V. J. and Melvill Jones, G. (1979). *Mammalian Vestibular Physiology.* Plenum Press, New York.

Chapter 6 Hearing input

References:

von Békésy, G. (1960). *Experiments in Hearing,* Trans. & Ed. E. G. Wever. McGraw-Hill, New York.

van Bergeijk, W. A. (1962). Variation on a theme of Békésy: a model of binaural interaction. *Journal of the Acoustical Society of America* **34**, 1431–1437.

Brodal, A. (1981). *Neurological Anatomy in Relation to Clinical Medicine.* Oxford University Press, Oxford.

Carpenter, M. B. (1976). *Human Neuroanatomy.* Williams & Wilkins, Baltimore.

Dadson, R. S. and King, J. H. (1952). A determination of the normal threshold of hearing and its relation to the standardisation of audiometers. *Journal of Laryngology and Otolaryngology* **66**, 366–378.

Katsuki, Y. (1961). Neural mechanisms of auditory sensation in cats. In *Sensory Communication*, Ed. W. A. Rosenblith. MIT Press, Boston.

Spoendlin, H. (1968). Ultrastructure and peripheral innervation pattern of the receptor in relation to first coding of the acoustic message. In *Hearing Mechanisms in Vertebrates.* CIBA Symposium. Eds A. V. S. de Renck and Julie Knight. Churchill, London.

Wood, A. (1930). *Sound Waves and Their Uses.* Blackie, Glasgow.

Reading

von Békésy, G. *Experiments in Hearing*, Trans. & Ed. E. G. Wever. McGraw-Hill, New York. (A gripping account of the elegant experiments that established the travelling-wave model of cochlear function).

Gelfand, S. A. (1981). *Hearing: an Introduction to Psychological and Physiological Acoustics.* Marcel Dekker, New York.

Gerber, S. E. (Ed.) (1974). *Introductory hearing science.* W. B. Saunders, Philadelphia.

Moore, B. C. J. (1977). *Introduction to the Psychology of Hearing.* Cambridge University Press, Cambridge.

Roederer, J. G. (1973). *Introduction to the Psychophysics of Music.* Springer, New York.

Chapter 7 Vision

References

Aguilar, M. and Stiles, W. S. (1954). Saturation of the rod mechanism at high levels of illumination. *Optica Acta* **1**, 59–65.

Baylor, D. A. and Fuortes, M. G. F. (1970). Electrical responses of single

cones in the retina of the turtle. *Journal of Physiology* **207**, 77–92.
Blakemore, C. B. and Rushton, W. A. H. (1965). The rod increment threshold during dark adaptation in normal and rod monochromat. *Journal of Physiology* **181**, 629–640.
Campbell, F. W. and Robson, J. G. (1968). Application of Fourier analysis to the visibility of gratings. *Journal of Physiology* **197**, 551–566.
Dowling, J. E. and Boycott, B. B. (1966). Organization of the primate retina: electron microscopy. *Proceedings of the Royal Society B.*, **166**, 80–111.
Fisher, R. F. (1973). Presbyopia and the changes with age in the human crystalline lens. *Journal of Physiology* **228**, 765–779.
Hubel, D. H. and Wiesel, T. N. (1962). Receptive fields, binocular interaction and functional architecture in the cat's visual cortex. *Journal of Physiology* **160**, 106–154.
Hubel, D. H. and Wiesel, T. N. (1965). Receptive fields and functional architecture in two non-striate visual areas (18 and 19) of the cat. *Journal of Neurophysiology* **28**, 229–289.
Kaneko, A. (1970). Physiological and morphological identification of horizontal, bipolar and amacrine cells in goldfish. *Journal of Physiology* **207**, 623–633.
Marks, W. B., Dobelle, W. H. and MacNichol, E. F. (1964). Visual pigments of single primate cones. *Science* **143**, 1181.
van Nes, F. L. and Bouman, M. A. (1967). Spatial modulation transfer in the human eye. *Journal of the Optical Society of America* **57**, 401–406.
Normann, R. A. and Perelman, I. (1979). The effects of background illumination on the photoresponses of red and green cones. *Journal of Physiology* **286**, 491–507.
Rushton, W. A. H. (1965). Visual adaptation (The Ferrier Lecture, 1962). *Proceedings of the Royal Society B* **162**, 20–46.
Rushton, W. A. H. and Powell, D. S. (1972). The rhodopsin content and the visual threshold of human rods. *Vision Research* **12**, 1073–1081.

Reading

Cornsweet, T. N. (1970). *Visual Perception.* Academic Press, New York. (Particularly intelligent accounts of colour vision and visual acuity).
Davson, H. (1980). *Physiology of the Eye.* Churchill Livingstone, Edinburgh.
Gombrich, E. H. (1982). *The Image and the Eye*. Phaidon, London. (For those interested in the visual arts: an incisive discussion of the relation between pictorial representation and visual perception).
Howard, I. P. (1982). *Human Visual Orientation.* Wiley, Chichester.
Moses, R. A. (Ed.) (1981). *Adler's Physiology of the Eye.* C. V. Mosby, St. Louis. (A useful source for information about the structure and function of ocular tissues, rather than vision).
Robinson, J. O. (1972). *The Psychology of Visual Illusion.* Hutchinson, London.
Rodieck, R. W. (1973). *The Vertebrate Retina.* Freeman, San Francisco.
Walls, G. L. (1963). *The Vertebrate Eye.* Hafner, New York. (Mostly on comparative aspects).

Chapter 8 Smell & taste

References:

Amoore, J. E. (1963). Stereochemical theory of olfaction. *Nature* **198**, 271–272.

Benjamin, R. M. (1963). Some thalamic and cortical mechanisms of taste. In *Olfaction and Taste,* Ed. Y. Zotterman, pp. 309–329. Pergamon, London.

Carpenter, M. B. (1976). *Human Neuroanatomy.* Williams & Wilkins, Baltimore.

Cohen, M. J., Magiwara, S. and Zotterman, Y. (1955). The response spectrum of taste fibres in the cat: a single fibre analysis. *Acta physiologica Scandinavica* **33**, 316–332.

Davies, J. T. (1969). The 'penetrating and puncturing' theory of odor: types and intensities of odors. *Journal of Colloid Interface Science* **29**, 296–304.

Davies, J. T. and Taylor, F. H. (1959). The role of adsorption and molecular morphology in olfaction: the calculation of olfactory thresholds. *Biological Bulletin of the Marine Biology Laboratory, Woods Hole* **117**, 222–238.

Gesteland, R. C., Lettvin, J. Y. and Pitts, W. H. (1965). Chemical transmission in the nose of the frog. *Journal of Physiology* **181**, 525–559.

Kessel, R. G. and Kardon, R. H. (1979). *Tissues and Organs: a Text-atlas of Scanning Electron Microscopy.* Freeman, San Francisco.

Kimura, K. and Beidler, L. M. (1961). Microelectrode studies of taste receptors of rat and hampster. *Journal of Cellular and Comparative Physiology* **58**, 131–139.

Moncrieff, R. W. (1967). *The Chemical Senses.* Leonard Hill, London.

Tanabe, T., Iino, M. and Takagi, S. F. (1975). Discrimination of odours in olfactory bulb, pyriform-amygdaloid areas, and orbitofrontal cortex of monkey. *Journal of Neurophysiology* **38**, 1284–1296.

Reading

Amoore, J. E. (1964). Current status of the steric theory of odour. *Annals of the New York Academy of Science* **116**, 456.

Davies, J. T. (1965). A theory of the quality of odours. *Journal of Theoretical Biology* **18**, 1.

Doty, R. L. (Ed.) (1976). *Mammalian Olfaction, Reproductive Processes and Behaviour.* Academic Press, New York.

Keller, Helen (1903). *The Story of my Life.* Hodder and Stoughton, London. (A fascinating account of the subjective world of someone blind and deaf from an early age, with unusual olfactory powers).

Moncrieff, R. W. (1967). *The Chemical Senses.* Leonard Hill, London. (Little physiology, but useful accounts of the phenomenology of smell and taste, particularly the relation between chemical structure and sensation).

Stoddart, D. M. (1980). *The Ecology of Vertebrate Olfaction.* Chapman and Hall, London.

Wolstenholme, G. E. W. and Knight, Julie (Eds.) (1970). *Taste and Smell in Vertebrates.* Churchill, London.

Wright, R. H. (1964). *The Science of Smell.* Allen and Unwin, London.

Chapter 9 Types of motor control

References

Rashbass, C. and Westheimer, G. (1961). Disjunctive eye movements. *Journal of Physiology* **159**, 339–360.

Reading

Milsum, J. H. (1966). *Biological Control Systems Analysis.* McGraw-Hill, New York. (An introduction, for the mathematically inclined, to feedback and other types of control system).

Powers, W. T. (1974). *Behaviour: the Control of Perception.* Wildwood House, London. (A thoughtful discussion of general aspects of motor control).

Stelmach, G. E. (Ed.) (1976). *Motor Control: Issues and Trends.* Academic Press, New York.

Chapter 10 The spinal level

References:

Bell, G. H., Davidson, J. N. and Scarborough, H. (1961). *Textbook of Physiology and Biochemistry.* Livingstone, Edinburgh.

Eldred, E., Granit, R. and Merton, P. A. (1953). Supraspinal control of muscle spindles and its significance. *Journal of Physiology* **122**, 496–523.

Liddell, E. G. T. and Sherrington, C. S. (1924). Reflexes in response to stretch (myotatic reflexes). *Proceedings of the Royal Society B* **96**, 212–242.

Marsden, C. D., Merton, P. A. and Morton, H. B. (1972). Servo action in human voluntary movement. *Nature* **238**, 140–143.

Matthews, P. B. C. (1972). *Mammalian Muscle Receptors and their Central Actions.* Edward Arnold, London.

Scheibel, M. E. and Scheibel, A. B. (1960). Spinal motoneurones, interneurones and Renshaw cells: a Golgi study. *Archives Italiennes de Biologie* **104**, 328–353.

Vallbo, Å. B. (1971). Muscle spindle response at the onset of isometric voluntary contractions in Man: time difference between fusimotor and skeletomotor effects. *Journal of Physiology* **218**, 405–431.

Reading

Creed, R. S., Denny-Brown, D., Eccles, J. C., Liddell, E. G. T. and Sherrington, C.S. (1932). *Reflex Activity of the Spinal Cord. Oxford University Press, Oxford.*

Matthews, P. B. C. (1972). *Mammalian Muscle Receptors and their Central Actions.* Edward Arnold, London.

Sherrington, C. S. (1906). *The Integrative Action of the Nervous System.* Yale University Press, New Haven, Conn. (1906, reprinted 1961).

Taylor, A. and Prochazka, A. (Eds.) (1981). *Muscle Receptors and Movement.* Macmillan, London.

Chapter 11 The control of posture

References

de Kleijn, A. (1921). Tonische Labyrinth- und Halsreflexe auf die Augen. *Pflügers Archiv* **186**, 82–97.
Martin, J. Purdon (1967). *The Basal Ganglia and Posture.* Pitman, London.
Rademaker, G. G. J. (1935). *Réactions labyrinthiques et équilibre. L'ataxie labyrinthique.* Masson, Paris.
Walsh, E. G. (1964). *Physiology of the Nervous System.* Longman, Harlow.
Witkin, H. A. (1949). Perception of body position and of the position of the visual field. *Psychology Monograph No. 302,* **3**, 1–46.

Reading

Carpenter, R. H. S. (1978). *Movements of the Eyes.* Pion, London.
Reason, J. T. and Brand, J. J. (1975). *Motion Sickness.* Academic Press, London.
Roberts, T. D. M. (1967). *The Neurophysiology of Postural Mechanisms.* Butterworths, London.

Chapter 12 Higher levels of motor control

References

Holmes, Gordon (1922). Clinical symptoms of cerebellar disease and their interpretation. (Croonian Lectures). *Lancet* **203**, 59–65, 111–115.
Netter, F. H. (1962). *The CIBA Collection of Medical Illustrations. Vol. 1: Nervous System.* CIBA, Basle.
Penfield, W. and Rasmussen, T. (1950). *The Cerebral Cortex of Man: a Clinical Study of Localisation of Function.* Macmillan, New York.
Rosen, I. and Asanuma, H. (1972). Peripheral afferent inputs to the forelimb area of the monkey motor cortex: input-output relations. *Experimental Brain Research* **14**, 257–273.
Woolsey, C. N. (1958). Organisation of somatic sensory and motor areas of the cerebral cortex. In *Biological and Biochemical Bases of Behaviour,* Eds H. F. Harlow and C. N. Woolsey. Univ. of Wisconsin Press, Wisconsin.

Reading

Brooks, V. B. and Stoney, S. D. Jr. (1971). Motor mechanisms: the role of the pyramidal system in motor control. *Annual Review of Physiology* **33**, 237–391.
Denny-Brown, D. (1962). *The Basal Ganglia.* Oxford University Press, Oxford. (A useful account of the classical signs of basal ganglia disorders).
Fox, C. A. and Snider, R. S. (1967). The cerebellum. *Progress in Brain Research,* **25**. Elsevier, Amsterdam.
Lance, J. W. and Mcleod, J. G. (1975). *A Physiological Approach to Clinical Neurology.* Butterworths, London. (Clear descriptions and illustrations of clinical motor signs).

Marr, D. (1969). A theory of cerebellar cortex. *Journal of Physiology* **202**, 437–470.

Martin, J. Purdon (1967). *The Basal Ganglia and Posture.* Pitman, London.

Palay, S. L. and Chan-Palay, V. (1974). *Cerebellar Cortex.* Springer, Berlin. (With outstanding micrographs and diagrams).

Phillips, C. G. and Porter, R. (1977). *Corticospinal Neurones: their Role in Movement.* Academic Press, London.

Schmitt, F. O., Worden, F. G., Adelman, G. and Dennis, S. G. (1981). *The Organisation of the Cerebral Cortex.* MIT Press, Cambridge, Mass.

Chapter 13 Analysis and storage of information by the cerebral cortex

References

Carpenter, M. B. (1976). *Human Neuroanatomy.* Williams and Wilkins, Baltimore.

Cobb, S. (1946). *Borderlands of Psychiatry.* Harvard Univ. Press, Harvard.

Critchley, Macdonald (1971). *The Parietal Lobes,* Hafner, New York.

Freeman, W. and Watts, J. W. (1948). The thalamic projection to the frontal lobe. *Research Publication of the Association for Nervous and Mental Diseases* **27**, 200–209.

Gazzaniga, M. S. (1967). The split brain in Man. *Scientific American* August 1967; reprinted in *Physiological Psychology*, Freeman, San Francisco, (1971).

Geschwind, N. and Levitsky, W. (1968). Human brain: left-right asymmetries in temporal speech region. *Science* **161**, 186–187.

Lassen, N. A., Ingvar, D. H. and Skinhój, E. (1978). Brain function and blood flow. *Scientific American* October 1978, 50–59.

Maclean, P. D. (1975). An ongoing analysis of hippocampal inputs and outputs: microelectrode and neuroanatomical findings in squirrel monkeys. In *The Hippocampus,* Eds R. L. Isaacson and K. H. Pribram, Vol. I. Plenum, New York.

Ochs, S. (1965). *Elements of Neurophysiology.* Wiley, New York.

Penfield, W. (1967). *The Excitable Cortex in Conscious Man.* Liverpool University Press, Liverpool.

Zangwill, O. L. (1967). Speech and the minor hemisphere. *Acta Neurologica Belgica* **67**, 1013–1020.

Reading

Brain, Lord (1975). *Speech Disorders.* Butterworth, London.

Critchley, Macdonald (1971). *The Parietal Lobes.* Hafner, New York.

Isaacson, R. L. and Pribram, K. H. (1975). *The Hippocampus.* Plenum, New York.

Mark, Richard (1974). *Memory and Nerve Cell Connections.* Clarendon, Oxford. (A lucid review of neuronal theories of memory).

Olds, J. (1970). The behaviour of hippocampal neurons during conditioning experiments. In *The Neural Control of Behaviour,* Eds R. E. Whalen,

R. F. Thompson, M. Verzeano and N. M. Weinberger. Academic Press, New York.

Renzi, Ennio de (1982). *Disorders of Space Exploration and Cognition.* Wiley, Chichester.

Walsh, K. W. (1978). *Neuropsychology: a Clinical Approach.* Churchill Livingstone, Edinburgh.

Chapter 14 The highest levels of control

References

Dement, W. and Kleitman, N. (1957). Cyclic variations in EEG during sleep and their relation to eye movements, body motility and dreaming. *Electroencephalography and Clinical Neurophysiology* **9**, 673–690.

Engel, G. L. and Schmale, A. H. (1972). Conservation-withdrawal: a primary regulatory process for organismic homoeostasis. In *Physiology, Emotion and Psychosomatic Illness.* CIBA Symposium 8. Elsevier, Amsterdam.

Fonberg, E. (1972). Control of emotional behaviour through the hypothalamus and amygdaloid complex. In *Physiology, Emotion and Psychosomatic Illness.* CIBA Symposium 8. Elsevier, Amsterdam.

Gould, P. and White, R. (1974). *Mental Maps.* Penguin, Harmondsworth.

Penfield, W. and Jasper, H. H. (1954). *Epilepsy and the Functional Anatomy of the Human Brain.* Churchill, London.

Roche (publ.) (1968). *Concepts of Sleep.* Roche Products Ltd, London.

Scheibel, M. E. and Scheibel, A. B. (1957). Structural substrates for integrative patterns in the brain stem reticular core. In *The Reticular Formation of the Brain*, Eds H. H. Jasper, L. D. Proctor, R. S. Knighton, W. C. Woshay and R. T. Costello. Churchill, London.

Simon, H. (1981). *The Sciences of the Artificial.* MIT Press, Cambridge, Mass.

Reading

Hebb, D. O. (1954). The problem of consciousness and introspection. In *Brain Mechanisms and Consciousness,* Eds E. D. Adrian, F. Bremer, H. H. Jasper and J. F. Delafresnaye. Blackwell, Oxford.

Kleitman, N. (1939). *Sleep and Wakefulness.* University of Chicago, Chicago. (The classical account of the phenomenology of sleep).

Kubie, L. S. (1954). Psychiatric and psychoanalytic considerations of the problem of consciousness. In *Brain Mechanisms and Consciousness,* Eds E. D. Adrian et al. Blackwell, Oxford.

O'Keefe, J. and Nadel, L. (1978). *The Hippocampus as a Cognitive Map.* Clarendon, Oxford.

Orem, J. and Bernes, C. D. (Eds) (1980). *Physiology in Sleep.* Academic Press, New York.

Walshe, F. M. R. (1972). The neurophysiological approach to the problem of consciousness. In *Scientific Foundations of Neurology,* Eds Macdonald Critchley, J. L. O'Leary and B. Jennett. Heinemann, London.

Index

Aberrations of eye 144, 145, 171
Acalculia 287
Acceleration
  angular 112
  linear 109
Accidental attributes of an object 188
Accommodation
  of lens, 143, 146, 160, 184
  of nerve 41–2, 52
Acetylcholine 2, 61, 66
Action potential 8, 19–46
  biphasic 37
  compositon of external medium and 28
  compound 35
  conduction velocity 33–5
  initiation 51–5
  mechanism of propagations 19–35
  monophasic 28, 36
  repetitive 52
Activation system, reticular 196, 311–15
Acuity
  auditory 129
  somatosensory 90–93
  visual 166–73
Adaptation
  bleaching 163
  chromatic 181
  dark 141–2, 146, 161–6
  field 161–3
  functions of 55–8
  general mechanisms of 49–51
  in muscle spindle 102
  in Pacinian corpuscle 49
  in vestibular system 110, 113
  of joint receptors 106
  olfactory 200
  visual 141–2, 154, 161–6, 173
Adiadochokinesis 268, 272
Adrenaline
  as neurotransmitter 66
  emotional release 307
Aerial perspective 185
After-image 165, 182
Ageing and accommodation 143
Agnosia 283–4
Agraphia 285
Akinesia 274
Albedo 139, 174, 182, 188
Alcohol
  effects on cerebellum 269
  effects on vestibular apparatus 253
Alexia 285
All-or-nothing law 23, 42–4
Alpha
  fibres 38, 234–40
  -gamma co-activation 234, 239
  waves 312
Amacrine cells 149, 154
Amnesia
  anterograde 296
  retrograde 296, 301
Amusia 287
Ampulla 111
Amygdala 6, 195, 196, 291, 292, 313
Anatomy of brain 5–7, 10
Angular acceleration, velocity 112
Anomalies of colour vision 183
Anopia, Anopsia 157
Anosmia 205
Anterior
  commissure 5, 193, 194
  corticospinal tract 227–9
  spinothalamic tract 88
Anterograde degeneration 10
  amnesia 296
Anterolateral
  chordotomy 89
  system 88
Anxiety and frontal lobe 281
Aphasia 284–9
Appetite
  and hypothalamus 315
  sodium 205
Apraxia 283, 284
Aqueduct, cerebral 5, 6
Aqueous humour 142
Arch of Corti 124, 125
Archicerebellum 264
Archicortex 292, 293
Argyll Robertson pupil 147
Arousal 307, 309
Ascending reticular activating system 310, 311, 314
Associated memory 270–73, 297–302
Associated movements 274
Association
  and motivation 196
  and recognition 190
  areas of cerebral cortex 277
  fibres of cortex 6, 258
Astereognosis 97, 283
Astigmatism 144
Asymmetry of brain 287–291
Asynergia 268
Ataxia 268
Athetosis 275
Attenuation 116
Audiometric curve 116
Audition 115–37
Auditory meatus 122
Automatic gain control 58, 161–3
Autonomic
  effects of emotion 196
  ganglia 3, 85
Axo-axonic synapses 74
Axone 8
  effect of diameter 22, 33
  electrical properties of 20–22, 33–35
  hillock 68

mechanism of conduction 19–33
myelination 9, 34

Babinski sign 222
Ballismus 272
Ballistic control 213–16, 272
Barbiturates 75
Basal ganglia 258, 273–6
Basilar membrane 124
Basket cells 267
Bechterew nystagmus 252
Behaviour
control of 303, 307
olfaction and 194–8
tropistic 303–6
Betz cells 259, 260
Binocular
cells in visual cortex 160, 186
vision and depth perception 185–7
Bipolar cells of retina 8, 44, 149, 154
Bleaching or retinal pigment 152, 163–6
Blind spot 147, 148
Body-image 283
Bowman's glands 192
Bradykinesia 275
Brain
development and evolution of 1–7
general topography of 5–7, 10
left-right asymmetry 287–91
Brainstem
and consciousness 318
and decerebrate rigidity 229
visual responses 160
Brightness 140
Broca's
aphasia 285
area 285
Bud, taste 205, 206
Buttress reaction 243

Cable properties of axon 20–22, 33–5
Calcium 63, 78
Caloric vestibular stimulation 112, 248
Capacitance
of neural membrane 33
of post-synaptic cell 60
Caudate nucleus 6, 273, 274
Central aphasia 286
Centre of gravity and posture 241
Centres
of satiety 315
pleasure 196, 316
speech 285–7
Cerebellum 5, 6, 7, 228, 264–73
afferent fibres 105, 109, 266–7
and co-ordination of visual and vestibular information 252, 272
and motion sickness 253
and parametric feedback 239, 272
and prism reversal 272
celltypes 265
disorders of 268–70
nuclei 265
theories of action 270–73
Cerebral
aqueduct 5,6
blood-flow 15, 288
cortex (*see* Cortex)
dominance 287–91
hemispheres 5
Channels
colour 178
ionic
and resting potential 25
in EPSP 65
in IPSP 70–71
in visual receptors 152
role in action potential 24, 29–33
voltage-dependent 29–33
Chemoreceptors 191, 206
Chiasm, optic 156
Chloride ions, role in inhibition 73
Cholinesterase 62, 65
Chorda tympani 207
Chorea 275
Choroid of eye 142, 149
Chroma 178
Chromatic
abberration 145, 171
adaptation 181
Chromophore 150
Ciliary muscle 142, 146
Cingulate gyrus 292, 316
Circuit
equivalent 20
local, nerve conduction 20–24
Circumvallate papillae 205
Civetone 197
Clarke's column 105
Clasp-knife reflex 106, 232
Classification
dangers of 84
of colours 179
of cutaneous stimuli 82–4
of nerve fibres 38
of odours 201
Climbing fibres 266
Cochlea 124–131
Cochlear
microphonics 125, 129
nerve 131
nuclei 131, 136
Coding
of auditory signals 128–31
of nervous information 44–6, 58
of olfactory information 199
Cold
paradoxical 95
receptors 94
Colliculi 6, 7, 228
inferior 130, 132
superior 160, 228
Colour
anomalies 183
blindness 182–3, (colour plate)
complementary 179, 182
contrast 182, (colour plate)
matching 178
mixing 177–81
opponent cells 157
triangle 179, (colour plate), 201
vision 139, 142, 176–83
Columns
Clarke's 105
in motor cortex 261
in somatosensory cortex 96
in visual cortex 158, 159
Coma 312
Comparator 214, 235, 251
Complementary colours 179, 182
Complex cells 158, 159
Conchae 191, 192
Conditioning, Pavlovian 298
Conduction, nervous 19–35
anoxia and 95
passive 20–22
saltatory 34
speed of 33–8
Cones 141, 148, 150–52, 172, 176, 178
Confabulation 297
Conjunctiva 142
Conscious control of movement 259, 317
and cerebellum 268
Consciousness 291, 316–19
Conservation 307–9
Constant-field equation 26
Constructional apraxia 284
Context, sensory 271
Contrast 167, 169, 171, 176
colour 182
simultaneous 173
Convergence
of eyes 146, 187, 215
of neural pathways 211, 224, 311
Convulsants 71
Cornea 142
Corpus callosum 6, 290
Corpus striatum 7, 273, 274
Corpuscle, Pacinian 47–51, 84
Corresponding points of retina 186
Cortex
auditory 132
cerebellar 264–73
cerebral 6, 7, 96, 257–63

cingulate 292
entorhinal 196, 291
frontal 280–82
hippocampal 291–5
motor 228, 256–63, 277-80
olfactory 206
parietal 282–7
periamygdaloid 196, 206
prepyriform 196, 200
somatosensory 86, 96, 258
temporal 291–7
visual 157–61, 188, 211, 293
Corti, arch of 124, 125
Corticospinal tract 228--9, 239, 259–63
Cribriform plate 192
Crista 111
Cuneate nucleus 86, 105
Cuneocerebellar tract 105
Cupula 111
Current
flow in action potential 20–24, 32
generator 49
synaptic 67
Cutaneous receptors, adaptation in 49–51, 55, 89, 93, 95
Cut-off frequency 169, 172, 176
Cylindrical lens 144

Dale's hypothesis 68, 71
Dark adaptation 141–2, 161–6
Dark light 161
Dash-pot 50, 51, 103, 113
Decerebrate
preparation 229, 235
rigidity 229, 234
Decibel 116, 117
Decomposition of movement 269
Decorticate preparation 229
Degeneration, nerve 10
Degrees of freedom in colour vision 176
Deiter's nucleus 228
Delayed reaction test 282
Dendrites, 8, 299
Dendritic spines 59, 299
Denervation sensitivity 79
Densitometry, retinal 152, 164, 182
Dentate nucleus 265
Deodorants 201
Deoxyglucose 15
Depression
emotional 308
post-tetanic 77
Depth perception, visual 184–7
Dermatomes 84–5
Descending tracts 225–31
Desynchronization of EEG 312
Detection
disparity 186
line 159
of visual targets 166–71
Determinism 319
Deuteranopia 182
Development
of brain 3--7
of synaptic connections 78, 189
Diencephalon 7
Diffraction in eye 146, 171
Dioptres 142
Direction perception
auditory 133–7
olfactory 194
visual 183
Disc, optic 148
Discrimination
between auditory frequencies 129
between stimulus intensities 57
colour 176–83
spatial 90, 166–73
two-point discrimination test 93
Disparity, retinal
and vergence movements 214
cortical detectors 186
use in depth perception 185--7
Distance perception
auditory 133
visual 184–7
Dominance
in cerebral hemispheres 287--91
ocular, in visual cortex 158
Dopamine 66, 275
Dorsal column nuclei 86, 94,105
Dorsal roots
distribution to skin 84
ganglia 5, 8, 84
Drawing
and parietal lobe damage 283, 284
by children 286
in 'split-brain' patient 290
Dreaming 313
Dummy head recording 136
Dynamic
gamma fibres 101, 103
range 57, 117, 140
vestibular reflexes 245, 247
vestibulo-ocular reflexes 247
Dysarthria 285
Dyslexia 285
Dysmetria 268

Ear, structure 122--6
Early receptor potential (ERP) 153
Efference copy
and knowledge of eye position 250
in internal feedback 217
in postural control 255
Efferent fibres to sense organs 103, 109, 129
Effort, sense of 263
Elastic element
in muscle spindle 103
in Pacinian corpuscle 51
in semicircular canals 113
Electrical properties of axone 20–22, 33–5
Electroconvulsive therapy (ECT) 301
Electroencephalogram (EEG) 312
Electromyogram 236, 238
Electro-olfactogram (EOG) 198
Electroretinogram (ERG) 153
Emboliform nucleus 265
Emmetropia 143
Emotional
behaviour 306–12
content of speech 122
effects on autonomic nervous system 307, 316
feelings 195, 307
Encapsulated sensory endings 47–55, 85--9, 106
Encephalization 3, 220
Endolymph 109, 124, 129
Endorphins 97
Endplate, motor 61–4
Enkephalins 97
Entorhinal cortex 196, 291
Epilepsy 256, 290, 295, 311
Equilibrium potential 24
Equivalent background 165
Equivalent circuit 20
Error
in control systems 214, 235
refractive 143, 147
Essential attributes of an object 188
Eustachian tube 124
Evolution
and olfaction 194
of brain 1–7
of cerebellum 264
of cerebral cortex 277, 292
of colour sense 180
of limbic system 292
of vestibular apparatus 107
Excitatory postsynaptic potential (EPSP)
generation of 64--8
summation of 67
Expressive aphasia 285
Extensor thrust 243
Extrapyramidal pathways 259
Eye
movements
and sleep 313
miniature 163
nystagmus 247, 250
saccades 160, 217, 228
vergence 146, 215
position, knowledge of 240, 250--51
structure 142–50

Facilitation 77
Far point 143
Fastigial nucleus 265
Feedback
 and motor control 212–19, 234–40
 in basal ganglia 273
 in cerebellum 271
 internal 217, 273
 parametric 216, 239, 272
 positive 39
Feedback, negative
 in action potential 39
 in control of muscle 214, 231–40
 in voltage clamp 29
 inhibition 71
Feet, role in upright posture 241–5, 255
Field adaptation 161–3
Fila olfactaria 192
Filiform papillae 205
Filter
 as model for recognition 188
 energy 47, 53, 103, 112
 high-pass 49, 51, 169, 174
 second, in cochlea 129
Final common path 224
Firing frequency 45, 52, 68
Flaccid paralysis 229, 260
Flexion reflex 225
Flicker fusion frequency 174, 176
Focussing by the eye 142–6
Follow-up servo 235
Force
 control of 235, 239, 262
 role in cutaneous sensation 263
Forebrain 5
Formants 121
Fornix 6, 292, 295
Fourier analysis and synthesis 119–22
 by the cochlea 126–8
Fovea centralis 148
Free will 319
Frequency
 analysis by cochlea 126–31
 code of action potentials 44–6, 58
 cut-off 170, 172, 176
 firing 45, 52, 68
 fundamental 119
 of sound 115
 spatial 169, 174
 temporal 176
Frontal leucotomy 281
Frontal lobe 280–82
Fundamental frequency of a sound 119
Fungiform papillae 205, 206
Fusimotor fibres 38, 101, 231–2

Gage, Phineas 280, 281, 318
Gain
 control in retina 161–3
 of stretch reflex 237
  and motor cortex 262
Gamma
 -amino butyric acid 66, 71, 75
 fibres 38, 101, 231–40
Ganglion
 cell, retinal 149, 154, 173
 dorsal root 5
 spiral 125, 128
 sympathetic 3
Gating
 by presynaptic inhibition 76
 in lateral geniculate 157
 of pain 97, 98
Generator current and potential 49
Geniculate ganglion 207
Glare 171
Glial cells 76
Globus pallidus 273, 274
Globose nucleus 265
Glomeruli
 cerebellar 267
 olfactory 193, 194
Glycine 66, 71
Golgi
 apparatus 300
 cells of cerebellum 267
 silver stain 10
 tendon organ 100, 104, 231, 262
Gracile nucleus 86
Granule cells
 in cerebellum 8, 265
 in olfactory bulb 193, 194
Grasping
 and motor cortex 261, 262
 sensory aspects 89
Gratings, as visual targets 168, 172
Gravity
 centre of, and posture 241
 direction of 109, 245
Guided systems 214–16
Gustation (*see* Taste)

Habituation 56, 78, 295
Hair cells
 cochlear 124–6
 olfactory 192
 vestibular 108–9
Hair follicle, innervation of 86, 89
Handedness 288, 289
Harmonics 119
Head
 position, sense of 245, 251, 255
 righting reflexes 245
 tilt 109, 245
Hearing 115–37
Helicotrema 124, 127
Hemianopia 157
Henning's prism 201, 202
Hermann grid 173
Heteronymous hemianopia 157
Hierarchical oraganization 220, 275, 278
High-pass filter 49, 51
Hindbrain 5
Hippocampus 196, 291–7, 304, 310
Histamine 96
Holism 13
Homonymous hemianopia 157
Homunculus 87, 88, 257, 262
Hopping reaction 244
Horizontal cells 149, 154, 194
Horopter 186
Horseradish peroxidase 10
Hue 178
Humour, aqueous and vitreous 142
Hypercomplex cells 159
Hypermetropia 144
Hypertropism 315
Hypothalamus 6, 97, 196, 291, 304, 315
 and emotion 315–16
Hypotonia 260, 268

Idiots savants 287
Illuminance 139
Illusion
 Müller-Lyer 185
 Ponzo 185
 waterfall 183
Image
 after- 165
 body- 283
 retinal, qualtiy of 166–73
 -sharpening 91
 stabilized 163
Impedance matching in ear 123
Impulse, nervous (*see* Action potential)
Inactivation, sodium 40
Increment threshold 161
Incus 122
Inferior
 colliculus 7, 130, 132, 228
 glossopharyngeal ganglion 207
 olive 259, 266
Inflow theory 250
Infra-red 138
 theory of olfaction 203
Inhibition
 current 73
 feedback 94
 feedforward 93
 lateral (*see* Lateral inhibition)
 presynaptic 73–6
 reciprocal 68
 remote 74
 synaptic 68–76
 voltage 72

Inhibitory postsynaptic
potential (IPSP) 68–71
Initiation
impulses 51–5
movement 234–40, 276
Inner ear 124–5
Innervation
of cochlear hair cells 128
of muscle 231
of skin 84–9, 94
reciprocal 68, 81
Intelligence 279, 281, 286, 287
apparent 212
Intensity
coding of 56
of sound 115, 133
Intention tremor 268
Interaural sound differences 133
Internal capsule 86, 229, 274
Internal feedback 217
in cerebellum 272
Interneurones 2
in cerebellum 267
in cerebral cortex 258
in olfactory bulb 267
in retina 149, 154–6
in spinal cord 69, 72
Interpositus nucleus 265
Intrafusal fibres 100
Ionic
distribution across
membranes 24, 27
permeability channels (*see* Channels)
Iris (*see* Pupil)
Itch 83, 96

Jacksonian epilepsy 256
Jargon 287
Joint receptors 106–7, 253

Kinocilium 108, 124
Kluver–Bucy syndrome 315
Korsakov syndrome 296, 301
Krause end-bulb 86

Labyrinth 107–14
Landoldt C chart 170
Language
cerebral localization of 285, 288
disorders of 285
Larynx 120
corticospinal tract 227–9
geniculate nucleus (LGN) 157, 279
inhibition 14, 91–4
in cerebellum 267
in cerebral cortex 258
in cochlea 130
in olfactory bulb 194
in skin 91
in spinal cord 224
in vision 173–4
lemniscus 132
line organ 107
olfactory tract 194, 196
reticulospinal tract 226, 227
spinothalamic tract 88
ventricles 5
vestibular nucleus 228, 229
Learning
brain lesions and 295–7
cerebellar 270–73
in motor system 216–17, 263, 269–70
in visual system 252
possible mechanisms 297–302
Left-handedness 288, 289
Lemniscal system 87
Lemniscus
lateral 132
medial 86
Lens
of eye 142
spectacles 144
Lesions, interpretations of
effects 16–18, 220–23, 275, 278
Leucine, tritiated 10
Light
adaptation 141–2, 161–6
measurement 139
visible spectrum 138
Limbic system 7, 273, 291, 292
olfaction and 195, 196
Linear
acceleration 109
perspective 185
Load, compensation for 214, 237–9
Localization
cutaneous 90
of function in brain 15–18
olfactory 194
sound 133–7
visual 183–7
Locus ceruleus 313
Long-term memory (LTM) 300
Low-frequency cut
spatial 174
temporal 176
Lower motor neurone 259, 260
Luminance 139

Macrosmatic 192
Macula
lutea, of eye 148
of otolith organs 109
Malleus 122, 123
Mammillary bodies 291
Manipulation 229, 261
Maps
collicular 160
motivational 303–6, 310
motor 256, 260, 262
Mass action 17, 279
Mechanoreceptors
in cochlea 124–5
in muscle 100–106
in skin 47, 85–9
in tendons 104
Medial
forebrain bundle 196, 292
geniculate nucleus 132, 279
lemniscus 86
reticulospinal tract 226, 227
Medulla 5, 226
pyramids of 228
raphe nucleus 98, 313
Meissner's corpuscle 84, 89
Membrane
axonal 9, 20–22
basilar 124
Reissner's 124
tectotonial 124
tympanic 122
Memory
and olfaction 197
and RNA 300
disorders of 291, 296
long-term (LTM) 300
mechanisms 297–302
short-term (STM) 300
types of 297
Merkel's disc 84
Mesencephalon 5
Mesopic vision 140
Microelectrodes 15
Microphonic potential, cochlear 125, 129
Microsmatic 191
Midbrain 5
Middle ear 123–4
Miniature
endplate potentials 63
eye movements 103
Mitral cells 192
Modalities, sensory
of skin 82–4
of taste 205
Modulation, amplitude and
frequency 45
Monochromat, rod 164, 182
Monosynaptic reflex (*see* Stretch reflex)
Mossy fibres 267
Moth lure 197, 204
Motion perception, visual 159, 160, 183, 249–50
Motion sickness 252
Motivation 276, 303–6
olfaction and 195
Motivational maps 303–6
Motor
aphasia 284, 285
cortex (*see* Cortex)
homunculus 257
maps 256, 260, 262
programs 213–19, 239
unit 224
Motor control 210–12, 256–76
and efference copy 217
ballistic 212
guided 214
heirachical 220–23
prediction in 217
programs for 213

servo-hypothesis 234--9
Motor neurones 8, 224--31
'lower' 259, 260
membrane adaptations in 53
synaptic potentials in 64--6, 68–72
'upper' 259, 260
Movement
control (*see* Motor control)
parallax 184
sense of 249–53
sensitivity, somatosensory 96
sensitivity, visual 159, 160, 183, 249
Müller-Lyer illusion 185
Muscle
gradation of contraction 224
innervation of 224, 231
proprioceptors 100–106
spindles 100, 231–4, 240, 267
stapedius 124
tensor tympani 124
tone 229, 234, 308
Musical instruments 120
Mydriasis 146
Myelin 9, 34
Myopia 144
Myotatic reflex 233

Near point 143
reflex 146
Neck reflexes 253--5
Need 276, 310
Negative feedback (*see* Feedback)
Neglect 283
Neocerebellum 264
Neocortex 291, 292
Neostriatum 273
Nernst equation 25
Nerve
chorda tympani 207
cochlear 131
conduction 19–38
growth factor 79
net 2
olfactory 192
optic 147
trigeminal 206
vestibular 108
Nerve fibres (*see also* Axon)
classification 38
Neural
codes 44–6, 58
plate, tube 4
Neuromuscular junction
denervation sensitivity 79
mechanism 61–4
Neurones 1, 8–10, 19–46, 47–81
summation in 67–8
Neurotransmitters 65, 71
Nociceptors 83, 95
Node of Ranvier 9
Noise
in axons 56
in motor system 213, 215
retinal 161
signal-to-noise ratio 56
thermal, and olfaction 204
Noradrenaline
and sleep 313, 314
as neurotransmitter 66
Nose, internal structure 191, 192
Nucleus (*see* names of specific nuclei)
gigantocellularis 226, 258
solitarius 206
Nystagmus
Bechterew 252
optokinetic 250
per-rotatory 248
positional 253
post-rotatory 248
quick phase 247
slow phase 247
vestibular 247

Ocular dominance columus 158
Olfactory
bulb 192, 200
cilia 192
cleft 191
effects on behaviour 197
epithelium 191
pathways 196, 292
receptors 192, 198
sensitivity 200
theories 202–5
tract 194
tubercle 196
Olive
inferior 259, 266
superior 129, 132, 136
Ophthalmoscope 147
Opiates 97
Opsin 150, 165
Optic
chiasm 156
disc 148
nerve 147, 149
radiation 157
tract 156
Optics of eye 142–6, 166, 171
Optokinetic nystagmus 250
Ossicles of middle ear 122
Otoconia 109
Otolith organs 107, 109–110, 245
Outflow theory 250
Oval window 123

Pacinian corpuscle 47–58, 84
Pain 97–9, 205
and frontal lobe 281
receptors 95–6
referred 96
Paleocerebellum 264
Paleostriatum 273
Panum's fusional area 186
Papillae 205, 206
Paradoxical
cold 95
sleep 313
Parallax, movement 184
Parellel fibres 267
Paralysis
agitans 274
flaccid 229, 260
in stroke 229
spastic 229
Parametric feedback 216, 272
in control of muscle 239
Paresis 260
Parietal
cortex 282–3
operculum 207
Parkinsonism 274
Pattern-recognition 187, 210, 270
Patterns of activity
in cochlear units 126-8
in skin 83
olfactory 199
Pavlovian conditioning 298
Peptides as transmitters 66
Periamygdaloid cortex 196, 206, 291
Periaqueductal gray 97
Periglomerular cells 194
Perilymph 109, 124, 127, 129
Permeability, ionic
and resting potential 24–9
during action potential 29
in receptors 47
in synapses 63–6, 70, 75
Per-rotatory nystagmus 248
Perspective 185
Phantom contours 190
Phase
-locking 130
of sound waves 118, 133
Pheromones 197
Photometry 139–40
Photopigment of retina 150–52, 163, 179
Photons 138
Photopic vision 140, 146, 148, 172
Phrenology 17
Picrotoxin 71
Pigment
epithelium 149
olfactory 192, 204
retinal 150–52, 163
Pinna 122, 133
Pitch 115, 131
Pituitary 196, 292
Place theory of hearing 126
Placing reactions 244, 262
Plasticity 14, 17, 297
in cerebellum 270–73
of cerebral cortex 17, 263
of vestibulo-ocular reflex 252
Pleasure centres 196, 316
Pointspread function 166

Pons 6, 7, 160, 226, 259, 267
Ponzo illusion 185
Position sense
  of eye 183, 250
  of head 109–10, 245, 251, 255
  of limbs 106
Positional nystagmus 253
Positive supporting reaction 243, 245
Posterior spinocerebellar tract 105
Post-rotatory nystagmus 248
Post-tetanic depression 77
Postural
  sway reaction 243
  vestibular reflexes 245–8
Posture 241–55
  and basal ganglia 275
  and cerebellum 268
  criteria for upright 241
  in depression 308
  vestibular contributions 245–7
  visual contribution 248–50
Potassium
  and action potential 28, 31
  and glial cells 76
  an resting potential 24
  role in inhibition 70, 73
Potential
  action (*see* Action potential)
  cochlear microphonic 125, 129
  early receptor (ERP) 153
  endplate (EPP) 62–4
  equilibrium 24
  excitatory postsynaptic (EPSP) 65
  generator 49
  inhibitory postsynaptic 68
  miniature endplate 63
  resting 24–9
  reversal 65, 70
Potentiation 77
Poverty of movement 274
Predictions
  in motivation 303
  in motor control 217–19
Prepyriform cortex 196, 200, 291
Presbyopia 143
Pressure receptors 47–51, 84
  and posture 242
  and stretch reflex 237, 263
Presynaptic inhibition 73–6
Primary
  afferent depolarization (PAD) 74
  colours 178
  odours, lack of 201
Prism
  Henning's 201, 202
  reversal 252, 272
Programs, motor 213–23, 239
Proprioceptors
  Golgi tendon organs 104
  in neck 253
  joint receptors 106
  muscle spindles 100
  vestibular apparatus 107–14
Protanopia 182
Pulvinar nucleus of thalamus 279, 282
Pupil
  Argyll Robertson 147
  control of 146, 160, 165
  effect on visual optics 146, 171
  emotional influence 307
Purkinje cell, cerebellum 8, 265, 270, 299
Putamen 273, 274
Pyramidal
  cells of hippocampus 294
    motor cortex 257–9
  tract 228–9, 260, 262
Pyramids of the medulla 228

Quick phase of nystagmus 247

Radiation
  electromagnetic 138
  optic 157
Range, dynamic
  auditory 117
  visual 140
Range of accomodation 143
Ranvier, node of 9
Raphe nucleus
  and pain 98
  and sleep 313
Rapid eye-movement (REM) sleep 313
Rate sensitivity
  as function of adaptation 58
  in muscle spindle 102
Rebound 268
Receptive fields
  binocular 186
  in motor cortex 261
  in temporal cortex 160
  in visual cortex 157–60
  of bipolar cells 154
  of ganglion cells 154, 169
  of hippocampal cells 294, 310
  of lateral geniculate cells 157
  of Purkinje cells 267
  somatosensory 89–91
Receptors 47–58
  adaptation in 49–51
  auditory 124–6
  cold 94–5
  cutaneous 84–89
  joint 106–7
  mechanical 47
  olfactory 191, 198
  pain 95
  retinal 141, 150–54
  stretch (*see* Muscle spindles)
  taste 205, 207, 208
  temperature 94
  vestibular 107–14
Reciprocal innervation 68, 81
Recognition
  disorders 283
  of sounds 120–22
  of visual targets 187–90
Recording methods 15
Recruitment 44, 131
Red nucleus 228, 258, 265, 273
Red-green blindness 182
Reductionism 13
Redundancy 56, 92
Referred pain 96
Reflex
  clasp-knife 106, 232
  conditioned 298
  flexion 225
  head righting 245
  monosynaptic 64, 105, 225
  myotatic 233
  near 146
  neck 253–5
  pupillary 146
  righting 244
  spinal 244
  stretch 105, 225
  tendon jerk 233
  vestibular 245
  vestibulo-ocular 245
  wiping 220
  withdrawal 225
Refractive
  error 143, 147
  index of parts of eye 142
  power of eye 142
Refractory period 40–41
Regeneration
  of tectal afferents in frog 80
  of visual pigment 151, 152
Reissner's membrane 124
Release 16, 17, 221, 229, 260
REM sleep 313
Renshaw cells 71, 224
Resolution, visual 167–71
Resonance
  in speech production 121
  of outer ear 122
  theory of basilar membrane 126
Resting potential 24–9
Reticular formation
  activating system 196, 311–15
  and decerebate rigidity 229
  medullary 226, 265
  nucleus gigantocellularis 226, 258
  pontine 226, 265
  raphe nucleus 98, 313
Reticular lamina 124
Reticulospinal tract 226, 227
Retina 142, 147–9
Retinal (retinene) 150–52

Retrograde amnesia 296, 301
  degeneration of neurones 10
Reversal potential 49, 65, 70
Rhinencephalon 197, 291
Rhodopsin 132
Rigidity
  and basal ganglia disorders 274
  decerebrated 229, 234
RNA and learning 300
Rods 141, 150–52
Rotation
  reflex effects of 247
  vestibular response to 111
Round window 124
Rubrospinal tract 227, 228
Ruffini end-bulb 86, 89

Saccadic eye movements 160, 217, 228
Saccule 107–10
Salt taste 205, 208
Saltatory conduction 34
Satiety centre 315
Saturation
  in colour-space 179
  of retinal receptors 153, 162
Scala
  media 124
  tympani 124
  vestibuli 124
Scanning speech 268
Schwann cell 9
Sclera 142
Scotopic vision 140, 146
Secondary motivators 195
Semicircular canals 107, 111–14, 247
Sensory aphasia 284–7
  context 271
  homunculus 87
  modality 82–4
  receptors (*see* Receptor)
Septal nuclei 195, 196, 291, 292
Serotonin
  and sleep 313
  as neurotransmitter 66
  in pain transmission 98
Servo
  -assistance 239–40
  hypothesis, simple 234–9
Shadow
  and depth perception 185
  in auditory localization 133
Shock
  electroconvulsive 301
  emotional effects of 307
  spinal 220
Short-term memory (STM) 300
Sight (*see* Vision)
Signal-to-noise ratio 56
Silent areas 277
Sinusoidal
  gratings 168
  sound-waves 115
Simple cells of visual cortex 157
  servo hypothesis 234–9
Skin, receptors in 82–4
Sleep 307, 312–16
Slow phase of nystagmus 247
Slow-wave sleep 312
Smell 191–202
Snellen chart 170
Sniffing 191, 201
Sodium
  channels 25
    refractoriness 40
  pump 26
Somatosensory
  cortical areas 86, 258
  modalities 82–4
  system 82–99
Sound 115–22
  localization 133–7
  timbre 120
Space
  colour 179, (colour plate)
  constant, of axon 21
  orientation of head in 245, 251, 255
Spasticity 229, 260
Spatial
  agnosia 283
  discrimination 90
  frequency 169, 174
  localization of sound 133–7
  pattern of activity in cochlea 126–28
  resolution 168
  summation 67
Spectacles 143
Spectrum
  of visible lights 138, (colour plate)
  of a sound 118–22
Speech
  disorders 285
  localization of cerebral areas 285–6, 288
  mechanism 120
  scanning 268
  spectrum 121
Spherical aberration 145, 171
Spinal
  preparation 220, 229
  reflexes 220, 224
  shock 220
Spinal cord 3, 224–40
  ascending tracts 86–9
  descending tracts 225–31
  hemisection of 88
Spindle, muscle (*see* Muscle spindle)
Spines, dendritic 59, 299
Spinocerebellar tract, posterior 105
Spino-olivary tract 105
Spinothalamic tracts 87, 88
Spiral ganglion 125, 128
'Split-brain' preparation 290
Stability
  in standing 241–5
  of axon 39
Stabilized image 163
Stains for neural tissue 10, 65
Standing mechanisms 241–5
Stapedius 124
Stapes 123
Stars, visibility of 167
Static
  gamma fibres 101, 103
  vestibulo-ocular reflexes 245
Stepping reaction 244
Stereocilia 108, 124
Stereophony 136
Stimulation 15, 97, 211
  of motor cortex 261, 263
  of temporal cortex 294
  of visual cortex 294
Stretch receptors (*see* Muscle spindles)
Stretch reflex 105, 232–4
  and decerebrate rigidity 231, 234
  gain 237, 262
Striate cortex (*see* Visual cortex)
Striatum, corpus 273
Stroke 229, 285
Strychnine 71
Stuttering 285, 289
Subiculum 294, 295
Substantia nigra 273–5
Subthalamus 273, 275
Summation
  olfactory 202
  spatial, in neurones 67–8, 73
  temporal 67–8
  vestibulo-visual 249, 251–3, 255
Superior
  colliculus 7, 160, 228
  olive 129, 132, 136
Synapses 2, 9, 58–73
  axo-axonic 74
  chemical development 78, 189
  electrical 59, 109
  inhibitory 68–72
  interactions between 9, 66, 73
  memory and 298–300
  structure 58
  summation 66–8, 73

Tactile agnosia 97, 283
Taste 205–7
  bitter 205, 208
  buds 205, 206
  mechanism of 207
  modalities of 205, 207
  receptor responses 207
  sour 205, 207
  sweet 205, 208
Tectorial membrane 124
Tectospinal tract 227, 228

Tectum (*see* Colliculi)
Temperature
  and conduction velocity 35
  receptors 94
Temporal lobe 291–7
  stimulation of 294
  visual cells in 160
Temporal summation in neurones 66
Tendon
  jerk 233
  organ, Golgi 100, 104, 231, 239, 240, 262
Tension, muscle
  control of 237–40, 262
  receptors for 104
Tensor tympani 124
Texture
  cutaneous 83, 89, 263
  gradient 185
Thalamus 6, 7, 279
  and pain 97
  and taste 206, 207
  nuclei of
    anterior 292
    centromedian (CM) 258, 274
    dorsomedial (DM) 316
    intralaminar 88, 258, 310
    lateral geniculate 157
    lateral posterior 279, 282
    medial geniculate 132
    pulvinar 279
    ventral posterolateral (VPL) 86, 106, 207, 282
    ventro-anterior (VA) 258, 274
    ventrobasal (VB) 99
    ventrolateral (VL) 258, 265, 274
Thermoreceptors 94–6
Thiamine deficiency 296
Threshold
  for action potential 23, 38–42, 51
  for sound intensity 116, 130
  for vision 140, 152, 161–6, 169, 171
  increment 161
  olfactory 200, 204
Tickle 83, 96
Timbre 120
Time constant of axonal membrane 34
Time-differences, auditory perception of 134
Time-intensity trade 137
Tone, muscle
  and decerebration 229
  and depression 308
  and muscle spindles 234
  and vestibular system 229
Tongue
  and taste 205
  in speech 121
Tonic
  neck reflex 253
  stretch reflex 233
  vestibular reflexes 245
Touch 82–99
Tracts (see under appropriate name)
Transduction 47–51
  in cochlea 124, 129
  in Pacinian corpuscle 47–51
  olfactory 198, 202
  vestibular 108
  visual 152
Transmitters
  at neuromuscular junctions 61–4
  evidence for 66, 71
Trapezoid body 132
Travelling wave 127
Tremor
  at rest 274
  intention 268
  of eye muscles 163
Triangle, colour 179, (colour plate), 201
Trichromacy 177
Trigeminal nerve 206
Triple response 146
Tritanopia 182, (colour plate)
Tropisms 303
Tuning curves 130, 132
two-point discrimination test 13, 283
Tympanic membrane 122

Upper motor neurone 259, 260
Utricle 107, 109–10

Velocity
  of nervous conduction 33–5
  sensitivity of semicircular canals 113
Ventral horn cells (*see* Motor neurones)
Ventricles of brain 5, 6
Vergence movements as example of guided system 215
Vertical, sense of 248
Vesicles, synaptic 61, 77
Vestibular
  apparatus 44, 107–14
  contribution to upright posture 245–7
  influence on decerebrate rigidity 229
  nuclei 109, 226, 265
  nystagmus 247
  projections to spinal cord 227
  reflexes 245
Vestibulo-cerebellum 253, 264, 268
Vestibulo-ocular reflex (VOR)
  dynamic 247
  plasticity 252, 272
  static 245
Vestibulospinal tract 227
Vibration
  and stretch reflex 106
  sensitivity of Pacinian corpuscle 60
Viscera, sensory afferents from 84, 96
Viscous element
  in muscle spindle 103
  in Pacinian corpuscle 51
  in semicircular canals 113
Vision 138–90
Visual
  acuity 166–73
  adaptation 141–2, 154, 161–6, 173
  agnosia 283
  contrast 167, 169, 171, 176
  contributions to posture 248–50
  control of eye movements 250
  cortex 157, 188, 211, 293
  pathways 156
  pigment 150--52
  resolution 167–71
Vitamin A 150
Vitreous humour 142
Voice 120–22
Voltage clamp 29
Voluntary movement 259, 317
Vowels 121

Waking-sleeping cycle 313, 314
Walking
  by newborn child 222, 223
  in cerebellar disorders 268
  mechanisms 220
Wallerian degeneration 11
Warmth receptors 94–5
Water, as a taste stimulus 207
Waterfall illusion 183
Wavelength
  of sound waves 115
  of visible lights 138
    and diffraction 171
    discrimination 176–80
Weber–Fechner relationship 161
Weight, sense of 263
Wernicke's area 285, 288, 291
White matter 5
Wiping reflex 220
Withdrawal
  behaviour 307–9
  reflex 225
Word deafness 285

Xenon, radioactive 15, 288